BEING THE PERSON YOUR DOG THINKS YOU ARE

ALSO BY JIM DAVIES

NONFICTION

Riveted: The Science of Why Jokes Make Us Laugh, Movies Make Us Cry, and Religion Makes Us Feel One with the Universe

Imagination: The Science of Your Mind's Greatest Power

SERIALIZED FICTION

Eve Pixiedrowner and the Micean Council (at alteredrealitymag.com)

PODCASTING

Minding the Brain (mindingthebrainpodcast.com)

You can follow Jim Davies at @AuthorJimDavies on Facebook and at @DrJimDavies on Twitter.

BEING THE PERSON YOUR DOG THINKS YOU ARE

The Science of a Better You

JIM DAVIES

PEGASUS BOOKS
NEW YORK LONDON

BEING THE PERSON YOUR DOG THINKS YOU ARE

Pegasus Books, Ltd.
148 West 37th Street, 13th Floor
New York, NY 10018

Copyright © 2021 by Jim Davies

First Pegasus Books cloth edition February 2021

Interior design by Maria Fernandez

All rights reserved. No part of this book may be reproduced in whole or in part without written permission from the publisher, except by reviewers who may quote brief excerpts in connection with a review in a newspaper, magazine, or electronic publication; nor may any part of this book be reproduced, stored in a retrieval system, or transmitted in any form or by any means electronic, mechanical, photocopying, recording, or other, without written permission from the publisher.

Library of Congress Cataloging-in-Publication Data is available.

ISBN: 978-1-64313-650-9

10 9 8 7 6 5 4 3 2 1

Printed in the United States of America
Distributed by Simon & Schuster
www.pegasusbooks.com

To Darren McKee

CONTENTS

Front Matters   xi

Front Unmatters and Front Antimatters   xiii

PART I: ABOUT YOU   1

1. **Productivity**   3

    An Optimizer   3

    Personal Productivity   4

    Sharpening the Saw vs. Cutting   5

    Your Mind, Your Time, Your Attention   10

    The Effects of Distraction   13

    Why Are People Sabotaging Themselves?   18

2. **Hacking Your Brain, Hacking Your Life**   26

    How to Hack Your Cognitive System   30

    How to Hack Your Habit System   40

    How to Hack Your Reward System   51

    Working Without Distraction: I'm Here to Make an Hour Out of Fifteen Minutes   59

    The Project List   62

    The Half-Hours Method   66

    The Half-Hours Method: Master Class   81

Competing Methods	82
How to Read More Effectively	86
Idea Capture	96
How to Sleep	105
Helping Yourself by Helping Others	109
3. Happiness	112
Your Life Satisfaction	123
Habituation	128
Happiness Is In-born	138
Happiness Comes from Within	139
Happiness Comes from Without	144
Hang Out with Friends and Loved Ones	156
A (Good) Marriage and (Maybe) Children	161
Happiness and Memory	166
There Is More to a Good Life than Being Happy	171
PART II: IT'S NOT ABOUT YOU	**177**
4. What We Think Is Right and Wrong	179
Morality vs. Selfishness	185
The Expanding Circle	187
The Social Group	188
The Out-Group and Animals	189
Anything That Can Suffer	193
Morality, Personality, and Emotion	196
The Good of the Many vs. the Rights of the Few	201
Your Identity as a Charitable Person	211

	Doing Bad Is More Powerful than Doing Good	213
	Your Brain on Morals	215
5.	What's Actually Right and Wrong	228
6.	A Work in Progress	238
	Having and Not Having Children: Moral Issues	240
	Ethical Purchases	245
	Think Globally, Act Globally	247
7.	The (Numerical) Value of Human Life	254
8.	Choosing a Career	269
9.	Measuring Good Done	280
10.	Animals	284
	Is Life Suffering?	305
	Helping Animals	315
11.	Comparing Human to Animal Suffering	330
12.	Environmental Morality	333
13.	Choosing Charities	341
14.	How to Mobilize People to Be Good	353
	How People Give, How People Donate	353
	Are Humans Good for the World?	363
15.	When Giving Gives Back	364
	Where Does It All End? How Much Is Too Much?	366
	The Only Local Activism that Makes Sense	374
	Acknowledgments	379
	Endnotes	381
	Index	419

FRONT MATTERS

Most of the work cited here is from psychology literature. If I mention a scientist or other scholar without their occupation, assume they are a psychologist. I will tag others as neuroscientists, economists, and so on as needed.

There's some neuroscience in this book. Like all popular accounts of neuroscience, what's presented here is a simplification. The brain is really, really, really complex. I mean *mad* complex. Some people, even some who have read Eminem's lyrics, say it's the most complex object in the known universe. If you haven't studied neuroscience, you have no idea how deep the rabbit hole goes.

My colleague and podcast cohost, neuroscientist Kim Hellemans, told me that it's common in undergraduate education to teach one thing to the students in their first and second years to only tell them later that what they were taught before isn't quite true—the reality is much more complex. But if you try to teach what's actually going on in the brain right away, with no simplification, it's hard to get off the ground at all with any idea of what's going on. Useful fictions become the scaffolding for an understanding of the deeper ideas that will be better understood as one climbs it.

To understand what neuroscientists actually think about how the brain works requires reading primary material—articles in neuroscience journals. Unfortunately these papers are all but incomprehensible to anyone without a graduate degree in neuroscience (my graduate degrees are in psychology and computer science, so I have trouble with many of them, too).

Furthermore, there are enormous controversies in neuroscience, and many of the things I will say about the brain will be in conflict with what some neuroscientists believe. I can guarantee you that for everything I say about the brain *some* neuroscientist agrees with me, but rather than burdening you with endless hedges and qualifications, I'm just going to tell you now that many things I say about the brain are still debated but the account I'm giving you here is plausible by the standards of at least some qualified neuroscientists.

Technically, we don't encounter this in scientific psychology, except in the most complex areas. When I simplify a psychology theory, I will say so. But rest assured that *everything* I say about the brain is a simplification. If you want more detail, I encourage you to use the references I cite as a treasure map and start digging.

FRONT UNMATTERS AND FRONT ANTIMATTERS

Whether or not it's completely true, we strongly feel that dogs look at us with love and awe. Our dogs make us feel like we are important and deserving of love, even when we don't feel that from people in our lives. I like the vague, demigod feeling I get from a dog's attention. It's something to live up to. But a dog's love is big on heart and short on specifics. This book is about the specifics.

In this book I'm going to talk about what you can do to make yourself better in three major areas that people care about: personal productivity, happiness, and moral goodness.

I'm a scientist, and I'm going to shed light on these issues and our ultimate goal of improving both ourselves and the world around us through a scientific lens. Science is often numerical in nature, and the wonderful thing about numbers it that you can look at two of them and know *which one is bigger than the other*. Isn't that exciting?

It is, when you consider that many qualitative analyses don't enjoy this benefit. If having a bit more income makes you happier, and spending a bit more time with your friends makes you happier, too, without numbers you can't tell their relative importance.

Even more fundamentally, maybe one of them doesn't really do any good at all. You can read all kinds of things telling you that this or that will make your life better, but without numbers it's hard to tell if it actually does. If it does, it *matters*. And some things matter much more than others.

Numbers can tell you if something is ineffectual, or has an effect so small that it's not worth concerning yourself with—these things *unmatter*.

Then there are things that are supposed to help, but actually hurt. These things *antimatter*. Of course, the word "antimatter" has another meaning, in physics, but I'm using it differently here because there is no word in English that means "something that appears to make things better but actually makes things worse."

Sometimes there are things that matter a little bit. Sometimes it might make sense to focus on these things. This is the idea behind the idiom that "every little bit helps." But unfortunately, focusing on things that matter only a little bit has a hidden cost. The resources (energy, money, attention, time, or whatever else) you're putting into things that matter only little bit are not going toward something that matters more. Furthermore, everyone has in their minds a bunch of equilibria that they are trying to maintain, that we can think of as psychological thermostats. For example, you don't want to be too hungry or too full, so you eat when the hunger thermostat gets too low, and you stop eating when it gets too high. We also have these thermostats for happiness, productivity, and morality.

Let's take a look at climate change as an example. Nowadays lots of people acknowledge that climate change is a real problem and that something should be done about it. One way that people often talk about how they can help reduce climate change is to reduce their energy consumption. And indeed, in many places, reducing your energy consumption does have an effect on reducing climate change. But how much can you really do for the climate by changing the way you live in your own household? Governmental laws and regulations have the potential to have far bigger effects, because they will effectively force many, many people into changing their energy consumption habits. So maybe you should advocate for policy changes instead of taking short showers.

You might be thinking: why not do what you can in your own home, and support policies in your government as well? Unfortunately, because of the nature of these equilibria that we are constantly maintaining in our brains, focus on one thing will crowd out the other. One study introduced people to a potential carbon tax as well as a suggested governmental intervention to then help people protect the climate through their own actions. Merely exposing them to the "nudge" idea reduced their support for a carbon tax.[1] There seems to be a trade-off going on. Focusing on something that matters very little, even though it does actually matter, can have negative side effects. People already feel they've done "something." This is how things that appear at first glance to matter might in practice antimatter, as they draw attentional and time resources away from more important things.

This effect is more relevant for morality than for the other things I'm going to be talking about in this book. Because when it comes to your own productivity and happiness, it is precisely those things you do in your life that matter the most. (What are you going to do, support a sadness tax?) But morality involves more than just you. It involves the entire universe, until the end of time. You can find calculations, errata, and other materials at http://www.jimdavies.org/science-of-better/.

These cutesy words that I'm using very unconventionally are intended to draw interest, but also to frame the whole book. Science can help guide us to focus on the things that matter, ignore things that unmatter, and oppose things that antimatter.

Let's get started.

PART I
ABOUT YOU

1
PRODUCTIVITY

An Optimizer

I want my life to be as good as it can be.

You might think that everybody's like this. I don't think that many people would come out and say they want their life to be worse, but there is a particular personality type that actively tries to make things as good as they can be, always searching through options in search of something better. These people are optimizers.

What does it mean to have a good life? We'll get to that. But for now, let's assume that optimizers actively try to make their life better, whatever a better life means to them.

My name is Jim, and I'm an optimizer.[1]

I'll give you an example of what I mean. At the time of this writing, I use an iPhone 6 for my cell phone. iPhones have virtual pages of app icons. Some people leave the app icons where they first appeared. But if you're optimizing icon location, you will move them around.

One kind of optimizing is to move the app icons around so that the apps you use pretty often are on the front page, and the ones you use the most often go in the "dock," where they are visible no matter which page you're looking at. So if you play *Hearthstone* a lot, it would make sense to put it on your front page.

My optimization of this is a bit different. I put on the dock and front page not those apps I use most often, but the apps that I *aspire* to use most often. If I'm hooked on a game that wastes a lot of my time, if I can't bring myself to delete it completely, I'll bury it in a folder on page three. I do this because I know that if you add some cognitive speed bumps to a task—make it just a little bit more of a hassle to do, you'll do it less often.

Throughout this book I'll refer to what matters, what unmatters, and what antimatters. When you're trying to optimize, it might *feel* like everything is equally important, but when examined more closely, and more scientifically, you find that some things matter much more than others. Figuring this out requires thinking in terms of the magnitude of something's impact. This isn't always easy to do, and requires modern tools, data, and ways of thinking that aren't natural. If you decide things with your gut, or intuition, you won't optimize very well. Emotions can often draw us to things that antimatter, so using data can help us stay away from having our emotions override reason to detrimental effect.

There is lots and lots of advice you can read on these topics already, but most of it is not based on any science at all. I know because I've looked at most of it in researching for this book. I'll give my opinion on what science says matters and what unmatters, but the main takeaway is *how* to think about these things. With new data, different values, or better reasoning, you might disagree or change your mind about what matters and what unmatters. That's great, as long as you're using science, rationality, and data to inform your decision. Although I do draw conclusions in this book, they are all preliminary.

It's the way of thinking I want to communicate: using science, respecting magnitudes, and to never stop reflecting on how you're living your life.

Personal Productivity

Productivity is a word that gets thrown around without a lot of reflection on what it really means. The root of the word invokes the notion of some "product," which, in its more crass meaning, is some commodity that can be aggressively sold to people. Its more benign meaning suggests something that is produced. A nicer notion, but it still has connotations of creating widgets of some sort or other.

For many professions, the widget idea isn't too far off. If you're a professor like me, your widgets are books, papers, and competent students. If you're a painter, you want to produce paintings. If you run a theater company, you want to produce shows—which are, appropriately, called "productions."

But sometimes productivity is less tangible. If you're a defense lawyer, being productive means handling many clients. If you're a social worker, your productivity is helping people help themselves with their lives. If you're an athlete, productivity might mean getting better at your sport.

Every endeavor has its own definition of what productivity means, and what non-productivity means. I'm using the word "endeavor" because your career or your job might not be what you care most about optimizing. You might do blue-collar work for money, but are a novelist when you're not working. You might have different standards for productivity for different endeavors in your life.

Whatever you are trying to do in your life, there are activities you engage in that interfere with those goals. How much time should you spend directly pursuing your most important goals? How much time should you rest? What counts as busy work and what makes real progress? Let's see what the science says.

Sharpening the Saw versus Cutting

Sometimes I like to think of my life as a saw. Being really productive, to me, is like using the saw to cut wood and build things. But you can't cut wood very well with a dull saw, so I have to spend time sharpening the saw, too. Sharpening the saw isn't productive, exactly, but it makes it so you're more effective when you do cut something. Sharpening the saw is improving yourself, and cutting with the saw is effecting change in the world.

This part of the book is about productivity, which is different from making yourself happy. If you're lucky, being productive makes you happy, but when I talk about being *productive* here, simply making yourself happy doesn't cut it. You have to effect change in the world beyond your own mind. So even if you like to get high and play video games all day, even if it makes you happy, it's not being productive.

Let's look at reading. Reading is a wonderful thing, providing the reader with new information about the world, new ways to approach things, the perspectives of other people, and stories and examples the creative mind can use for a lifetime. It's also fun. With all of these benefits, some people believe that reading books is, all by itself, productive, but I don't see it that way. I read because I want to *do* something with that information. Reading is not inherently productive, it only prepares you to be productive later on. Reading is sharpening the saw, not cutting.

To take an extreme example, imagine a person who does little else with their life other than read books. Even if they're reading great books, they're not using what they've learned from those books to do any good. They spend their whole life sharpening the saw and never cutting anything. There is an important caveat here—if the person *enjoys* reading, there is some happiness gain, which is important, but a fairly small contribution to the world, given how much wisdom they must have gained from a lifetime of reading!

Reading, education, training, and even networking are all ways to sharpen the saw. It makes sense that younger people should spend more of their time sharpening the saw than cutting anything. For many activities, young saws aren't yet sharp enough to cut much very effectively. Even when they do produce things, they're not all that good, for the most part, and that's okay. They are producing things for practice. They have to write many, many terrible essays, or play the violin many, many times, or draw horses over and over before they create something anybody else can appreciate.

But as you grow older, you should spend more and more of your time cutting and less time sharpening. Again, an extreme example is helpful: suppose we have a ninety-year-old man who is too old to travel. He doesn't know anybody who speaks German, but has always thought he should learn it. Finally, at ninety, he starts learning German. Your first instinct might be to say "good for him!"

But from a productivity point of view it makes no sense to sharpen the saw of someone when they're not going to have an opportunity to use that skill for cutting anything. There is nothing inherently good about learning German, or any other language. If that old man has sharpened his calligraphy skill for thirty years, he should be producing works of calligraphy, not learning new skills. The saw is sharp enough, and an older productivity optimizer needs try to cut as much as possible before they die. I feel this acutely, and sometimes try to estimate how many books I'll be able to write before I run out of time. Remember, we're talking only about productivity here, not happiness, which we will discuss later.

Near the end of one's life, one might lose the ability or drive to do much cutting with the saw. This is a time of retirement from productivity (which might or might not coincide with a retirement from one's occupation). I look at this time as one of pure appreciation. When this happens to me, my mission, if you can call it that, will be to enjoy myself and appreciate the world around me, hopefully without making it much worse in the process. A person in productivity retirement should only do saw-sharpening activities that they find enjoyable. That might be reading or learning German, but that's not because it's productive, but just because those activities happen to be fun to do.

You can further break down sharpening the saw into two categories: the first one I'm going to call "maintenance." Maintenance activities are those that help you maintain your general well-being. These include things like meditation, time with friends and family, writing in a journal, exercising, and so on. In terms of health and happiness, socializing with people you care about is about the most important thing you can do. It's one of the most important maintenance-sharpening activities. Objectively speaking, socializing should take priority over exercise. So keep that in mind, even if you're the kind of person who prefers exercise to socializing. Keeping yourself happy and healthy is an obvious prerequisite for optimizing your productivity, so at least some amount of this saw-sharpening is necessary.

But there's another kind of saw-sharpening that you can do, and that includes improving skills that you can use later in life. Things that you want to get better at, like learning to write, fixing a bicycle, or washing a horse.

How much sharpening versus cutting you should do every day should be based on an estimate of optimizing the amount of (and quality of) cutting that you will do over the course of your entire life.[2] Crudely put, this means that the best thing you should do at any given time will be affected by how much time you have left to live.

We can make an analogy with eating at restaurants. Suppose you have no plans to move out of the city you're in. If you're in the mood, you might want to explore and find a new restaurant. You might not like it, but it might be a new favorite. It's risky. But imagine that you are moving out of this city in two weeks. The benefit of exploration is much lower. You would probably try to hit all of the restaurants you already know you love in the remaining time that you have. The situation is analogous to sharpening and cutting. Sharpening is preparing for the future, and it makes less sense the less future you have to take advantage of that preparation.

I think it's good to think about all the ways of sharpening the saw that are important to you. Recently I did it for myself, and I'll describe my thinking process to give you an idea of how it works. There are several things that I want to get better at, and they include drawing, math and statistics, computer programming, writing fiction, dancing, playing guitar, cardistry, and calligraphy. I also have a lot of saw-sharpening activities that I have abandoned, like martial arts, swing dancing, and comedy improvisation.

Of the things on this list the most important ones to me are drawing, math, programming, and writing. The benefit of three of these (drawing, programming, and writing) is that I'm already good enough at those things that while I sharpen the saw I'm also cutting with the saw. That is, the drawings that I make will be useful for something (I'm illustrating a serialized fiction story in *Altered Reality Magazine*, so these drawings will be published), the programming practice that I do will probably contribute to my scientific progress and productivity, and the writing that I do I might very well be able to publish someday. Producing is also practicing, so I simultaneously cut and sharpen. That makes my inner optimizer very happy.

I want to learn more math so that I will be better at understanding scientific papers and have more mathematical sophistication with my science down the road. But the process of learning math, at this point, is not going to be anything that's productive for the world, or leading to publication. It's more like traditional formal education; pure saw-sharpening without producing anything useful along the way.

With this list of your most important ways to sharpen the saw you can try to think about what kind of schedule you want for yourself, so you can make sure you get to them pretty regularly. The bottleneck is your limited time and energy. How do you prioritize doing meditation, exercising, journaling, practicing drawing, practicing writing, learning programming, all while doing your actual job, commuting to said job, sleeping, hanging out with your friends, and having a bit of leisure time to watch new *Star Wars* content? There's no easy answer to this. If you work forty hours a week you're not going to have a lot of free time to dedicate to all of these other things. If you have to take care of young children, your time is extremely limited.

There are two ways to deal with it. One is simply to prioritize. Of the ambitions that you might have to make yourself a better or more skilled person, or to make yourself happier, some ambitions might have to be simply dropped. If there's something that you really want to get to every day, then you can give that priority, but something else might have to fall by the wayside. (I miss you, martial arts!)

The other thing you can do is to schedule different things on different days. This allows you to do more things, at the expense of slowed progress. If you want to do it this way, it's best to follow the standard cyclical notions of time: we have the hours in the day, the days of the week, and the days of the month, because it's simpler to keep track of. If there is one thing you want to practice more than anything in the world, you should try to make time every single day to practice it. If you have *seven* things that are very important to you to do, you can dedicate perhaps one half hour every day of the week to doing a different thing. Drawing on Monday, learning programming on Tuesday, and so on.

If you have *thirty* things that you want to do, then you might want a bit of time dedicated on a different day of the month to each of those activities. Although you might scoff at the idea of having *thirty* activities that you might be interested in doing, one thing that could probably be applied to everyone is reinforcing social connections with people you care about. If you can come up with a list of thirty people you most want to stay in touch with, it might be worth dedicating a half hour a day every day to making sure you've reached out to some particular person for each day of the month (you get to rest on the 31st). If today is Tammy, you send her a little note, or give her a call if you haven't corresponded with her in a while, expressing gratitude for her, or asking how she is. But if we're talking about general skill-building activities, one half hour a month spent in practice is very little time, and improvement will be very slow.

Another interesting finding from the psychology of expertise is that not all practice is equal. When I was young, I took piano lessons. My teacher told me to practice a half an hour a day. I tried to keep to this, but in retrospect I see that I did it poorly: I spent the half an hour playing the songs I was already good at playing. I was enjoying the good feeling associated with doing something well. What I wasn't doing was *deliberate practice*:

working on the things I was bad at. The most efficient way to practice, in terms of time, is to practice what you are bad at, not what you're good at. Unfortunately, this takes more effort, energy, and is less fun—which is why I didn't do it. But to the extent that you care about getting better over having fun, you should engage more in deliberate practice.

For me, I really want to get better at drawing. So when I practice drawing, I try to draw things that are going to turn out terrible, because I need to work on my weaknesses. I also play the guitar, but I decided that I don't really care about getting really good. I play guitar *only* to have fun, and getting better at it, though it is happening, isn't very important to me. So when I "practice" the guitar, I allow myself to play songs I already am pretty good at. I'm just happy I can play some Paul Simon songs. (It feels good not to optimize *everything*!)

Keep in mind that some endeavors *matter* more than others. This is important to think about when you prioritize your list of saw-sharpening activities. You might find yourself with a goal to complete a difficult video game, or get better at playing pool. I understand the draw of these activities, but it's important to reflect that the skills you gain when you get better at video games or pool are limited, and you might want to prioritize other skills more. I'm a sucker for this, and have to keep myself in check. There's a game I like to play—*Hearthstone*—and it's so easy to get caught up in the ambition of improving your rank, getting better decks, etc. It *feels* productive when I make progress on these things, but it all unmatters. I'm climbing ladders that don't lead anywhere.

For purposes of this book, doing anything productive is cutting with the saw. If we are going to optimize productivity, we need to cut effectively. We need to do it at the right times, for the right lengths of time. Likewise, we need to cut where it's most efficient and effective.

Your Mind, Your Time, Your Attention

One important aspect of being productive is being able to completely focus on what you're doing. Cal Newport calls it "deep work," and defines it as "professional activities performed in a state of distraction-free concentration that push your cognitive capabilities to their limit. These efforts create new value, improve your skill, and are hard to replicate."[3]

Working without distraction seems to be getting increasingly difficult. Computer technologies are so advanced that a large and growing number of people on Earth regularly carry around supercomputers in their pockets. (I'm talking about smart watches and smartphones.) Email, web surfing, social media, and video games all vie for our attention, and many of us are unable to resist. Part of this is because there are lots of very smart people designing these systems to optimize your paying attention to them. Companies have created applications that, in many cases, are diabolically designed to draw attention to themselves. Design ethicist Tristan Harris said, "There are a thousand people on the other side of the screen whose job it is to break down the self-regulation you have."[4] These smart people are working hard to make technologies designed to keep you from working hard.

Things are constantly happening in our environments, and we ignore most of them. The ones that are likely to become distractions are ones that are surprising or salient. Something can be salient for many reasons. You might get excited when your phone makes a *ping* sound—it's distracting because you know you got a text message, and you're used to being rewarded by reading texts. A similar sound from a truck backing up outside is not as distracting because it doesn't mean anything to you: it is a signal of something that is irrelevant to your life.

In 2012, the average "knowledge worker" spent 60 percent of their workweek dealing with email and searching the Internet.[5] Young people have a reputation for always being on their phones. A study of students showed that most of them weren't even able to get through ten minutes without checking some kind of device, be it a phone, tablet, or e-reader. Some students switch tasks every two minutes![6]

Chris Bailey, author of *The Productivity Project*, confesses to how he used to start his day: "After I woke up, I would immediately reach for my phone and then mindlessly bounce around between my favorite apps in a stimulation-fueled feedback loop for about thirty minutes, continuously bouncing around between Twitter, email, Facebook, Instagram, and several news websites until I snapped out of my trance."[7]

The reputation that younger people have with constant interaction with technology is true: the younger someone is, the more likely they are to multitask with texting, music, television, and other technologies. But careful studies show they aren't any *better* at doing it.[8]

Distractions come in a few forms. Some distractions are forced upon you. If someone walks into your office to tell you that vampire rappers are the original sucker MCs, then you either have to deal with them or risk social rudeness trying to get them out. Either way, you are distracted.

YOU HAVE TOO MANY THINGS TO DO TO WORRY ABOUT OTHERS' LAMENESS: A LESSON FROM THE 2 LIVE CREW

One source of distraction that many people struggle with is the desire to complain about and correct other people's actions. There have been many times in my life when I've wasted far too much time on fruitless Internet debates. Every time I read a bit of newspaper (which isn't often), I have to resist the urge to get online and rant about what garbage it is, with a point-by-point analysis of what the paper is doing wrong. For the most part, actually doing this would unmatter. I could fill my life complaining about other people's products, and never produce anything of my own.

I found solace in the most unlikely of places: a track from the 2 Live Crew's album *As Nasty As They Wanna Be* called "I Ain't Bullshittin'." This strange track seems to be a drunken rant about some other MC, but it contains a line that basically says that this MC should create his own music, rather than criticizing the 2 Live Crew.[9]

For decades, this line has helped keep me focused. Every time I feel the temptation to spend my mental energy and time criticizing others, I remind myself of all of positive things I can do in this world, and that every minute I spend criticizing others—and even *thinking* about criticizing others—is time taken away from what I can accomplish in this world.

As for the album in general, I can't recommend it unless you have a high tolerance for misogyny. Beats are sick, though.

Incoming phone calls are distracting, even if you don't answer the call. You can turn off your phone's sound notification and vibration, but many people are loath to do this because phone calls are sometimes important. You should know that on many phones there is a function that allows a call to come through even if the phone is on silent if the caller is in your contact list, *and* they call twice in a row. This allows you to have your phone quiet a lot of the time—at the very least, when you're sleeping, but allowing emergency communications to come through.

Similarly, text messages are external interruptions, but they are the kind that you can turn off. That is, when you want to concentrate, you can make your phone silent. (And by silent I mean you are not aware of any notifications. Phone quiet, not vibrating, screen down.) In general, notifications on your phone that interrupt you should be kept to an absolute minimum.

The same goes for email. Reading email puts your mind into a stressed mode, and cutting yourself off from it makes you more relaxed.[10] Just about everybody should turn off notifications for email, and check it only periodically.

THE EFFECTS OF DISTRACTION

So how bad is it, really, if you check your email in the middle of working on a paper? Science shows us that it's pretty bad.

The most obvious problem with checking your social media (and I'm including email in this concept) is that while you're doing it you're not doing something else. This is the "opportunity cost." The cost of what you're doing isn't just time, money, and other resources you invest to do it, it's also the cost of not doing all of the other, better things you could be doing with your time.

So if you play an hour of video games, that's an hour you aren't being productive.

But it's worse than that. Because when you switch from one task to another, say, from manipulating a spreadsheet to playing *Mario Kart*, or back again, you are just a little bit dumber for a short time. If you're switching a lot, you're just plain dumber a lot of the time. There are many studies that show that if you're rapidly switching from one task to another, you suffer performance deficits in at least one of the tasks.[11]

Multitasking

Doing more than one thing at a time is often called "multitasking." Sometimes this is true multitasking, such as when you're jogging while listening to an audiobook. But in general, when you multitask, you're worse at all the things you're doing. For example, talking on the phone while driving makes your driving worse. (Laws that allow hands-free cell phone use in the car don't make any sense, because it's the *talking* that distracts you, not whether or not you're using your hands.)[12] Because we crave constant stimulation, multitasking "rewards" our brains with novelty, creating a constant dopamine rush, effectively training your brain to enjoy being distracted. At the same time, it causes stress (increased cortisol) and increased adrenaline (putting you in fight-or-flight mode).[13] Though it can be exciting, it is not good for your long-term prospects.

But most of the time when people talk about multitasking they're really talking about rapid task switching. If you're texting while watching television, for example, what you're really doing is paying attention to the show, then the text conversation, and back to the show. While you're typing out a text, you miss what's going on in the show.

I often listen to audiobook novels while I walk my dog. This isn't a problem, though, for two reasons. First, even if my "performance" drops for dog-walking and novel-listening, it unmatters. The stakes are really low. Second, these tasks don't interfere with each other very much because they use different parts of my mind. Dog-walking involves watching my dog and navigating my physical environment, and listening to a novel involves verbal comprehension and imagination. Interference is worse when two tasks are competing for the same functions in your mind—listening to (and comprehending) an audiobook while you read a magazine is impossible, for example. When the tasks are similar you get much higher costs. When you're watching TV and texting, the two tasks are visual and verbal, and when your mind is occupied with one, it has no resources left to process the other. Experiments in which people did a visual task while engaging in a voice chat experienced less of a performance drop than they did for tasks that drew upon the same kind of thinking.[14]

There are exceptions, times when multitasking is okay. Listening to instrumental music while you study or do computer programming can sometimes help.[15] Doodling during a boring lecture helps you retain more of the lecture. This seems to be because when an important task is really boring, your mind actively tries to search for something else to do. If you're listening to a boring lecture, your mind might wander, looking for a more interesting thing to think about. In one study, the students who were instructed not to doodle often detached completely from the lecture, getting nothing out of it. But the doodling students were less bored. With part of their attention on the lecture, and part of it on doodling, they were better able to attend to the lecture content.[16] Now we're talking doodling here, not making some masterpiece. It has to be relatively mindless or you'll be too cognitively engaged in the drawing to have any attention left for the lecture.

So if it's a boring, cognitively engaging task, you can do another task at the same time, as long as it's *not* cognitively engaging—walking, instrumental music, doodling. But if your main task is cognitively demanding and engages your interest, you would be best off not doing anything else at the same time.

For noncognitive tasks, multitasking is a good way to optimize your life. Let's take exercise, which I personally find so boring that I can't get myself to do it for very long. My solution has been to play squash, which is a game that I find fun (also it's a game I can play year-round). But if you are into weight lifting, or running, or some other repetitive, non-game exercise, listening to music helps you enjoy it more, and might even make you exercise harder or longer.[17] If you have a long commute, listening to podcasts or audiobooks might help pass the time, and you get to read a lot of great books. Since I started listening to audiobooks, in addition to traditional reading, my book consumption has tripled. (Audiobooks are not as bad as having phone conversations in the car because they are not interactive and can be turned off pretty easily or safely ignored when driving gets hairy.)[18]

So if you're going to multitask, it should be with tasks that don't use the same parts of your mind, because switching tasks has a cost. The trick to it is not to multitask anything that really requires *thinking*. Don't listen to the radio while you write a report. Don't have the TV on while you're doing your taxes, but listen to music while you exercise.

Task-Switching Costs

Scientists have put numbers to these costs, in terms of time. So what is the task-switching cost? Measurements range from 200 milliseconds to 25 minutes.

Wha—?

With a range this big, we need to dig a little deeper into the research. The 200-millisecond measure comes from psychologists studying people doing tasks on a computer screen. For example, they might show people a series of faces, some happy, some sad, some male, some female, and instruct them to *identify the gender* of the face as fast as possible. The task switch is to start *identifying the facial expression* instead. In this study, switching from gender-identifying to expression-identifying cost about 200 milliseconds (one-fifth of a second) in response time.[19]

The 25-minute measure, on the other hand, came from real-world, observational studies of interruptions at the office. You might be working on a budget, and a coworker comes by your cubicle to tell you about how they binge-watched *Fringe* over the weekend. This kind of thing happens all the time, and the U.S. economy suffers an estimated loss of $650 billion a year from these distractions.[20] Computer scientist Gloria Mark found that when people got interrupted like this, it often took them a while to get back on track—often about twenty-five minutes.[21] It often takes over a minute to even remember what you were doing before you got interrupted, and sometimes people never got back to what they were doing at all.[22]

We often are vaguely aware that distractions and interruptions make us less productive, but the scary thing is that a lot of times we don't. Often people think they are just fine at multitasking, and are oblivious that their performance suffers.

Can You Get Better at Multitasking?

You have to wonder what the long-term effects of multitasking are. On the one hand, frequent multitaskers get a lot of *practice* multitasking, so maybe they get better at it. On the other hand, perhaps all that training is rendering them less able to focus when they need to, because they are "addicted" to the dopamine rush of the constant novelty associated with task switching. Really, these are separate, independent questions: does habitual multitasking make you better at multitasking, and does habitual multitasking make you worse at focusing?

Multitasking seems to be difficult for two underlying cognitive reasons. The first is that when you change to a new task, your mind has to prepare to do it. Think of your mind like a desk. When you're working on your taxes, you put the stuff on your desk that you'll need—your tax forms from work, your donation receipts, and so on. When you want to do some painting, you clear all that stuff off and get out your painting stuff. Your working memory is a bit like this. When you switch from one task to another, this "advance preparation" means activating in your mind the right representations, responses, and internal processing mechanisms you'll need. Because they are somewhat different from the previous task, it takes some time to prepare them, causing a time delay, and sometimes a performance reduction as well. The other problem is that after you switch to the new task, the activation of all of those things related to the *previous* task are still active and potentially interfering. These take time to become less active, in a process known as "passive decay." It's the combination of these two factors that generate the lion's share of our task-switching problems.

Advance preparation is related to fluid intelligence, and heavy multitaskers seem to be a little better at this aspect of multitasking, though we do not yet know if they got this way through practice. The studies to date are correlational, and don't shed light on causation.

Passive decay, in contrast, is unrelated to intelligence and seems to be independent of how much multitasking a person does, which suggests that it doesn't get any better with training. This is a large enough factor, though, that we see in some studies there seem to be no differences between heavy and light multitaskers.[23] Unfortunately, the science on this isn't conclusive. Another study shows that frequent multitaskers have higher task-switching costs and are worse at ignoring irrelevant information than low-multitaskers.[24] But again these studies don't show whether multitasking causes poor task-switching, or that people with poor cognitive control can't help but be distracted more often, and multitask more.

Adam Gazzaley created a video game called *NeuroRacer* to test to see if practice at multitasking could make people better at it. Specifically, the game required you to drive while attending to some distracting signs while ignoring others. Playing this game improved people's multitasking abilities.[25]

In sum, the science of getting better at multitasking is inconclusive. We know that frequent multitaskers are worse at it, suggesting that you probably are not going to get much better at multitasking by doing it more. In the modern world, so many people are so distracted that the ability to concentrate is like a superpower. Learn to harness it, stop multitasking, and you will have a competitive advantage.

Work at the Office or Work at Home?

One way to get more concentrated work is have your door shut at work, but that's only possible if you're lucky enough to have a door (see sidebar "What a Workplace Should Look Like"). But another option is to work from home, or telework. Many of us were able to experiment with telework during the 2020 coronavirus pandemic. You'll likely experience predictable benefits and drawbacks when you spend more than half of your workweek teleworking.[26] This is assuming that you can concentrate at home, don't have to do your own child care, have a good Internet connection, etc.—something not everyone had access to during the pandemic. Further, many people have only laptops at home, which have smaller displays. Using a large monitor helps your productivity because you don't have to constantly switch between windows—you can just move your eyes.[27]

On the bright side, at home you'll have less work-life conflict, greater job satisfaction, less stress from meetings and interruptions, less exposure to office politics. Telework also reduces commuting, which means more time for work as well.[28] You'd think telecommuting would save energy, but this is doubtful.[29]

On the downside, you'll have lower quality relationships with your coworkers. A study of scientists showed that teams have more impact than people working alone, and that collaborators had the most impact when they were physically close together—preferably in the same building.[30]

Now, communications technologies have gotten so good that you can work from home and really stay connected to the people at work. The problem with this is that the more intensely one engages with this, the fewer benefits one derives from teleworking in the first place! One gets more interruptions, and gets involved with discussions they might otherwise have been able to avoid.[31]

WHY ARE PEOPLE SABOTAGING THEMSELVES?

In a typical work environment, people average about two to three minutes on a task before switching to something else. This is fairly rapid task-switching, and performance suffers from it. The younger generation checks its mobile phones every fifteen minutes, and about 75 percent of young adults sleep with their phones nearby, *with vibration or the ringer on* so as not to miss nighttime alerts![32] This is a terrible idea.

There are two important parts of removing distraction from your life. The first is the removal of external interruptions—other people, and your devices, alerting and interrupting you. The second is the removal of internal distractions, which is your own mind sabotaging your concentration by giving in to temptation to check your phone, daydream, or otherwise screw around when you should be working. Most interruptions are of the internal kind. A study of workers found that over half of the interruptions involved "checking in" with social media or something else, with no alert or notification prompting them to do so.[33]

So if multitasking is distracting, and impairs productivity and performance, then why are people doing it so much?

When you ask people why they multitask, many say they believe it helps them. Without their beliefs in the right place, they have no motive to make a change. Hopefully this book will fix this for you. People think they're good at multitasking. Young people believe they can successfully juggle six forms of media simultaneously.[34] But in truth, heavy multitaskers are worse at multitasking than light multitaskers.[35] This is really important: *you cannot trust your feelings about your productivity when it comes to multitasking*. But even if you know it's bad, it's hard to stop.

Sometimes people switch tasks because they get frustrated with what they are doing. They don't know the way forward, and that doesn't feel good, so their minds start looking for something that is more rewarding. Switching tasks can feel amazing, and gives the *feeling* of productivity because you're responding to so much.[36] You get a notification, and your brain anticipates that it might get a reward. It might be a message from someone you like, or a cool picture. This anticipation can build if you don't check the notification, causing you to obsess about checking in, causing further internal distraction from what you're trying to concentrate on.[37] So you give in, and check the notification. Your curiosity is satisfied, but you've distracted yourself from what you were doing, and also made the checking more of a habit, making it harder to resist phone-checking in the future.

Unconsciously, we discount future rewards, like eventually getting some major project done. Switching to another task feels rewarding, and we can get that reward right now. This is how we can constantly switch between tasks, feel great and productive, and yet never get anything accomplished.[38]

THE IMPORTANCE OF WRITING

As an academic scientist, much of my productivity is based on writing. I run experiments, I make computer programs, I have meetings and discussions with people, but eventually these are in service of creating scientific papers and books, which means writing. Other people do other things, like dance or painting. There are lots of productive activities.

But there seems to be something special about writing.

I have a writing assignment in the class I teach, which has over 1,000 students in it. Why would I put myself through this?

Writing can help people deal with emotional issues in their lives. Jamie Pennebaker ran an experiment where people wrote about upsetting, traumatic experiences they had in their lives—preferably ones that they haven't told anybody about. The people who did this went to the doctor fewer times over the course of the next year.

And the kind of writing mattered—those who just complained about the trauma, or ranted with anger didn't feel any better. Catharsis doesn't always work, it can just remind you of what makes you upset and make you upset all over again. The biggest benefit went to people who tried to make sense of what happened, and had increasing insight over the course of a few days. He found that dancing and singing about emotions didn't work. It had to be creating words, through writing or talking into a recorder. If you're dealing with a traumatic event in your past, you might want to write for fifteen minutes a day about it, trying to figure out why it happened and what benefit you could get from it.[39]

Writing is hard for people, and I can't find any research looking into why that is. If you read books about writing fiction, for example, it seems that a third of every book is about motivational strategies to get your ass in the seat to actually write, and the rest is about how to make the writing actually good. In contrast, books about painting do not dedicate pages to trying to get you to paint. But people have a strange block about writing. This is a shame, because writing is so good for you.

> When it comes to writing fiction, there is also this strange idea that it's purely a matter of talent. Compare this to, for example, playing the violin. If someone picks up a violin for the first time, maybe watches a video about how to play, and then tries to play a song, they don't record that first attempt and then try to peddle it to record labels. People understand that learning to play the violin, like drawing, is difficult and takes a lot of practice before you're good at it. They understand that for a while your violin playing will be poor.
>
> But with writing, you'll often see people decide at some point in their adult life to try writing, say, a short story or a novel, and then show it to friends, hoping to be told that it's brilliant and should be published. Of course it's terrible, just like one's first playing of the violin is terrible. But people will conclude from its terribleness that they are bad writers, and never write again.
>
> Writing takes practice, just like violin playing, but for some reason I don't know, people don't appreciate that.

If you want to optimize your productivity, you have to fight this instinct.

People are tempted by multiple, competing needs all day long. The most common desires that people get include wanting sex, sleep, or food. But stopping work, including checking email and social media, wanting to watch television or listen to music, are also very common distractions. This was found in a study where they gave people beepers that randomly went off, and asked people what they were thinking about.[40]

To understand this, let's back up and look at why anybody does anything: people have various brain functions simultaneously trying to control the body and what the conscious mind focuses on. Each of these works in a different way—you might think of them as valuing different things. Often, the behaviors being pushed by those brain areas are different. It's like a board meeting, where the board members disagree on what the company should do. But instead of reasoning, each just shouts as loudly as it can. The board member (or brain function) who shouts the loudest determines what the company (or your body) does.

We can look at multitasking as an attempt to gratify multiple needs. The varying strengths of those needs determine what we turn our attention to at any given moment. Over time, a social "need to connect" might grow in one's mind, making turning attention to chatting with someone or checking in on social media more likely. When this temptation is indulged, the need is satisfied, and becomes low-powered enough that other needs (such as a cognitive one, such as curiosity about a book you're reading) end up determining behavior in the next moment.

Suppose a student is studying because of a cognitive need to understand class material. As the studying progresses over time, other needs grow in strength. The need for social engagement is held at bay for only so long, but when it gets stronger than the cognitive need, the student checks their social media accounts to see if they got any "likes" recently. They got a few, so they are satisfied and return to studying.

Or perhaps the studying is boring, or depressing, and the student has an emotional need to be happier. The student turns on music, which improves mood. The emotional need is satisfied, and the student studies while music is playing, even though it renders the studying less effective. A study by Zheng Wang shows that this happens, but the emotional need is subconscious. That is, students don't turn on the music *because* they will be emotionally satisfied, but they end up getting emotionally satisfied anyway.[41]

Crucially, these self-interrupting behaviors are reinforced by the gratifications they cause. If listening to music while studying makes the student happier, then they are more likely to listen to music in the future during study time. This is conditioning.

It's also important to realize that there is often a mismatch between gratifications sought and those actually obtained. That is, sometimes we engage in a behavior to satisfy some need, but that behavior actually doesn't pay off the way we want it to. This is clear in cases of uncertainty—a person might attempt to initiate a conversation with a stranger on a bus because of a social need for human connection, but if that stranger blows him off, his need is not satisfied. Or a person might check Facebook to get a mood boost, but come away feeling worse because they attend to someone else's life that looks so much better.

Another way to think about our constant task-switching is in terms of foraging for information. Just like an animal will spend some time at a location until it seems to be exhausted of food before going off to another location, we too gather information at a "location," like Instagram, and then feel the urge to switch to another source of information, like television or email. The instincts that guide this behavior evolved in a world where switching locations had a higher cost. But in today's world, we can have a hundred information locations *on our phone*. The cost to switch from one to the other is so low that we overestimate the utility of switching. We also get bored more quickly than we used to with the information we're getting from one source. Unfortunately, our instincts lead us astray: we often are better off staying at one particular information source than switching. If you follow your feelings, you are likely to go down the rabbit hole, jumping from app to app on your phone, wasting the day away, not getting anything done, feeling productive the whole time, but feeling bad afterward![42]

WHAT A WORKSPACE SHOULD LOOK LIKE

Helena Jahncke ran a study in which she simulated two open-plan office work environments: high-noise and low-noise. Participants in the high-noise environment rated themselves as being more tired and less motivated to work than those in the low-noise condition. Noise impairs cognition.[43]

Open-plan offices reduce privacy and are much louder. They allow easier interaction, more creativity, and are more flexible in terms of change. They also are more stimulating, so they can be better for routine, boring tasks.[44] But whether these benefits are smaller than the penalties is controversial.[45]

A study by Craig Knight showed that worker performance was better in "enriched" offices than in sparse, minimally decorated offices. Performance was best when people could decorate their workspace however they liked, but even top-down placement of plants and art made for a more productive workplace.[46]

Procrastination

Try to be the kind of person who doesn't need to have others remind you to pursue your own goals. Part of being like this is avoiding procrastination.

Procrastination is what we call it when your immediate needs for gratification win control of your behaviors at the expense of your longer-term goals. Let's take an example that many people have to deal with: doing their taxes. Let's assume that you are going to do your taxes at some point, it's just a matter of when. There is a span of a few months before the spring deadline when you have all the information you need to do them. If you don't enjoy doing your taxes, then it's an easy thing to procrastinate. Even cleaning your house might seem more promising, in terms of expected reward, than doing your taxes, at every given moment.

We feel the temptation for procrastination when we expect to feel worse doing what we *should* do than some other, more fun thing you could be doing instead. Which means that, ultimately, procrastination is about emotion. The things you're likely to do when procrastinating, are, by definition, temptations: things that feel good but don't move you toward your greater values and life goals.

Thirty-one percent of people admitted to procrastinating at least an hour a day, and another 26 percent said they wasted more than two hours a day. And these are just the people willing to admit it. My colleague Tim Pychyl found that university students wasted about a third of their waking hours procrastinating.[47] So if you have a problem with procrastination, join the club.

Tasks are much more fun when you think they can be finished, offer novel challenges or things to learn along the way, and you get feedback according to how well you're doing while working on it. Tasks you are tempted to procrastinate are those that are boring, frustrating, ambiguous, difficult, or lacking in meaning or a good reward structure.

We procrastinate because it feels good. But it's a classic short-term gain with a much worse long-term payoff, like eating junk food now and gaining weight later. That's why you should try to eliminate procrastination from your mental diet. If you want to abandon a task, do it for the right reasons.

Not every problem with getting things done is due to procrastination—some people have trouble starting projects, and some have trouble finishing them.[48] In these cases, a coach, either formal or informal, can help with these specific problems.

PERFECTIONISM

Perfectionism—the classic answer to the dreaded interview question "What is your greatest weakness?"—actually has two different kinds, one good, one bad.[49]

The good perfectionist simply has high standards, a striving for excellence, and persistence that encourages better performance.

But perfectionism gone too far is associated with anxiety, depression, stress, and test-taking anxiety. When I am considering taking on graduate students, I see perfectionism as a big red flag. I attribute some of my modest success to my non-perfectionism. When I was in graduate school, I would write crappy drafts of things, turn them in to my supervisor, who would tear them apart, then I'd write another draft, and go through two or three iterations of this while the students sitting next to me were still "perfecting" their first draft. In many projects, not just writing, you risk excessive polishing of parts that will be cut from the final product, which antimatters. It's rearranging deck chairs on the *Titanic*.

The bad kind of perfectionists probably get less done because they keep polishing things past the point when the improvement of the product for a given amount work gets negligibly small. Do you want to make one perfect project, or three really good ones?

I tell my students that their job is to give me shitty first drafts.

The perfectionists react to this with surprise and horror.

2

HACKING YOUR BRAIN, HACKING YOUR LIFE

The mind and brain are incredibly complex, but in this book I'm going to simplify things just a bit into a structure that is mostly true and easy to understand. Let's ask a fundamental question: what is happening, in your mind and brain, that determines what you do at every moment? Why do people do anything?

The basal ganglia part of your brain controls action. It is your procedural memory. Think of it as storing the codes for everything you can do. This includes everything from simple physical actions like reaching for peanut butter to complex actions like break dancing, and also includes purely mental procedures, such as the steps required to do long division in your head.

Interestingly, all of the things you could do are constantly trying to happen—the basal ganglia actually works by suppressing all of these actions in the supplementary motor area, and selectively allowing just one at a time to happen by stopping the suppression.[1] When you're cognitively taxed, your habits, encoded in the dorsolateral striatum (DLS) part of your basal ganglia, can take control of your body with the habits stored there.[2] Habits form because of repetition and conditioning (associating the action with something pleasant or unpleasant). In this way the basal ganglia can control what you do all by itself. This is the habit system.

Let's review the systems involved in how your mind chooses what to do next, and come up with cute names for them. These systems compete for control of your body. If your brain is functioning properly, each of these systems will be in control at the right time.

The Habit System tries to get you to do what you're used to doing, and isn't particularly goal-directed. One of the marvelous things about your mind is that it can automatize things. This is how you can learn to talk to a friend while driving, or tie your shoes while thinking about what to have for lunch. There's nothing inherently good or bad about habits in general—you can have good ones or bad ones. The important thing to understand is that in many situations, you have habits waiting in the wings to take over, and your basal ganglia will be pushing you toward engaging in those habits. If your cognitive system is otherwise distracted, you are more likely to engage in habitual action. What I casually refer to as "habits" can be genetic instincts or learned repetitive behaviors. Habits are less sensitive to reward in any particular context—that is, you might engage in habit without your mind considering very carefully whether it would be appropriate in this particular situation. The environment cues the habitual behavior, but it's less like a deliberative decision.[3]

The Cognitive System uses something variously called willpower, self-control, executive control, and impulse control. It tries to get you to behave in ways that are in line with your long-term goals and "better judgments." They involve conscious, slow thinking and, sometimes, reasoning and goals. It's primarily located in the cortex, and it has a function to suppress subcortical drives when needed. In fact, most of the prefrontal cortex's connections to the rest of the brain are inhibitory—there to prevent the rest of the brain from doing what it normally would.[4] When you say "I ate more than I wanted to," you are usually referring to your whole self with the first "I" and your cognitive system with second "I." The cognitive system also triggers the basal ganglia, and with sufficient input, you can override the habits that the basal ganglia would engage in if left to its own devices. In the scientific literature, the cognitive system is often referred to as the new brain, System 2, or the rational system. Motivations originating with your cognitive system can feel less emotional. You might start engaging in a behavior because of some expected outcome, or some goal you have, but after it is repeated over and over it gets turned over to the habit system, and thereafter requires the cognitive system to step in to stop from doing it.

The Reward System tries to get you to do things that avoid bad feelings and facilitate good feelings, where "feelings" refer to any valanced (pleasant or unpleasant) mental states. When you're trying to choose between eating raw vegetables and potato chips, you can feel your reward system making its opinion strongly! Again, eating either one involves input to the basal ganglia, but the reward system sends information from a different part of the brain than the cognitive system. The cognitive system is evolutionarily newer and in the neocortex, and the reward system is older, shared with many other species, and is from the emotional areas in the middle of the brain—the mesolimbic system.

The Mind-Wandering System goes into effect when the task you're currently engaged is boring or not particularly cognitively demanding. Neuroscientists call this the "default mode network," and it allows people to refocus on longer-term tasks important to their cognitive system (among other things). It is also responsible for a lot of anxious thoughts. There seems to be a trade-off between the cognitive and mind-wandering systems, such that if one is active, the other is less active.[5]

Let's make these different origins of action clear with two examples. The first is when the different parts of your brain are trying to get you to do different things.

Suppose Julie is out walking her dog, and she hasn't yet planned dinner. She tends to take the same route every day. Today is also the day when everyone has put out the recycling to be picked up. She notices a can that someone has tossed on the sidewalk. There is also a cute puppy across the street. Now we can think about all of the different parts of her brain that might have different opinions on what she should do next: keep walking, pick up the can, go pet the puppy across the street, or maybe plan dinner?

The habit system encourages her to continue walking around the block, because that's what she does every day.

The cognitive, rational part of her brain, in the dorsolateral prefrontal cortex, thinks that she should pick up the can and throw it into the recycle bin. This might accompany a feeling of moral obligation, but it's not like she's expecting to get a rush of pleasure from it. She "knows" that she should do it, but isn't primarily driven to by an expectation of pleasure, or to substantially decrease some negative emotion such as dread or anxiety.

The reward center of her brain thinks that she should vary from her normal route to go meet the cute puppy, which is expected to be pleasurable.

The mind-wandering system is trying to get her to think about what to have for dinner, because she forgot to shop, and now the lines at the store

CHAPTER 2: HACKING YOUR BRAIN, HACKING YOUR LIFE

are going to be long. If she engages in this activity, her habit system will take over control of her body, and she will continue to walk her normal route. Only her mind will be elsewhere.

All of these (and possibly more) factors are relevant to which thing Julie actually does. All of these drives have relative strengths in her mind, and the one that ends up the strongest, in that moment, will be the one that she actually does. In Julie's example, we can see how the basic motivational systems in her mind might be trying to get her to do different things. Some of these systems can, at any given time, be dormant—in fact, they might be literally asleep for a few moments. Different parts of our brain can display neural indicators of sleep while "you" are awake.[6]

In Julie's example, we can see how the different brain functions are working toward different behaviors. In my next example, we can see how the *same action* might be motivated by the different motivational systems. Let's suppose you grab a piece of chocolate, put it in your mouth, and eat it.

First, **habit**. The basal ganglia responds to perceptions of the environment, and can do things out of habit. If you have just eaten several chocolates, for example, your habit system might make you take one more. This is more likely to happen if the other things that can affect the basal ganglia (such as your attention and cognitive system) are otherwise occupied. You might have experienced eating a bag of chocolates or something while watching a movie, and at some point reach in to find that you'd eaten them all without even realizing it. That's habit.

Second, conscious intention through the **cognitive system**. You might be about to go on a long hike, and reason that you need the sugar for energy. You decide to eat chocolate, not because you expect it to taste delicious, or because you have an urge (you might not even like chocolate), but because you believe it will help you with a longer-term goal. Your frontal cortex activates the actions in your basal ganglia, and the chocolate gets eaten.

Third, **compulsion**. As we'll talk about in more detail later, the reward system has several components, two of which are liking and wanting. Compulsions are from the wanting part. You might be hooked on eating chocolate. You are not eating it with a particular expectation of pleasure, but you feel an urge to eat. There might be a small amount of tension or anxiety that is relieved by eating it. This is mesolimbic, and possibly dopaminergic.

Fourth, **pleasure-seeking**. The second part of reward is liking. You see chocolate, and expect to get pleasure from eating it, so you reach for it and eat it. Your cortex might have been recruited to justify this action (e.g., "I went to the gym, so I deserve it!") but the original motivation is the seeking of pleasure. This is also mesolimbic, and more opioid.

Because wanting and liking are often felt together, I'm going to group them as reward—but they can be separated, particularly in addictions.

So far I've ignored learning, which is another story entirely. Conditioning and repetition can cause habits and compulsions to form, but later, when the actions are taken, conditioning and the immediate context is less relevant to whether the action is taken or not (in cases of habit or compulsion).

I just want to remind you that this is a simplification. I'm not talking about drives like hunger and the need for sleep, for example. The whole thing is really complicated. But now that we've got four systems with cute names, we can think about what you can do to each of them to get yourself doing more of what you want, to make your life more like you want it to be.

How to Hack Your Cognitive System

When you are trying to do something important, and your phone pings a notification sound, your brain has a choice to make. Do you stay on task, or do you check to see what news your phone brings? I say "your brain" and not "you" because often what you end up doing isn't the result of a deliberate choice at all. If you pick up your phone, you might not have consciously chosen to do so. You're reacting without even considering. It might happen so fast that your cognitive system doesn't have time to even weigh in.

Even if you know about how damaging constant interruptions are, and wish to not give in to them, it can be hard to avoid them. You remember you have a notification that hasn't been checked. It sits there, taking up valuable working memory in your mind, causing a mild anxiety. Eventually, the tension gets so distracting that you just have to check it. This falls under a larger problem of "weakness of will." Cognitive science can help us look under the hood a bit and see what weakness of will actually means in the mind.

I had a friend who was doing consulting for a snack food company. He told me that when he'd go to a meeting, there would be a big pile of these snacks in the middle of the table. At the start of the meeting, everyone would avoid them. But as the meeting dragged on, everyone's fingers would start itching toward the center, and eventually everybody was eating snacks and would go into a sugar coma. They'd all feel regret.

This is a classic example of weakness of the will. What's curious about it is that we might describe it as doing something we didn't want to do. But if we didn't want to do it, how could it have happened? Philosophers call actions that are against your better judgment "akratic actions."

Your "better judgments" are a function of your cortical brain regions, associated with reasoning, memory, and belief. The part of your brain that actually initiates action, the supplementary motor area, decides what to do based on the judgments of the cortical brain areas, but others as well, such as the reward system, the emotional system, and the habits (and related behavioral inclinations) stored in your basal ganglia. These subcortical systems do not have representations of judgments as we normally think about them. The supplementary motor area takes information for all of these areas and "decides" what to do. And sometimes the subcortical "opinions" on what to do win the day.[7]

These subcortical areas are much older than your cortical areas. They're also faster. People are about a fifth of a second faster at judging the tastiness of food than judging healthfulness.[8] We share versions of the emotional system, reward system, and basal ganglia with lizards, who have no "better judgments" at all.[9] Rather than scratching our heads, wondering why we act against our better judgments, we should be thankful that we even have better judgments at all! For millions of years, our ancestors didn't.

These subcortical inputs are very powerful. They respond to immediate gratification, whereas your cortical areas respond relatively strongly to long-term reward.[10] Even a ten-minute delay makes something feel like a long-term reward.

What we call discipline, self-control, or willpower is best described, neurally, as the relative strength of your cortical areas to suppress your subcortical areas when it comes to action selection. It is a self-initiated suppression of impulses in the service of longer-term goals.[11] We need discipline because when we're doing something hard, our mind tries to find something else that's more rewarding.

Acting badly isn't always caused by a weakness of will. Sometimes you can be "irresolute." Let's say you have an intention to avoid cake at a party. You get there, and you find yourself thinking of reasons to eat the cake:

you haven't had cake in a long time, you deserve it, you went on a run that day, it would rude to the host not to have any, the cake was baked by poor people who need the economic support, etc. Eventually you actually *believe* it's okay to eat the cake. You've talked yourself into it! You can see how being smarter allows you to think of *more* justifications to eat the cake. Thanks, prefrontal cortex.

Some say that this is different from a weakness of will. Rather than giving in to temptation and doing what you know is wrong, it's your old brain recruiting the rationalization powers of the new brain to work against itself by changing its mind about what's right and wrong.

People with more willpower have a lot of advantages. They are better able to deal with stress, adversity, and conflict. They are happier, healthier, make more money, maintain better and longer-lasting relationships, and are more successful in their careers. Willpower is more important to grades than intelligence. Oh, and they live longer, so they can enjoy these things for a longer time.[12]

When you're under stress, it takes more willpower to do what you want to do, so one way to take better advantage of the willpower you have is to reduce stress, both in the moment and in your life in general. Getting enough sleep is the easiest and most pleasurable way to do this.

Another way to increase your willpower in the moment is to forgive yourself when you slip up and give in to temptation. Just saying to yourself, "It's okay, everybody makes mistakes, I'll try to do better next time," increases your chances of doing exactly that in the future. Beating up on yourself actually reduces your drive to be better.[13] You've heard of the golden rule? Here's my platinum rule: Do unto yourself as you would do unto the people you love.

Can You Increase Your General Willpower?

Maybe. Some evidence suggests that meditation can improve it.[14] Research on meditation is difficult because it's hard to have a placebo group (it's hard to make someone think they're meditating but actually not be).[15] Aside from meditation, simply being *mindful* as you do things at least gives you the opportunity to exercise your willpower, because being mindful reduces the chance of falling back on habit (though, as we'll see later, you can hack your habit system so that this isn't such a bad thing).

CHAPTER 2: HACKING YOUR BRAIN, HACKING YOUR LIFE

Your willpower system is often in conflict with the other systems. Here's a metaphor I like to use. Think of your body like a car. The habit system is a silent cab driver going where she normally goes, your willpower system is an adult in the passenger seat opining on all the places the car needs to go, and the reward system is a hungry kid whining in the back seat. When the willpower/passenger is telling the driving where to go, there's a decent chance the driver will go that way. But if the willpower/passenger is quiet, because she's thinking about something else, the reward/kid will make the habit/driver go where he wants it to go. If the reward/kid is quiet too, the habit/driver will go wherever she's used to going. The more familiar the terrain is, the more power the habit/driver has—sometimes to the point of ignoring the other two people in the car altogether. In a new place, the habit/driver doesn't know what to do, because there are no triggers for the installed habits, and more readily takes instruction from the others in the car.

This was shown in a study of students who transferred from one university to another. The new environment broke habits to some extent, and intentions were able to gain control of the student. If the student had been exercising at the old university, would she exercise at the new university? It depended on her *intention*, not so much habit. Students who wanted to exercise did, and those who didn't want to stopped. The environment failed to trigger their old habits, putting the students back under more cognitive control.[16]

Cognitive control is also increased by doing physical exercise, and the benefits are even greater if the exercise is cognitively engaging (like tennis) versus cognitively passive (like running on a treadmill). Even doing a brief bout of exercise has beneficial effects on your cognitive control system immediately after.[17]

You're already hacking your cortical system by reading this book. You are using your cognitive system's values to hack the other sources of motivation in you. You need to want the right things, and knowing what to want means having good values, and knowing how the world works well enough to know the difference between what matters, what unmatters, and what antimatters. Using conscious intention and willpower has limited power, so you don't want to blow it on intentions that do no good or actually hurt. Reading and getting advice from wise people you want to emulate installs values and conscious strategies that can be recruited by your willpower system.

In this section, I'll talk about conscious strategies and mindsets that can help you get your willpower system working better.

The Six-Second Rule

It feels like there's something magical about six seconds. Example: when you think "I should do some push-ups," your mind will start working on an excuse not to. If you wait more than six seconds, you'll probably come up with something. So when something like the sudden desire to do push-ups comes to mind, try to do it within six seconds of thinking about it.

I hurt the muscles in my back and hips a while back from playing squash. My physical therapist gave me a bunch of exercises to do every day. It took me about half an hour to do all of them. To help get myself to do it, I put on an audiobook, and did them one by one, as indicated by a stack of index cards. I like the stack because it's not easy to see how many are left, and because I can only look at one at a time, I'm not as tempted to skip ahead. If it were just a list on a piece of paper, my eyes would scan past the next exercise to the ones lower in the list, hoping to find one that looked more pleasant. Anyway, some of them, like plank, were very onerous for me. I find that when I turn the card over, and see something like plank, my heart sinks, and I have to get started doing it within six seconds. If I don't, I'm toast, because I'm smart enough to be able to rationalize a reason why I shouldn't have to do it. Although this works for me, I have not been able to find any research on anything like this six-second rule, so this advice is anecdotal, but perhaps will help you avoid the common pitfall of talking yourself out of something you don't *feel* like doing, but should.

Your Inner Voice

Have you ever been thinking of doing something, and some inner voice tells you that it's not going to work? Or that you're unattractive? That people will figure out that you aren't good enough? That nobody likes you? That you're in over your depth?

Since it's hard to get away from, you'd hope that your inner voice was soothing and encouraging, and for some people it is. But for many people, their inner voice is a little shit, undermining their confidence, causing self-doubt or self-loathing, and on balance holding their lives back.

In general, my inner voice isn't so bad, but I have troubles with anger. When I get angry at something, I get a mad, complaining voice in my head that crowds out other, more happy and productive thoughts.

Here is one strategy to deal with it: when my inner voice gets oppressive, I try to picture it as a one-foot-tall cartoon creature that is comically angry—I actually picture the Calcifer character from the film *Howl's Moving Castle*. When he has control over my mind, I picture him on my lap. I gently pick him up (in my imagination) and place him beside me. I immediately feel a little better. When I notice the angry thoughts showing up again, I "notice" that, like a persistent two-year-old, he has started climbing onto my lap, and I gently put him back on the ground. Picturing him over there, and not on me, screaming and complaining, gives me some distance from it. Again, this is only my personal experience, and I do not know of research showing that this works for negative emotions in general. However, there is research showing that visualizing pain outside of your body helps reduce the suffering associated with it. (I review this literature in my previous book, *Imagination*.)

A negative inner voice can diminish your optimism. Optimistic people generally think the future will turn out well, explain their lives in a positive light, are better at coping with difficulty, and have fewer depressive and other medical symptoms.[18] Although you can be more or less optimistic at different times, some people are more or less optimistic in general. Psychologists call this "dispositional optimism."

One reason optimists have these benefits is because they have excellent ways to cope with problems in their lives. They face them rather than avoid them, focus on what they can change, and make better future plans.[19]

Part of optimism is your self-esteem. Everyone has good and bad aspects of themselves, and you can't keep more than a few of them in mind at any moment. What are those things going to be, the good things or the bad things? Whether we are aware of it or not, we have stories in our heads describing our lives. Is it a heroic story, or a story of failed ambitions?

By changing the story, you can change your attitude toward yourself and foster a more optimistic explanatory style. We can see the powerful effects of narrative by looking at interventions for people who have undergone trauma. One method often used is to have people talk to a therapist as soon as possible about the traumatic event, airing their feelings about it. Unfortunately, this well-intentioned intervention antimatters. What it does is it freezes the feelings in place. The victims often relive the trauma, and this solidifies their feelings about it during therapy, permanently encoding the event as horrible in their memories. Although they report later that it helped, they're wrong. Careful studies show that the intervention leaves them worse off.

What works better was a method pioneered by James Pennebaker. A while after the event, the victim writes their deepest thoughts and feelings about the experience for four consecutive nights (about fifteen minutes each night). It's important that it's a while later so that they have a little distance from the event. It's also important that they reflect on the meaning of what happened when considering their whole lives. They need to "step back" and look at the event in a more detached way, as though they were watching themselves in a movie, and write about why the feelings that they experienced had come about, and what it means in their lives. This method is effective. They contextualize the event in a way that's more meaningful, and it doesn't bother them so much.[20]

AMBITION

Your self-image is affected by how well you are measuring up to your own standards. And in this, there's a danger with having ambitions. If you have high expectations of yourself, and then fall short of those dreams, you can get depressed. People who commit suicide often aren't in terrible circumstances, but rather are not living up to the standards in their heads.[21] The Theravada and Zen Buddhist solution to this problem is to not have any ambition.[22] Research supports the notion that ambition for *extrinsic* goals (money, looking good, and social recognition) leads to diminished happiness. Contrary to Buddhist teachings, however, ambition for goals like self-acceptance, community, and physical health are associated with better feelings of well-being.[23] Effects on your happiness aside, ambitious people are more successful: they are better educated, have more prestige in their occupation, and tend to make more money.[24]

Feeling terrible about the state of your life can be a real drag. One way that I get out of this problem is with a mindset I discovered while playing a video game called *Redneck Rampage*.

Let me first say that the term *redneck* is prejudicial, and the game is pretty offensive in terms of how it depicts lower-income people. It's full of unkind stereotypes. This was the 1990s, when sensitivity to this issue (with respect to the poor) was only starting to percolate into public awareness. Anyway, in the game, you play a low-income farmer with a thick Southern accent who is defending his land from alien invaders. You run around your property, shooting aliens, yelling "yee ha!," and collecting pork rinds to restore your health. I was in graduate school, struggling like hell to make sense of my research. And when I played this game, I had a minor revelation.

Here I was, playing a role. The role is of a very poor person, with a life I would never want. But am I (or is my character) wallowing in it? Am I thinking about all of the bad decisions in my life, and the strokes of bad luck that led me to this point? No, I was shooting aliens and doing the job that needed to be done, making the best of what I had, taking action. My revelation was that at any moment in your life, you might find yourself in a bad situation. But agonizing about how you got into that situation can hold you back. What you need to do is look at your situation, figure out the best thing to do, and do it, the past be damned, whether it's looking for a better job, getting out of a bad relationship, or shooting the aliens that threaten your prize pig.

I've actually used this mindset at several points in my life. I imagined that instead of me being me, with all my self-doubt and regret, paralyzed and overwhelmed with my life, that I was just a video game player, dropped into this terrible situation. What would that game player do? Not agonize. Just get to work. From its humble beginnings of a ridiculous first-person shooter game, I have gotten a surprising amount of mental support from this attitude over the years. It helps keep me from being paralyzed by whatever overwhelming situation I find myself in.

Even when we know what we need to do next, it's often hard to get ourselves to actually do it. When faced with something we should do but don't expect we will enjoy, our minds are very good at looking for other things that are more reliably rewarding. Look at the sidebar "Non-Productivity Scenario: Elaine." When Elaine is faced with working on her boring project, the idea of cleaning up her desk pops into her mind. Elaine is tempted by something less important but easier to make progress on. Over the next few minutes, cleaning the desk is more rewarding: She will make clear progress and see the results very soon. You might be the kind of person who never has a desk as clean as it is when you have a job to do you really don't want to work on.

NON-PRODUCTIVITY SCENARIO: ELAINE

Elaine has a very important task she needs to do, but she really doesn't want to do it. It's boring, and the way forward isn't clear. She sits down to work on it but feels so bored and so full of dread about the project that she finds other things that need doing. She answers emails, gets coffee, and organizes her desk. The day passes, and at bedtime she's made no progress and feels even more stressed-out and ineffective. She has trouble sleeping, and wakes the next morning feeling even more anxious and groggy.

Faced with this choice, it becomes a matter of willpower. Elaine will try to use willpower to force herself to do what she *should* do, rather than something that will be more rewarding in the short term. This is hard. When people do tasks designed to tax their willpower, they report difficulty, fatigue, and the requirement of high levels of effort. They don't like it.[25] But it's important, and a part of why disciplined people, people with a lot of willpower, do really well in this world. They can use their willpower in a brute-force effort and suppress impulses so that they can do the unpleasant thing that needs doing.

If you're one of those people so blessed with excess discipline that you're cursed with too much of it, you're a rare bird. It means that you have to make sure you use your discipline wisely. You could use it to work 70-hour weeks, for example. But this is a terrible idea. Economic research shows that for the average person, a 35- to 40-hour workweek is ideal for productivity. When you work more than that, productivity drops. If you consistently work 60-hour weeks, for example, you don't get any more done than if you worked 40-hour weeks.[26]

Can you run out of willpower? The concept that willpower was like a muscle was popular in the early 2010s. The theory was that, like a muscle, it got weaker the more you used it over the course of the day, but, also like a muscle, "exercising" it like this made it stronger in the long term. One study asked people to make a bunch of meaningless decisions, such as what kind of pen to use. They made poorer decisions on important matters later, as though using up some kind of cognitive fuel that got wasted on unimportant decisions. Trivial decisions take as much energy in your brain as important ones.[27] You can't keep it up all the time. These studies, however, which went under various monikers such as "ego depletion," fell out of favor in the social psychology replication crisis. A large, recent study, using twenty-three laboratories, failed to find any sign that your willpower decreases over the course of the day. But the theory's proponents have their reasons for doubting this attempted replication.[28]

The other aspect of willpower as a muscle, however, that training over time seems to make it stronger, has been supported by many studies. Exercise your willpower and you will gradually have more of it.[29] So you can increase your willpower by exercising it, and possibly through meditation.

You don't have to rely on your willpower—there are other things you can do. You can use mental tricks to avoid temptation, such as distracting yourself with *attention-deployment*.[30] You focus on something else, like making eye contact with your date instead of the cheesecakes on display in the background.

Sometimes, though, you should do the opposite. Sometimes focusing on the temptation is helpful. Suppose you're eating a bag of potato chips while watching a movie. If you're not careful, you might eat the whole bag before you catch yourself. In cases like this, actually paying attention to eating the chips (that is, being mindful of them) can help you eat fewer of them, because otherwise you are just eating one after the other on autopilot.[31] Furthermore, eating chips without paying attention to them is a real waste of calories—you're getting all the bad parts of eating chips without even enjoying them as much as you would if you were giving them your full attention. Savoring food allows you to enjoy it more, and also makes you eat less of it.[32] Some foods are just too fattening and delicious to warrant anything less than your undivided attention. If you're going to eat garbage, you should at the very least enjoy it!

In the brain, goal-directed behavior seems to be facilitated by the dorsomedial striatum, which is part of the basal ganglia. It takes into account expected benefits and drawbacks of this or that action. It is a competing circuit with the dorso*lateral* striatum part of the basal ganglia, which implements habits.[33]

Some people have naturally high willpower, and others have lower. The people with high willpower are lucky, as willpower might be the single most important trait for predicting long-term life outcomes.[34] Your willpower is like your bodyguard: a tough guy who protects you. But even people with excellent willpower should not rely too much on it, because using it requires a focused attention that you won't always have. Exercising willpower requires conscious effort, and is difficult, unpleasant, and causes fatigue.[35] And sometimes the bodyguard goes on a break.

In general you should try not to rely on your self-control if you don't have to. It's a last-ditch effort, difficult and generally unsustainable. I think it's helpful to think of having to exercise your willpower as a sign that your other systems have failed. If you are someone without much self-control, someone who gives in to impulses relatively easily, it's even more imperative that you don't rely on it.

Luckily, there are lots of other things you can do besides relying on willpower. Let's go over what those things are.

How to Hack Your Habit System

Sometimes people will deliberately buy small packs of candy rather than large ones, so they don't eat as much. This makes good scientific sense, as studies show that people eat less if food is put in smaller packages.[36] Why might this be? Once a package is opened, it's easy to just keep eating. When you have to open a new package, it's a salient cue that you're consuming. It breaks the grab-food-and-eat-it loop that your basal ganglia is running on mindless autopilot.[37] People tend to keep eating until food is gone, more or less ignoring their own hunger or satiety. One of the main theories of why French people are thinner than Americans is that they eat less due to smaller portion sizes.[38]

Habit is a major contributor to your behavior, more likely to take control when your conscious mind is focused on something else. You can use your cognitive system to override your habits, but this is not sustainable. Sometimes you have to consciously think about something other than what your body is doing: you might be giving a presentation about your company's marketing plan for next year, and not thinking about how many sips of latte you're drinking, or having a tough conversation with your daughter about her crappy boyfriend, and not focusing on driving, or wondering about why Qui-Gon Jinn didn't vanish when he died, even though he came back as a force ghost later, and not focusing on the peanut butter sandwich you're eating.

In moments like these, your body will act on habit.

We can see just how powerful habits are by looking at how durable they are in the face of other mental problems. People who suffer from an inability to form new memories can still learn new habits of surprising complexity. Charles Duhigg's book *The Power of Habit* describes Eugene, who was unable to create new episodic memories, and indeed could not remember anything from the past several decades. His wife was stressed because he would wander out of the house and get lost. But after taking him on the same walk every day, it became a habit for him, and when he wandered off again, he would just find himself going on the habitual route he'd taken with his wife over and over again, without knowing why, and without even consciously remembering ever going on that route before. Unless there was something different about how the route looked, like construction or something, he would just find himself back home again, much to his wife's relief.[39]

Habits Ignore Reward

I had a friend who was trying to avoid eating candy. He was at a restaurant, and at the end of the meal, was deeply engaged in a conversation. Then he looked down and saw the empty candy wrapper in his hands. He'd eaten the candy without pleasure, consciousness, or memory of the event. Such is the power of your automatic, habitual behaviors. When your goal-directed, cognitive brain functions are weakened, or otherwise occupied, habits are more likely to control you.[40]

People simply don't have the mental resources to behave deliberately all day long. Suppose you start craving a doughnut at 12:30 in the afternoon. You resist. Then at 1:15 you get another craving, which you also resist.

Then again 45 minutes later. By now, you're feeling pretty proud of yourself for successfully resisting eating the doughnut three times now. You almost deserve the doughnut for your efforts, right? You give in two more times before finally giving in and eating one at 3.

Forty percent of what people do is on autopilot, based on habit and what you're used to.[41] One way scientists can test to see if some action is the result of habit, rather than higher cognition or the reward system, is to see if doing the behavior is sensitive to changes in expected reward. That is, if you change how good or bad the expected outcome will be, if it is cognitively motivated, the probability of doing it will change. Not so much if it's a habit.[42] This suggests that the habit system bypasses the normal decision-making that weighs benefits and drawbacks to an action that we often engage in. When you want to drive from work to a doctor's appointment, but you end up driving home out of habit, this is what's happening. If you were considering the benefits and drawbacks, you would have realized that going home meant a waste of time, and that taking the route to the doctor is the best thing to do. But habit took over.

My writing of this book is a good case in point. I worked on it every morning of the week except Saturday. When it became a habit my experience was that I did not need make a *decision* every morning to work on the book. I just sat down and started working. My conscious mind was thinking about *what* to write, not *whether* to write. Ideas of doing anything else never even crossed my mind, freeing up my willpower for other things.

Even when you're paying *some* attention to what you're doing, your mind tends to look for reasons to engage in the habits you already have, making it harder to avoid bad habits even when you're thinking hard about it.[43]

Habits Have Big Effects Over Long Periods of Time

By definition, habits are things done over and over again. Because of this repetition, small changes can build up over time. For example, buying an expensive drink once in a while is no big deal, but if you do it every day, it can cost you thousands of dollars a year. Similarly, riding your bike to work once is nice, but doing it every day increases your lifespan. Habits are hard to change, but this has a good and bad side—good habits are hard to establish, but once they are there they work for you without effort, like interest in your bank account. With just a little habit maintenance, you can focus your limited attention on other issues in your life, trusting that what you'll do on autopilot is in line with improving your life.

This is why optimizing yourself requires curation of your habitual activities. First is to recognize the habits you already have. Then you can decide which ones to reinforce and which ones to replace. Think of them like apps on your phone, they take up memory, and some make your life better, and others make your life worse. Which habits should you uninstall?

In the moment, a habit is triggered by something perceived in your environment, and over time, habits change because of conditioning—reward and punishment from engaging in this or that behavior in this or that situation.

Changing Your Environment So the Right Habits Are Triggered

Habit cues tend to be company (I smoke when I hang out with the actors), locations (I buy a doughnut every time I'm at Dunkin' Donuts), emotional state (I eat when I get stressed), action immediately preceding (I wash my hands after I empty the cat's litterbox), and time (I eat lunch around noon).[44] You can remember these triggers with this mnemonic I made up: HABIT. H) Humans you're around. A) Activity. B) Bearings (your location). I) Internal state. T) Time of day.

There are two general classes of things you can do to hack your habits. The first is to alter your environment so that the habits you want are triggered more often, and the habits you don't want are triggered more rarely. For example, if you mindlessly eat an entire bag of chips while watching movies, then don't bring a bag of chips to the couch, or, better yet, don't buy the bag at all. If you drink too much when you hang out at bars with Pat, then hang out with Pat only during the day, and not at bars. In the short-term they don't actually affect your habits, they just keep them from being triggered at one particular time.

A study of trying to instill an exercise habit found that people stuck to their routines better when there was a specific cue to start, such as running as soon as they wake up, or going to the gym after work. A reward also helped (a beer, television, etc.).[45] Over time, exercise can be its own reward—you might feel physically good after exercise, or enjoy a sense of accomplishment. You then might crave exercise to get these feelings.[46]

Just figuring out what triggers your existing habits can be challenging. If you can't stop buying doughnuts when you are on break at work, what exactly is triggering this habit? Is it hunger? Is it boredom? Is it being tired? Is it a low mood because work doesn't allow you to take a nap in the afternoon? Identifying and monitoring these triggers, and changing reward and punishment of the actions you take takes sustained vigilance.

If you can engineer your environment to make you act (and not act) how you want, eventually you'll get into good habits.

Changing the Habit Itself

When you avoid a habit being triggered, you don't remove the habit from your mind. It's still there, lurking, ready to engage when the trigger comes back. So the other important thing to do is to try to change the habit itself. This is a longer-term strategy that takes sustained effort. You can alter your habit system by changing the punishments and rewards associated with habit behaviors. What makes this difficult is that many bad habits are kept in place because you get rewarded for behaviors that lead to immediate and visceral gratification.

To take an example in my life, I drink Vietnamese coffee every morning. It's got caffeine and a spoonful of sweetened condensed milk in it. Drinking sugary drinks causes a pleasure you don't have to wait for. We are conditioned much better when the reward or punishment comes immediately after the behavior. Furthermore, the pleasure is something very primal—we are wired for liking sweets. (In contrast with sugar, caffeine's reward takes about fifteen or twenty minutes to kick in.)

Breaking this habit would be challenging to do. The reward for *not* having a sugary drink every morning is what—better health in the long term, and perhaps a slight difference in my weight? Avoiding Vietnamese coffee on any given day simply does not create the same kind of feeling of reward that drinking it does. These far-future rewards are so uncertain and speculative that they are trapped in the cool, emotionless belief system in the cognitive part of my mind. The habit system barely even registers them. Instead of a great taste in your mouth, you have to settle for a belief that you're doing your future self a favor, and try to feel good from that. The satisfaction of the taste of sugar versus the satisfaction of a nice belief? You can see why changing habits is hard!

CHAPTER 2: HACKING YOUR BRAIN, HACKING YOUR LIFE

Similarly, watching television instead of exercising provides the immediate gratification of a primal urge to conserve energy, and our basic thirst for hearing good stories, where exercising is boring and strenuous in the short term, and you only feel good when it's all over, and experience health and happiness benefits down the road. Many new habits are hard to establish because they pit short-term gratification against long-term goals.

If you try to break a habit through sheer force of will, you are relying on your cognitive system, which has a lot of other jobs, too. When it's distracted, which it eventually will be, you will accidentally engage in the bad habit, reinforcing it again. This is why breaking habits is very hard for people.

So if you have trouble breaking bad habits, you are not alone. It's hard. In fact, you probably cannot ever completely remove bad habits from your brain. Studies show that the old habits are still in there. It sounds like an impossible task.

But there are ways to break bad habits. Rather than thinking of getting rid of bad habits, it can be helpful instead to think of replacing them. That is, instead of having Vietnamese coffee every morning, I can try to have green tea instead. Over time the tea habit will be stronger than the coffee one, even though the coffee one is still in my head, waiting to be triggered again. I use my willpower in the short term to force myself to drink green tea for a while, until the habit takes over. Then I can let my cognitive system relax a bit, and my habits will be in line with my goals for this issue.

This is very important, so I'll say it again: your old, bad habits will probably always be in your mind, so if you want to "break" them, you need to install better habits that are triggered by the same things, and work at those until the new habit is stronger than the old, bad one. Don't think of eliminating habits. Think about *replacing* them.

When to Replace Habits

Because habits are triggered by environmental situations, making a big change to your life, like moving, getting married, or getting a new job, can disrupt your habits, for better or for worse. If you have a desire to change habits, either by removing bad ones or installing good ones, a big life change is a good time to do it. Studies show that big context changes disrupt your habit system, making them vulnerable to facts, goals, and information. So, for example, it's easier to start riding your bike to work when you change apartments, become vegetarian when you change jobs, and finally kick that infidelity habit the next time you get married.[47]

Similarly, your normal habits can be disrupted by a change in context. When you travel, and you're staying at the hotel, it is sometimes hard to keep doing the good activities you usually engage in, like exercising or taking your daily medicine. The normal cues are not there, and you're apt to just do whatever. I struggle with this. If I'm trying to write every day, I have to use my willpower to get myself to do it on my laptop in the hotel room, because it doesn't come as naturally as it does at home. I'm not at my desk, I don't have my Vietnamese coffee, I'm working on a tiny laptop, etc. My mind starts looking for reasons why I don't need to do it, considering all of the other things I could be doing instead. This never happens at home anymore, where the habit is reliably triggered, and other options and rewards are not even calculated.

THE MORNING IS A GOOD TIME FOR GOOD HABITS

We all have nightly habits, like brushing our teeth or reading. But the mornings are particularly good times for good habits. I don't know of any evidence for this, but it seems that mornings are much more similar to each other than nights are. My nights are very variable. I might eat in; I might eat out. I might go to a party until late into the night; I might go to bed early after watching a movie. But my mornings are much more routine. Things don't come up. This means that the potential triggers for habits are more reliably there.

So if you want to try to install good habits, like taking daily pills, getting writing done, or exercising, you might have more luck trying to do them every morning rather than every night.

You want to make sure your habits are working toward your *considered* interests. The interests you muse about when you're thinking about your future. The interests that resonate with your deepest values—not the interests of immediate gratification. How can you change your habits? I like to think of my mind like a computer, and a strategy or habit as a piece of software. How can you install a habit into your system to make it more effective?

Using Willpower

You can use your willpower to get a habit into place. So, for example, if you want to get into the habit of running every morning, for the first month or so you're going to need to use willpower (and other methods) to establish a routine before you can benefit from your habit system. But once the habit is there, the conscious decision-making step of action is skipped. This is why good habits are so powerful—your conscious mind is less likely to get in your way. You don't question the habit as much, and you're better for it.

Relying on intentions alone tends not to work very well in the long-term. It suffers from the New Year's resolution problem, in that compliance is difficult as the months go on. It also leads to lowered mood, and preoccupation with your habits.[48] Part of the problem is that your willpower and self-control fluctuates day-to-day, and even during a single day. When you are focused on something else, a bad habit can take control. This is why good intentions and willpower should be supported by environmental changes that encourage good habits and eliminate bad ones; they don't tend to work all by themselves.

One easy way to try to help install habits is with implementation intentions. Think about some goal you have, and something you would need to do as a step toward achieving that goal. Then, form a policy, like, "Whenever I'm in that situation, I will do this." For a weight-loss goal it might be, "Whenever I get tempted to eat junk food, I will chew gum." Studies show that just setting these in your mind help make your unconscious mind work for you.[49] Over time this can be turned into a bona fide habit. This has been found to be quite effective for installing new habits, and less useful for removing bad ones.[50] Remember, it's more effective to try to replace bad habits than to remove them.

Higher-level decision-making does not directly affect habits. Habits change over time, through repetition and conditioning. This is why merely learning about something does not change them. The part of your brain that learned in a book that, say, drinking too much is bad for you, is not actually affecting your habit system. If you are thinking about some fact like that, your willpower system might override the habit in the moment, but it doesn't change the habit itself, it just overpowers the force of the habit in that one instance. When you're not consciously thinking about the fact, it won't have any effect at all. This is why you need to do more than learn about what to do and what not to do—you need to train your habit system to change. Good intentions tend to have much less effect when there is an opposing strong habit.[51]

HOW TO HACK COMPUTER SYSTEMS AND OTHER PEOPLE

The easiest way to make a change is to delegate good activities to someone (or something) else. You can have your paycheck automatically deposited into your bank account, and set up a system to automatically put a portion of it in a savings account to help you save for retirement. Simply having that money out of your checking account makes it less tempting to spend it. You can often set up a system to put a portion of your wages put into a retirement plan.

If you are giving to charity, you can have an automatic donation schedule set up, so you never need to question it.

These strategies work because they *bypass the weaknesses of your brain completely*, and allow you to do the right thing, for yourself or others, without having to worry about your brain screwing you over.

The flip side, which you need to be careful of, is that you can also subscribe to services that drain your bank account. This is why businesses love subscription services. People get charged monthly and forget about it. Periodically check to make sure that the things you're being regularly charged for on your credit card are things you actually want and use.

If you have (or can get) an assistant, you can have the assistant remind you at regular intervals to do the right thing. This is a part of why having a personal trainer is so effective.

CHAPTER 2: HACKING YOUR BRAIN, HACKING YOUR LIFE

At the same time, using your willpower is crucial because you need to care for your habits like they are your pets.

Another trick you can use to promote a new, good habit is to break it up into pieces, and gradually do more and more pieces every day. For example, suppose you would like to go to the gym every morning, but it sounds like a drag. Think about the steps you need to take to work out at the gym. For example, for the first three days you go out of your way to walk by the gym on your way to work—but don't go in. Or, if you drive, drive to the gym and park, then go to work. That's not too hard, is it? If it doesn't feel like a habit yet, keep doing it until it does—it might take ten days. When that habit is established, pack your gym clothes and change into them at the gym. Then change out of them and go to work. I know it sounds stupid, but putting on gym clothes is a lot easier than exercising, and easier to make a habit out of. Then, add a very, very light workout in those gym clothes, maybe just a walk. Eventually, work your way up to a full habit of exercise every morning. If you find yourself slipping, go back to the easier versions.

In practice, you'll probably want to work out at least a bit if you put on your gym clothes. But you shouldn't count on this—be wary of breaking promises to yourself. If you promise yourself that you will only put on your gym clothes and then take them off, you need to let yourself do exactly that or the next time you won't do any of it, because even though you are telling yourself that you don't need to work out, you know that the last time you guilted yourself into working out anyway, so you say screw the whole thing. You won't trust yourself to keep your own promises.

Doing things in groups also encourages behavior, both good and bad. You might have drinking buddies, which will make you drink more. Addiction interventions always try to keep addicts from hanging around the people they use with. But your social group can also be leveraged for better habits: you can have workout buddies, writing buddies, or art buddies. I find that when I'm part of a writing group, I write much more, because I want something to bring to the group. When I do, I get rewarded for doing so. For normally solitary activities, like reading books, writing, working out, or doing many forms of art, you can *make* them social to encourage the habit to form. When I was living in Kingston, Ontario, I had a friend who would call me up and say, "Want to hang out at a coffee shop? I don't feel like being social." We'd just sit together and read.

Imagine the difference in motivation for going for a morning run in these two situations: In the first, you try to get yourself to run alone every morning. In the second, a group of three friends knock on your door at 7:00 A.M. to get you for the group run. Which do you think would be easier to skip on a day when you're feeling tired, or it's a little cold?

THE NEUROSCIENCE OF PLEASURE

It's not necessary for you to understand how reward works in the brain, but this sidebar describes the basics of it, in case you're interested.

Pleasure is associated with naturally occurring substances in your brain, such as opioids, endocannabinoids, and GABA-benzodiazepine neurotransmitters.[52] But many of these chemicals are used widely in the brain for lots of things, so for pleasure, it matters *where*.

When people experience reward, there are several parts of the brain that are more active than usual. These are the neural correlates of reward. But this does not mean that all of those brain areas *are* reward. Reward is used by your brain for all sorts of things—you change future behavior patterns, habits, associate memories, and so on. Brain areas that work on these things might be predictably activated whenever there is reward, but they are a consequence of reward, not the reward itself.

Brain studies, mostly done on rats, suggest that pleasure itself is distributed across a number of "hot spots": the nucleus accumbens medial shell, the posterior ventral pallidum, two places in the deep brainstem region, and the parabrachial nucleus in the pons. The only *necessary* hot spot is the posterior ventral pallidum, and perhaps some surrounding areas. There is no single "pleasure center" of the brain.

It is possible that these hot spots encode a basic *form* of pleasure, but not the *feeling* of pleasure. That is, it might be that having pleasure need not involve a conscious feeling of pleasure. I know this sounds strange, but we have a couple of reasons to think it might work this way.

First, people seem to be able to engage in learning through conditioning (that is, using reward and punishment) without being aware of any pleasure or displeasure. Drug addicts, for example, are able to learn which action to take to administer microdoses of their drug that are so small that they can't feel anything. Similarly, other subliminal pains and pleasures can be used for learning, completely outside of conscious awareness. In short, you can learn through reward and punishment *without feeling anything*.

Second, the pleasure system is very, very old, evolutionarily, and it might be that conscious awareness (of anything) came later. This suggests that having pleasure involves the hot spots listed above, but actually *feeling* the pleasure might involve some other spot, a conscious gloss on the pleasure that makes it consciously feel good. We don't know what that spot is.[53]

How to Hack Your Reward System

There are two distinguishable feelings in the reward system: liking and wanting. But they are difficult to tease apart because most of the time they are activated at the same time.

The first is the liking system. This involves pleasure and pain. The reward system has a very important function, in humans and most other animals: by using pleasant and unpleasant experiences, it shapes what the organism does in the future. If something hurts, we learn to avoid it. If something feels good, we learn to approach. Pleasure and pain serve as an interface between sensations and goal-directed action.[54]

We also feel pleasure when we accomplish something, or expect to accomplish something. From an evolutionary perspective, this is why pleasure exists at all. But some recreational drugs are intrinsically pleasurable because they directly activate pleasure systems. They are a pleasure hack, a cheat code, giving reward when you haven't done anything useful!

Pleasure is a gloss that gets put over a sensation or some other mental state. The sensation itself isn't the reward.[55] For example, you might like the taste of ice cream, but at some point you will have eaten enough, become satiated, and being forced to eat more would make you feel disgust. It still tastes like ice cream, you just don't find it pleasurable anymore. So, if you like ice cream, it's not the taste of ice cream that's pleasurable, it's the combination of the taste of ice cream combined with your state of hunger, how much ice cream you've eaten lately, how much you've been thinking about food or ice cream, etc. In short, pleasure emerges when the stimulus is in a particular relationship with your body and mind.

The second is the wanting system, which motivates action. There were famous experiments where rats and humans were given control over a device that directly stimulated certain parts of their own brains. They activated this device over and over again. At the time, the researchers assumed that this must be the pleasure center. (The rats preferred to activate this device over sleeping and eating, and kept at it until they died of exhaustion and starvation.)[56]

But now we think it's more likely to be the motivation center, or "wanting." It's more like an urge, or a compulsion, but not particularly euphoric. While pleasure involves opioids and the like, the wanting system uses the neurotransmitter dopamine. Sometimes people experience irrational wanting, which is wanting something you don't even like or expect to like.[57] One example of this "wanting" is to pick your scabs, even though doing so hurts and makes you bleed. In this case, your mesolimbic "wanting" system tries to get you to pick the scab, and your cortical system (and perhaps even the liking part of your reward system) wants you to stop. These conflicting wants in your mind are generated from different brain areas, fighting for control of the behaviors stored in your basal ganglia. Addictive behaviors might be examples of this, too. Some addicts claim to want drugs, even though they don't particularly like them.[58]

You might have experienced the dissociation between liking and wanting while eating lots of food. I've had situations where I found myself compulsively eating long after I've stopped experiencing pleasure at eating (high wanting, low liking). I've also found myself full, and not wanting to eat, but the food tastes so good that I wish I were hungrier (high liking, low wanting).[59]

It's important to distinguish the "wanting" system we're talking about here from the kind of "wanting" involved with things like conscious goal-setting (sometimes this kind of want is called a "desire"). When you "want" to graduate from college, or get a better job, it's using the higher-level cognitive system in the neocortex, not the mesolimbic wanting system. It feels different, too: the wanting to graduate does not have the same immediate urge-like compulsion that wanting to eat more candy does.[60]

Now that we understand a bit about how your reward system works, how can you hack it? You do this by conditioning yourself: rewarding yourself for things your cortex wants you to do more of (avoid watching TV, attend to your friends more, work harder, and so on) and punish yourself for doing things your cortex doesn't want to do, even though your reward system might be jonesing for it.

Punishment

Punishing yourself is difficult to maintain. Some people suggest that you can try to break bad habits by, say, putting on a jersey of a sports team you don't like every time you indulge in the habit you're trying to break. But having to do work to punish yourself is probably not going to last, because, well, punishing yourself is intrinsically unrewarding. This is why aversion therapy is often done by someone—or something—other than the person being punished.

There's a bracelet you can buy called the Pavlok. You get a chrome extension that connects to it, and it delivers an electric shock to your wrist every time you spend more than some specified amount of time on social media sites. You can also shock yourself with it by pressing a button. Although this, too, might be hard to maintain, at least pressing a button requires negligible effort. There are testimonials touting its benefits, and science backing up using electric shock to reduce bad habits,[61] but the Pavlok itself has not been rigorously tested.

Given that conditioning happens even in very small creatures that most scholars don't believe to be conscious, perhaps there is an easy way out: conditioning yourself with unconscious punishment. The idea is this: you use some mechanism, perhaps like the Pavlok, but have it deliver a "punishment" too mild to be detected, such as an electric shock you can't even feel. It might be that part of your mind processes this stimulation, and perhaps even learns from it, all outside of conscious awareness. Could this possibly work? This is currently a matter of debate in psychology, with some studies finding effects of unconscious conditioning, and others failing to find it.[62]

In general, it's better to use reward than punishment to change yourself. So what is rewarding? Some rewards are intrinsic. A great example is the taste of sugar. Babies do not have to learn to like the taste of sugar. Humans generally love sweet foods without being told or trained to do so, and many other mammals also love it. Mammals seem to be wired to like it without any learning. For many people, foods with fat, sugar, protein, and salt are intrinsically rewarding.

Sex is also intrinsically pleasurable (for most animals), but the sex drive only comes online later in life—it's still genetic, but those genes are expressed later in development. We evolved to feel pleasure from these things because eating and having sex led to more offspring. Higher-order pleasures, like music, might use the same brain mechanisms.

Similarly, some things are naturally aversive. Pain is intrinsically aversive, and serves as punishment.[63] Other tastes, such as bitter things like coffee, are initially aversive to most people, and require learning to grow to like.

You have to learn to love extrinsic rewards. Money is a good example. There's nothing intrinsically rewarding about finding a bunch of dollar bills on the ground. The positive feelings are only there because you've come to associate money with status and the things it can buy.

In 2007, long-distance runner Dean Karnazes generally only ate very healthful foods—grilled salmon five nights a week. But when he wasn't training (sharpening the saw), but actually competing (cutting with the saw), he needed lots of calories. He would order an extra-large Hawaiian pizza to be delivered to a particular intersection on his route. He'd roll it up and eat it while running. Not only did it give him the calories he needed (supplemented with éclairs and cheesecake), but it gave him something to look forward to during the most grueling parts of the race.[64]

But rewarding yourself can be difficult if you generally never deprive yourself of anything. For example, if you eat whatever you want all the time, and get yourself a massage whenever you feel like it, and basically do whatever you feel like doing all the time, then there's no particular reward that will feel very special. One thing I do is deprive myself of something I really like, and only let myself have it after I do something I really *should* do. Although you might feel you're happier if you seek pleasure all the time, and never deny yourself anything, this puts you in a weird bind: there's nothing you can do to really treat yourself *especially* well when you want to celebrate or reward yourself.

For example, a few years ago I was hooked on a video game called *Hearthstone*. Playing it took about ten minutes, and when I was really hooked on it I played it about two or three times a day. At that time I was concerned that I wasn't getting enough exercise. So I made a rule: I couldn't play *Hearthstone* unless I had exercised that day, like if I took the stairs at work (I work on the 22nd floor). I can't really say that this reworked my reward system so that I found exercise fun (I eventually had to find a sport I liked to make that happen, and thus start a new habit), but I was so hooked on *Hearthstone* that I exercised every day for three weeks so I could play it!

What would be really great is if you could hack your reward system so that the unpleasant things you believe you should do become rewarding in themselves. A way to do this is to use classical conditioning, where you come to associate behaviors with things that happen to you. The classic experiment in classical conditioning involved salivating dogs and footsteps. Ivan Pavlov noticed that not only would dogs salivate when the food was brought in, but when they heard the footsteps of the experimenters. They had come to associate the footsteps with the behavior of salivation.

Rachel, a friend of mine, was hooked on sleeping pills. She was studying psychology at the time, so she tried to get off them using classical conditioning. At bedtime she would take her pill and then put on the first Harry Potter movie. She'd fall asleep. Over time, she'd start taking fewer pills, then a pill every other day. Eventually she was off of sleeping pills, and could not get thirty minutes into that movie without falling asleep. (Her real name isn't Rachel; she doesn't want everybody to know about her Harry Potter addiction.)

But can you make something unpleasant into something fun? One method that seems to work is temptation bundling, which does this by having people simultaneously doing "want" and "should" activities.[65] In the experiment, they had audiobooks put on iPods. As the experimenters put it, these books are "lowbrow, page-turner audio novels." For one group, they only allowed them to listen to these while exercising at the gym. This group went to the gym 51 percent more often than the control group. The deprivation here is key. The people wanted to hear the book, but could not do it unless they were exercising, which made them exercise more. I have a friend who does this with TV: she won't allow herself to watch it unless she's on her exercise bike at the same time.

It's not a panacea, though. A second group was merely *encouraged* to only listen at the gym. These people went to the gym at a rate statistically indistinguishable from the controls. Also, the effect decreased over time, especially after Thanksgiving: a decrease of .07 visits per week for the nine-week length of the study. Some habits are hard to keep.

But did it make people enjoy exercising more? Not really. The differences between the iPod group and the control groups were not significantly different when asked how much they enjoyed the exercise.

Temptation bundling is a kind of precommitment device, where you commit to a policy to reward or punish yourself depending on a future action. It works better than resisting temptation in the moment because the precommitment happens when you're relatively cool-headed. Once there is a policy in place, like I won't watch TV unless I'm exercising, or I can't play *Hearthstone* if I had Swiss cake rolls today, or the fact that you scheduled reviewing the budget for 9:30 A.M., means the decision is easier in the moment. It takes effort to vary from the plan, so you make your own laziness work for you.

It's hard to make yourself like something, but it's a lot easier to make yourself hate something you like. You do this by putting extrinsic punishments in place. A treatment for alcoholics is to take an Antabuse, which causes nausea and vomiting when you drink alcohol. It prevents you from drinking, and when you do drink, you quickly build negative associations with it.[66] If you eat a certain food when you are sick with the flu, it can ruin that food for you for years. So if you're addicted to eclairs, and wish you didn't like them anymore, eat a bunch of them while you have the flu and it just might ruin them for you.

Sometimes a desire can be so strong that it can turn into a motivation without any external trigger. As I write this book, I'm struggling with a mild bubble tea obsession. A store opened half a block away from my house. But lately, I find myself thinking about bubble tea, and how I might be able to get some, even when the store isn't so close. I recently went on vacation to Cartagena, Colombia, where there was no bubble tea, and found myself thinking about it, and looking forward to coming home so I could have some.

You have probably felt this way about something in your life, whether it's a video game, some food or drink, or a powerful crush on somebody. But most motivations are triggered by perceived opportunities. Your desire to eat ceviche, for example, might only become a motivation when you see it on a menu (there was a lot of ceviche in Cartagena).

As such, you can manipulate your environment so that you perceive more opportunities for things you *should* do, and fewer opportunities for things you are *tempted* to do. As you engineer your environment to hack your habits, you can also engineer it to hack your reward system.

You might be using your cognitive system to try to focus on an important task, but your reward system is seeking out distractions, which, by their nature, are novel and rewarding. But when you cut yourself off from those distractions your reward system isn't as tempted. Turning off your phone, stopping email notifications, being in a place where you can't easily snack, and interventions like that just make it easier for your cognitive system to do what it needs to do, as you've removed many triggers of your reward system.

On the campus of Google, there are free food and drink stations all over the place. Being the data-obsessed company they are, they tracked snacking and tried to see what they could do to make people eat more healthful foods and less food overall. They found that simple things, like having the snack counter distant from the drink counter, reduced consumption of food considerably. Just serving M&Ms in smaller containers reduced their consumption.[67] In one study, simply having a freezer door closed rather than open resulted in 5 percent rather than 16 percent of people serving themselves ice cream.[68]

This works, in part, because of how motivations are formed. If the candy is not in front of you, then you are less likely to turn your desire for candy into a motivation to get some. (This also works because of the reward system—adding friction to a task, making it harder to do, makes the entire project less rewarding because of the negative feelings associated with doing even a little extra work, like opening a cabinet, or having to walk down the street.)

The lesson here is that you have some control over your work and home environment, and you can engineer it so it's harder to do tempting things. You can keep the remote control in a room distant from the TV, and keep the batteries farther still, requiring more effort to even get the TV on. You can keep brownies frozen, so you have to defrost them to eat them. I have seen no studies showing how much effort is required to keep you from indulging, but personal experiments suggest that requiring twenty seconds of work is enough to reduce indulgence considerably.[69]

This can also be used to *encourage* behaviors you want to do more of. I recently got a guitar, and my guitar friends say you should keep it in its case. Apparently it's better for the guitar. But the case is ugly, it takes up room in my living room, and opening and closing the case is a pain. This calculus goes though my head every time I think of playing the guitar. If I only want to play for two minutes, it's not worth the hassle. So I play the guitar less. But I'd like to practice more. It's a good break for me, and you can't get better without practice. So I got something for the wall where I can hang the guitar. When I'm sitting on the couch, I can pull it into my arms in about four seconds. It's easy to take out and put away, so I end up doing so a lot more.

You want to make tempting things more expensive, in terms of time, allocation of effort, enduring of pain, monetary cost, etc., and the things you want to do more of frictionless and less expensive: easy, quick, cheap, and rewarding.

This is the *situation-selection* method of self-control: choosing to put yourself in situations that make the "should" activities relatively more inviting, and the "want" activities less inviting. A particularly powerful way to engineer your situation to your benefit is to surround yourself with the right people. Who are "the right people"? The people who are acting in ways your higher-level cortical processes think you should act, and, importantly, not indulging in the things your reward and pleasure systems would have you do that are counter to your goals. The power to be influenced by the people around you is so strong that scientists use the word "contagion." Suicide, obesity, drug use, smoking, and pregnancy are all socially contagious. So is doing hard, productive work.

Situation-modification is when you manipulate your environment to help you with your goals: not keeping junk food in the house, planning time with friends to ensure enough social interaction, placing your alarm clock across the room so you have to get up to turn it off, or putting social media–blocking software on your computer. You can reduce the amount of willpower you need to use by manipulating your environment. These methods are often more effective than exercising self-control because many temptations grow over time, but changing your environment at the beginning happens when temptation is low or not yet present.[70] Calling your spouse and asking them to hide the cookies is easier than resisting cookies right in front of you. You can opt to not have unhealthful foods in your house, or to put them in a place that's hard to get to. My aunt once kept snack cakes in her house for her daughter's lunchbox. But she found that she would eat them, so she put them in the freezer. She wasn't tempted to eat them because they were frozen, but the ones she put in her daughter's lunch would thaw by lunchtime. This worked until she discovered she could defrost the cakes in the microwave. Still, having to microwave the cake was a bit more friction than just pulling it off the shelf. Smokers will sometimes try to quit by depositing money with someone that they will forfeit if their urine test reveals nicotine.[71]

When I'm working at my desk, I get snacky. I'm not hungry, but I want to snack. I'm constantly distracted by this even when I don't indulge. And when I do indulge, I have lots of temptations: spoonfuls of peanut butter, dark chocolate, dark chocolate covered with spoonfuls of peanut butter, spoonfuls of peanut butter with a chunk of dark chocolate in it . . . So I found a snack that I can eat constantly without having to worry about it: raw cabbage. I cut up some raw cabbage into Dorito-sized chips and keep it by my desk. It's crunchy, so it feels substantial, but it's nutritious. This actually works. Does raw cabbage taste good? No, it doesn't. But because I'm eating compulsively (the wanting system) the pleasure doesn't matter so much. I'm eating mindlessly, and multitasking a bit (eating while working), so I don't notice the bitter flavor much.

These are ways you can take advantage of however lazy you might be. Make the things you really want to do easier, and make the temptations harder.

If you have your cognitive system's high-level judgments in the right place (that is, you want the right things), and your reward system is tweaked so that you find it rewarding to do the right thing, and your habit naturally leading you to do the right thing, then all of your brain areas will be working in concert, rather than at cross-purposes, to make you act better.

Let's get into some specifics about how can you hack your environment to make yourself more productive. How do you keep yourself from being distracted from what you need to do?

Working Without Distraction: I'm Here to Make an Hour Out of Fifteen Minutes

It just might be that your job requires you to do "shallow" work a lot of, or all of, the time. If you work at a call center, or as a CEO, for example, your job is mostly rapidly responding to external demands on your time, or where constant connectivity on the Internet is actually a part of the job.[72] But for many jobs, some level of concentration is beneficial. But keep in mind that your job isn't your whole life—you might have endeavors in your leisure time that would benefit from undistracted phases of productivity.

Cal Newport, in his inspiring book *Deep Work: Rules for Focused Success in a Distracted World*, describes four different ways to structure your time so that you get more focused attention.[73] This means *not* multitasking. At the very least, don't let your devices distract you with notifications (see the sidebar "Remove Electronic Distractions").

The first is the "monastic philosophy," which is a really extreme approach. You cut yourself off from the rest of the world, making yourself unreachable by phone or email for long periods of time—maybe forever. Some people are really productive using this method, such as computer scientist Donald Knuth and novelist Neal Stephenson. But most people's occupations don't allow this kind of hermit lifestyle.

The "bimodal" philosophy involves having parts of your life designated for deep versus shallow work. You might do shallow work for a month, then deep work for a month. Or half your day might be deep, and the rest of it shallow.

In the "rhythmic" philosophy you have particular parts of the day dedicated to deep and shallow work. Even if you have to check email every half hour, you can alternate fifteen minutes of concentration with fifteen minutes of interactivity in each half hour of your day.

Finally, in the "journalist" philosophy, you fit in concentrated work whenever you can. You sneak in half an hour here, ten minutes there, as time allows, but during those times you are absolutely dedicated to the task.

It seems that different people prefer different kinds of structuring of their day. There are lots of methods out there. I use a version of the rhythmic philosophy I call the "half-hours method," and since I've never seen it described before, I'll tell you how it works. But perhaps the most important things to take away are these: you need *some* kind of strategy for concentrating and getting work done without distraction, and you might need to *experiment* to know which one is right for you.

Even if my method doesn't work for you, keep searching until you find a method that does.

REMOVE ELECTRONIC DISTRACTIONS

Keep your email window closed or minimized so you can't see incoming emails. If you have a sound or other notification when a new email comes in, disable it.

For most people, though, their primary source of distraction is their smartphone. I'm shocked (and annoyed) at how many of my phone apps ask me for permission to send me notifications. I understand why they want to do it—by sending you notifications, you use their app more, which means more revenue for them. But keep in mind that you are giving that app permission interrupt you. When an app asks you for permission to send you notifications, you should translate the request into: "May I slow your progress toward your most important goals so my company can make more money?"

If you already have lots of notifications pushed to your phone, remove as many as you can stomach. Ask if those notifications have made your life better or worse. If you need to check on an app, you can schedule a time to do it, rather than the app deciding when you should give it your attention. At the very least, do not allow the apps to make noise or vibrate your phone. Take control of your mind back from your phone!

Even I have *some* notifications: texts and phone calls. That's it. Luckily for me, I don't get a lot of them. If you do, or can't resist getting notifications from lots of apps, then I recommend turning off your phone when you want to do focused work. I mean *powered off*. If you can't get yourself to do this, at least turn off the sound and vibration and put your phone out of sight in your pocket or purse, so you don't see the distractions. Some studies show that even having a phone visible, with no notifications, is distracting, like a bowl of candy you're constantly tempted to sample.[74]

The Project List

Key to the half-hours method is to have a project list. This is a prioritized list of everything you have to do in your life that takes more than 20 minutes.

It's important that you keep this list curated. What I mean by that is you attend to it frequently, and make sure it has *all* of your projects on it. Many people report that simply making this list gives them an enormous sense of relief. The reason is that all of the things you need to do are otherwise swimming around in your head, distracting you from what you're trying to work on. This is because if you don't "rehearse" thoughts, you are more likely to forget them.[75] So if there's something important you need to do, your mind rehearses it so you don't forget. But if your mind is rehearsing lots of things, it's distracting. This, I believe, is one of the reasons we get overwhelmed: we cannot work on any given thing, because we are distracted by all of the other things we know we have to do!

That said, I cannot find any scientific research on the concentration benefits associated with making a project list. One bit of research is suggestive: E. J. Masicampo ran a study where he had people begin to work on a problem, and then made them work on something else. The unfinished task interfered with the second task: they had intrusive thoughts. But when they were allowed to make a plan for how to accomplish the goal, the distracting thoughts went away.[76]

The project list is ordered in terms of urgency, importance, and whether it needs to be put on someone else's desk. The thing you most need to work on is at the top. This is important because although it's easy to prioritize your projects when you make a list, your mind isn't very good at doing it automatically.[77]

Projects include work commitments, but also ambitions in your private life and hobbies. That is, reorganizing the basement is just as much of a project as writing a proposal for your boss.

There is a question as to whether or not these things should be on the same list. For me, they absolutely should be, because, as a professor, my time is extremely flexible. I have a lot to do, but not a lot of constraints on exactly when I need to do them. So I can go shopping in the middle of the day, and work all night, if I want to. If your life is like this, then I recommend having a single project list for all of your endeavors.

PRIORITIZING YOUR PROJECTS

When looking at two projects, most of the time people have a pretty good feeling about which one is higher priority. When two tasks are so close in priority that you don't know which one to put first, just pick any order and don't worry about it. The closer they are in priority, the less it matters what order those two are in.

Importance is how much getting the project done will help your life. Urgency is how soon it has to get done. Even if finishing your home renovation is more important to you than filing your taxes, it's only the taxes that has a real deadline, so it might get higher priority as the deadline approaches. The other thing to keep in mind is whether you can get the project off of your desk and on to someone else's. Getting someone else to work on a project that is also yours is a great way to get things done without you having to do anything.

When you first make your project list, it might be overwhelmingly long, and it might take tedious hours to rank order every single thing. In cases like this, break down the task into something more manageable. Maybe the most important tasks get an A grade, less important tasks get a B, and so on. Maybe the F tasks should be deleted. Once you have them in this rough order, then you can order within the grade: look at only the A-grade tasks, and order them, then do it with the B tasks, etc.

Because deadlines loom, and your life changes in many ways, you need to curate your project list and make sure the order is still good. Having the tasks in a system that allows easy reordering of tasks makes this much simpler. I use Google Tasks.

But if your work-work is only done during work hours, and you don't do any non-work projects during that time, then it makes sense to have separate work and home project lists. What also might make sense is to have different lists for different physical locations. Often this means work and home, but perhaps you sometimes work from home, and sometimes work from work. At one point I had a "computer" list. All my computers are synched with Dropbox, so being on an Internet-connected computer constituted a place, as far as my project lists are concerned. I also had a "car" list for things to do when I was out and about in my car.

This list can be electronic or on paper. There are benefits to electronic lists: they can be in the cloud, and thus always with you, and the items on them are easier to reorder. I use Google Tasks, and with it I can create a task from an email—that is, from Gmail I can click a button and it becomes a task in Google Tasks that is linked to the email. This is useful because many things I have to do eventuate in sending someone an email saying it's done. It's nice to be able to click right to it, rather than saving the subject line of the thread or something, so I can search later for the relevant email—which is how I used to do it.

There is a benefit to having a paper list, though, because it makes it easier to keep the list from getting out of control in terms of size. If you rewrite your list every day, or even every week, you will drop unimportant things from the list simply because they aren't important enough to even write down again and again. Having a short project list is less stressful. My electronic list is way too long.

You might be thinking "great, I already have a to-do list." But please note that a project list is not a to-do list. A to-do list is full of next actions that you can do and cross off. Things you can get done. Projects are longer-term things that you *work on*. Each project has a next action, but it's not the next action that goes on the project list. For example, I have a project to write this book, but on any given day that might mean reading a paper, processing something I've already read, editing, trying to sell it, or simply writing. Some projects stay on the list for years—each of my first two books took me four years to write.

Each item on the project list has some next step associated with it. Sometimes what you need to work on next for the project will be obvious, and when it is you don't need to actually write the next step on your list. But sometimes it's helpful. So I might have a project item called "Write book," but if I want to remember that I need to edit the book to include what I've learned from an article by Jonathan Haidt, I might change the entry to "Write book: process Haidt2003.pdf." There is always a next step, and if you don't know what the next step is, the next step is to figure out what the next step should be.

Also, the order has to be *maintained*. When you add something to the project list, make sure it's in the right place on the list, in terms of urgency, importance, and whether working on it can get it off your desk and onto someone else's. The list needs to always be complete and ordered.

As you reflect on the things you have to do in your life, think about how you might want to engineer your life so that you are doing more things you like to do and are good for the world, and fewer things that waste time. You might want to put some thought into changing your lifestyle so that you have fewer maintenance tasks, for example. If you live in a condominium or an apartment, you pay extra money to avoid lawn mowing, gardening, shoveling snow, and so on. The bigger your house is, the more maintenance work there is to do. Would it be better for you to hire someone to do that stuff or consider downsizing?

Quantity vs. Quality

Sometimes people worry that the way we think about productivity is flawed in that it emphasizes quantity over quality. What's the point of writing 1,000 words a day if they aren't any good?

Let's compare Beethoven and Haydn. If you ask "Hay-who?," you're not alone. Joseph Haydn was a classical composer of considerable skill who wrote over 106 symphonies, and is sometimes known as the "father of the symphony." But he's not nearly as famous as Ludwig van Beethoven, who only wrote nine. What a layabout! Who was more productive?

Obviously, creating even one masterpiece is better than creating a bunch of just really good things. The question is, how do you maximize your chances of doing really great things?

Evidence suggests that the best way to get quality *is by generating quantity*. Beethoven seems to be an exception, because studies of great scientists, inventors, and artists show that the ones who produce the *best* stuff, are, on average, those who produce the *most* stuff. Most super-successful people produce an enormous amount of content, like painter Salvador Dalí, who produced hundreds of sketches before starting a painting. That is, for the most part, the probability of a given scientific paper or work of art turning out to be an impactful masterpiece is roughly the same for famous and non-famous people alike—the difference is that the famous people tend to produce so much more that they have more famous pieces.[78] This does not necessarily mean that you should try to produce quantity without any notion of quality—the causal connection is not clear, here. But it's something to be mindful of.

The Half-Hours Method

Now that you know what you have to do, the question is when do you work on what? Here is where you make your "half hours."

At the beginning of the day, take out a sheet of paper and down the left side write one half hour per line, starting with the next half hour and going all the way until you plan to go to bed. So if it's 7:15 A.M., make the first half hour 7:30 A.M. If you're up for sixteen hours per day, you will have around thirty half hours on the paper.

The next thing you do is to go to your calendar, and fill in all of your scheduled activities: meetings, social events, exercise classes, things like that. Be sure to include commute times. Also put in time for meals, cooking, and a nap, if you do naps.

You might then find yourself with shockingly few half hours left. What you do then is simple: go to the top of your project list, and put the first item there (the most important/urgent) into the first free half hour you have of the day. Then put the second item on the project list into the second free slot, and so on. Also put in breaks, time to exercise, and leisure activities. If you don't want to work after five, then fine, but schedule your half hours anyway—just fill time after five with the leisure activities you plan to do.

I know this sounds insane. You are probably thinking that you can't get any meaningful work done in a half an hour on a project. You probably think that your mind will be scattered, switching tasks so many times. And maybe it's not right for you. But I also want to say that no matter how certain you feel this is not right for you, you really don't know until you actually try it for about three days.

Here is why I think it works for me.

It's long enough to make significant progress. Even though you're only working for a half an hour on something, it's enough of a chunk of time to make significant progress on a project, provided that you actually work on the project for a half hour. That is, no checking email, social media, and no surfing the web. (Of course you can look up something very specific for the project at hand, but that's not "surfing.")

I will admit that there are some projects that require more than a half an hour at a time to be productive. Sometimes, when you're programming, you might need to leave the code in shape for someone else, and this might take more debugging time. Or if you're painting with acrylics, the setup and cleanup take so long that you will need more than a half hour for it to be worth your time. Basically, anything with a big setup or cleanup can get several consecutive half hours. But this holds for fewer projects than you think.

Writing, for example, is something that lots of people want or have to do, and most people think that they need a big chunk of free time to make any progress. And I know where they are coming from. At the beginning of a writing project, or if you haven't touched it for several weeks, you might spend forty-five minutes just getting your bearings in the project and figuring out what you even need to do next.

Robert Boice ran an experiment with three groups of professors who all wanted to write more. Writers wrote in two phases. In the spontaneous phase, they scheduled five writing sessions per week, but only wrote when they felt like it for ten days.

In the second phase, the first group stayed in this "when they felt like it" mode for an additional five days, and then for the twenty days after that, were asked to write daily, no matter what mood they were in. But there was no quota, no reward, and no punishment.

Another group was asked to write three pages per session in the second phase, and if they didn't, they would have to donate $15 to a despised organization. That is, a charity that they fundamentally disagreed with. They were in this phase for thirty days.

The last group was a control no-writing group, which did no writing for fifty days, but were asked to write down any creative, new ideas they had.

The results are startling. The no-writing group wrote about .1 or .2 pages per day (that means a page every five or ten days). The spontaneous writing produced .3 or .4 pages per day. The group that wrote whether they felt like it or not wrote .9 pages a day, and the group with the threat of punishment wrote 3.2 pages per day.

Regular writing also produced a lot more creative ideas. The no-writing group produced around .08 per day, the spontaneous about .26, the regular writing group .63, and the forced writing group 1.39.

You should also know that all of the participants in this study believed that they needed lots of uninterrupted time to get any writing done.[79] This study shows that when you force yourself to write, the pages (and ideas) will come. Don't wait for inspiration. A half an hour is enough time to get something significant done.

There are a few important lessons from Boice's study and others like it. I think the most important one is that these professors, who are basically professional writers, were absolutely wrong about what kind of time management system would work for them. They believed that they could only write productively if they had lots of time to do it, but when these same people were put in an experiment, they were shown to be wrong. The lesson I want you to take from this is that when you read advice, in this book or elsewhere, you might have an opinion, perhaps even accompanied by a feeling of great confidence, that it would or would not work for you. Don't be so sure.

Boice's study is legendary, but has been criticized on a few grounds: all of the participants were having trouble with their writing, and the successful ones had an enthusiastic, in-person coach—Boice himself. The author even admitted that after the study was over, the participants fell back into their bad habits. Many participants dropped out of the study because they found all of the listing and charting of their behavior onerous. Helen Sword surveyed over a hundred successful academic writers and found that only two wrote the way Boice suggested![80]

NON-PRODUCTIVITY SCENARIO: ERIKA

Erika has to work on a project, and that involves reading a very boring report. She thinks she doesn't have the mental energy to read this report, so she thinks about other things she has to do. She finds another important project, but then discovers that also happens to require reading as a next step. Although the reading looks much more palatable, she pauses: "If I'm going to be reading," she thinks to herself, "I might as well read this boring report." So she turns back to the report. But it's awful; she can't read it. Caught in a standstill, she ends up checking social media and getting nothing done. After an hour, she feels like a failure.

Returning to the half-hours method, although the first two days of this method might feel frantic, and like you're not getting anything done, by the third day you will realize that, when you start working on a project, you pick up right where you left off the day before. You don't need a lot of ramping-up time, because you were working on it so recently. That ramping-up time is required for projects your mind hasn't thought about for weeks. A half hour, when done every day, is long enough to keep it fresh enough in your mind to make progress on it without catching up.

At the same time, a half hour has the benefit of being a *short* amount of time. Everybody has to work on projects they dread. (See the sidebar "Non-Productivity Scenario: Erika.") Some are boring, others difficult, others involving disturbing information, and some have all three. If you feel that you have to spend hours doing a task like that, you will be likely to procrastinate working on it, because the idea of working on a dreaded task for so long is painful. But if you commit to yourself that you only have to work on it for a half an hour, it's much easier to do it. It's even better if you schedule something fun, like a project you like, or a break, as a reward for doing it. I often find that dreaded tasks get done in just a few days, only working half an hour a day, without much pain. No matter how bad it is, I can do it for half an hour.

Similarly, having *only a half an hour* to work on something you know is so important puts the fear of God into you to actually take advantage of what little time you have given yourself. Often, when I do my half hours, I find that I don't have enough half hours in the day to work on everything important to me. (This probably means I need to drop some projects . . . that's something I'm working on.) So when the half hour for this or that task comes up, I get started on productive work immediately. I think to myself, "I only have half an hour to make progress on this important thing!"

When my beloved was finishing up her master's degree, she'd left campus and was back at home. Away from the context of school, with its reward structure, she was finding it very difficult to get to work on her thesis. She complained about it, and I timidly asked her if she'd made her half hours. (Timidly, because giving advice in a spousal relationship should always be done with compassion and extreme care! It's not always easy to be married to someone obsessed with optimizing.) She reluctantly made her half hours. A little while later I tried to ask her a question, and was immediately shut down: "Jim, I only have ten minutes left in this half hour to work on this so I really need to concentrate." From a woman who hadn't worked on the thesis in three weeks, now she had a sense of urgency that ultimately got her thesis finished.[81]

The other benefit of the half an hour being so short is that it allows you to work on multiple projects, and possibly all of your important projects, every day. This is important because they are all kept fresh in your mind. You will see connections to the projects you are working on every day in the things you read and the people you talk to. If you only work on one thing day after day, your other projects are far from your mind, and you might not see the opportunities that arise, like asking a geology expert you meet at the dog park a question relevant to one of your projects.

If you've ever been stuck with a problem, only to have the solution pop into your head while you're taking a walk or a shower, then you have experienced the power of what creativity scientists call "incubation." The theory behind incubation is that when you let a difficult task go, and focus the conscious part of your mind on other things, your unconscious mind continues to work on the project. A review of many studies showed that this effect is real, and is particularly useful for tasks requiring creativity. Longer incubation times are better, and doing relaxing or low-demand cognitive tasks makes it work better as well.[82]

But incubation cannot work on a problem that you haven't been thinking about recently. If you think of your mind like a police station, certain tasks become "cold cases," projects that you never think about or make progress on. With the half hours method, you're working on all your most important projects every day, and so they are fresh in your mind and available for your unconscious mind to incubate. You don't want your mind to turn any really important projects into cold cases.

Working on your most important projects every day has another benefit: you end up starting work on projects early. Sometimes, when you begin work on a project, you quickly realize that the scope of it was bigger than you thought. You might need help from someone. You might need to gather materials. One of my professors in graduate school, Charles Isbell, said, "It's fine to wait until the last minute. The secret is knowing exactly when the last minute begins." But in the real world, this is always uncertain. You don't know how much time a project will take (and people always underestimate task completion times), and new things come up that interrupt your projects. You don't want to start something at the last minute only to realize that you need to ask someone for something, and who knows when they will get back to you.

Because I use this method, for the most part I don't have to worry about deadlines. Things just get done long before.

The idea of working on something different every half hour might sound very stressful to you. And I will grant you that the first two days of it will feel very frantic. But for me, it's much *less* stressful. Have you ever had a million things to do, and while you're trying to get work done, you are distracted by thoughts of all of the other things you need to do? That's your mind rehearsing. You keep thinking of the other things you need to work on so you don't forget them. But keeping your project list updated relieves this stress. You don't have to think about the other projects because you have written them down, allowing extreme focus and productivity.

NON-PRODUCTIVITY SCENARIO: TED

Ted has a million things to do. He goes to bed stressed, and wakes up happy only until the crushing weight of his responsibilities comes to mind. Facing this mammoth pile of tasks, he looks to completely escape. Instead of doing any work, he does something fun and distracting. He watches a good television show. It's engaging, funny, and most important, it allows him to not think about everything he has to do. He's happy—while he's watching. But when it's over, he's wasted even more time. At the last minute, he finally is forced to get to work, but his work is rushed, late, and of poor quality. Ted dimly reflects that he maybe should not have watched so much TV.

You also might be distracted by thoughts that maybe you should be working on something else. So you're trying to write a proposal, but you keep wondering if maybe you should be brainstorming your next steps on some other project. I find that when I'm feeling overwhelmed, what it really means is that I have so many things to do that the very thoughts of those things keep me from making progress on any of them.

Many executives have assistants who schedule their days for them. This way, they can feel undistracted, confident that someone else has done the worrying about what should be done when.[83] With the half-hours method, you are acting as your own executive assistant. Act like your assistant for a few minutes at the beginning of the day, and then trust your assistant for the rest of the day, relatively stress-free.

In the sidebar "Non-Productivity Scenario: Ted" we see one possible outcome of being overwhelmed: not doing anything, and doing something fun instead. By strictly managing your time, these ultimately non-enriching leisure activities are easier to resist.

But when you make your half hours for the day, you've done some important stress-*relieving* things: you have a plan in which you're going to make progress on all of your most important projects, and you're going to work on the most important ones first. This gives you a great feeling: that what is in that little slot next to your current half hour is the most important thing you should be working on right now. Freed from worry that you should be doing something else, you get things done.

Sometimes, when people first start trying to use this method, they plan half hours starting whenever—like, a half hour starting at 4:46. Once I spoke to someone who was trying to get started on something, but she kept procrastinating. She was supposed to start at 4:30, but she stalled, checked email, etc. Then at 4:40 she said, "Okay, the half hour starts now." Then she dilly-dallied again. This is one reason I endorse using the half hours on the clock. If you schedule 4:30–5:00 for working on your proposal, and you screw around and miss it, tough luck. You'll have to get to it tomorrow. This is a mild form of self-punishment that will help you get started working in the future. If you miss the half hour, it's gone, and you're on to other things.

After doing this method for over fifteen years, my internal clock is pretty good at knowing when the half hour is up. I have my computer desktop wallpaper change every half hour to remind me. If you find that you work through your half hour and fail to notice that it's up, you can set your watch to beep, or even set a timer.

Another trap is that sometimes people make their half hours, but then start on the first project, get on a roll, and just don't stop. They are thankful that the half hour plan got them started, but are happy to abandon it when they get on a roll.

I'm wary of this abuse of the method. Part of the power of the half hour commitment is that it's *only* a half hour. This helps you get started on difficult tasks. The promise to yourself that you only have to work on it for such a short time is motivating to get started. If you start going long, your mind stops thinking of it as "just" a half hour. It becomes "a half hour and then possibly more," which is a different commitment. So yes, even when I'm on a roll, I stop after a half hour. I'll pick up where I left off tomorrow, when I'll still be on that roll.

The other downside of abusing the half hour is that there are all of those other important projects you were supposed to be working on today but didn't. That said, I'm not a complete slave to this method. I will make exceptions once in a while. But after many years of experience with it, in general things go better if I stick to the plan.

I don't work on Saturdays or in the evenings, but I also plan my half hours for my leisure time. This might sound weird, but many people actually have trouble enjoying leisure time. It's too unstructured, there are too many choices, and people end up just watching TV or doing something that is very easy to start doing (as opposed to coordinating to hang out with people you care about, which makes you happier but requires more effort). Believe it or not, studies show that you are happier and less stressed by scheduling leisure time.[84] You can keep a list of things you know you like to do, and schedule your free time to try to get to them. Maybe calling a distant friend, reading, playing pool, going for a walk, and yes, even watching TV. You just want to be deliberate about it, so you can actually get to do the leisure activities you would like to get to, and have enough time spent with the people you care about.

Again, this is something you might want to experiment with. I have had one major exception to the half hour: writing nonfiction books like the one you're reading now. I often give myself an hour to work on it, first thing in the morning. There are two reasons for this: first, it's the most important thing in my life, so I want to dedicate more than a half an hour to it, and second, I enjoy it so much that I'm not tempted to procrastinate.

Sometimes I am under deadline and need to spend more than a half hour on a task in a given day. But even for this, I don't do the time consecutively. I will work on it for, say, three half hours, spaced out over the course of the workday. This absolutely doesn't work for some people. Some people have high task-switching costs, and it really helps for them to have longer, uninterrupted periods of concentration.

In general, I find that if I give myself more than a half hour—like, say, an hour—to work on something, I find myself more tempted to futz around, checking email and social media. After all, I have a whole hour. Then I find that fifteen minutes have passed, and I haven't gotten started. If I only have a half hour, I get right to work.

But you might be different. Is a half an hour the right amount of time for you? Maybe not. If you are not tempted to procrastinate, maybe a longer period will work for you. An analogous "hour method" might work better for you. For lots of people, those large chunks of time are few and far between, and they end up working on the project only rarely, losing all of the benefits of keeping the projects fresh in your mind. One study suggests that the average worker is maximally productive with a fifty-two-minute work stretch, followed by a fifteen-minute break, and repeating. But this is just an average. Computer scientist Cal Newport reports that many very successful people had widely varying stretches of work vs. break time.[85]

Even if you want to have two-hour blocks, you can use a method similar to the one I've described. Suppose your schedule looks something like this:

7:00 A.M.	wake
7:30 A.M.	breakfast
8:00 A.M.	work on project one
10:00 A.M.	coffee break
10:30 A.M.	work on project two
12:30 P.M.	lunch
1:30 P.M.	work on project three
3:30 P.M.	break (nap or coffee)
4:00 P.M.	work on project four
5:30 P.M.	stop working

Here you've got three two-hour blocks for uninterrupted work. The downside is that you only can keep four projects fresh in your mind. If you use the half-hours method, you might have as many as twelve half hours during a day of working hours (almost everyone will have far fewer, but let's look at the most generous example). This means that you could potentially work on twelve projects "at once." If your working size is one hour, it's half of that, and if you work in two-hour chunks, you only have three or four projects. If you have more projects than you have time slots, then they are not all on your plate at once, and you're working more serially—one, or maybe four, projects at a time. So another consideration is to choose the size of your working time based on how many projects you need to have on your plate at once. If you don't work on everything every day or two, you forget where you left off, and have a larger start-up cost when you get back to the project.

My point is that no matter what your perfect amount of time is for working on a project, you can use some version of the method described here. Experimentation will help you tweak the parameters to optimize what's best for you.

Project List: Master Class

In addition to my project list, I also keep a list of things I need to do that take less than a half an hour. I call this list TCOB, or "Taking Care Of Business." This has things like making a dentist appointment, sending someone a question over email, etc. The TCOB list is more like a traditional to-do list than the project list. Each day I try to schedule at least a half an hour for this list. I put "TCOB" as the task for whatever half hour, and when that time comes I just crank through the TCOB list.

I have yet another list called "Waiting On" for everything I'm waiting on from other people. Often I'll work on a project until I can't do anything else until someone else does something. Maybe I'm writing an article with someone and it's their turn to do an edit. I move the entry from the project list to Waiting On. Or if I submit a poem to a poetry magazine, I'll put an entry into Waiting On about it until I get a response. (With Google Tasks, it's easy to move a task from one list to another.) To process this list, I go through and nag everyone on it every month.

I keep another list for things to do when I don't have the mental energy to do what I had scheduled for myself. I call this list "Too Tired." On it are things that are not urgent, nor particularly important, and don't require much cognitive ability. Examples include unsubscribing from unwanted email lists, reading Star Wars comics, practicing music, and sketching.

When people think of what productivity really means, they think of getting things done. And sure, how to get things done is a big part of productivity, but what's even more important is choosing what things *should* be done. Having great time and energy management, delegation abilities, and so on do you and the world no good if the goals you are working toward aren't worth pursuing.

Sometimes you can *feel* productive when you're actually not. Being busy and responding constantly to things that come your way can put you in the zone. You feel like you're solving problems left and right, and it feels productive. But often, these responsive activities unmatter. You could respond quickly to things for a year and have little to show for it. (I respect that many have jobs that require constant responding—but you have many leisure hours to work with every week, too.)

Some goals matter. They make you happier and make the world a better place. Some goals unmatter, and have a negligible effect on you and the world. And some goals antimatter: somebody carrying out atrocities in an efficient manner certainly doesn't help the world.

Even if you have lots of goals that are laudable, there's the question of whether you have too much on your plate. What I mean by "on your plate" are goals that you plan to make progress on, say, a little bit every month. Here we run into real limits on your time, energy, and memory.

This means that your project list needs constant curation. The more projects you have, the longer it will take (in terms of days) to complete them. For most people, this amounts to cutting projects you're excited about. Personally, I have many things I'd love to do, but *never do*, simply because other things have taken priority. Some people have an idea in their head that any stopping of work on a project qualifies as "quitting," and is something to be avoided.

You should purge the notion immediately from your head that quitting is bad. If you finish every book you start, if you complete every project you begin, two bad things will happen. First, you'll finish a lot of projects that unmatter, and maybe even antimatter. You can't always know the value of a project before you begin working on it. Second, this policy will make you risk-averse. If you know that you'll always finish what you start, you're likely to avoid project ideas that you are not sure will be good. This means doing a lot of safe stuff, and avoiding risk. If, on the other hand, you practice "strategic quitting," you can feel free to attempt risky projects, because you know you can just dump it in the future if it turns out to be a dead end.[86]

Success requires grit and working through difficult parts of a project, but giving up on a project because it unmatters or antimatters is very different from giving up out of laziness.

Many projects are collaborative. If you have the kind of collaborative project that you work on, then you pass it to someone else to work on, this project might get some extra priority on your project list in terms of what you work on soon. Getting a project off your desk and onto someone else's is a great place for your project to be. I slightly prioritize getting the ball in someone else's court when I choose what to work on every day.

You can *make* some projects collaborative, or make them not your projects at all, by delegating. Most people think that delegation is only possible if you have people working for you in your job. But don't forget that you can often hire people, on your own dime, to do some of your work.

I'll start with something you're familiar with: cleaning your living space. You can do it yourself, or you can hire somebody to do it. Which should you pick? Lots of people do it themselves because they think it's cheaper. But is it?

To know for sure, you have to think of how much your time is worth. But your time's value is relative to the activity you're talking about. I hate weeding, for example. So I think to myself, how much would a stranger have to pay me to do an hour of weeding in their garden? At the time of this writing, probably about $200 Canadian (on a beautiful day). So this means it makes sense to me to pay someone else to do it in my own garden if it costs me less than $200 an hour to get it done. Now, this is relative to your wealth and income as much as it is how much you enjoy it. If you have a low income, or love gardening, you might weed a stranger's garden for less.

Take something you do in your life that isn't fun. Think of what you would need to be paid to do it for somebody else, and if you could get someone to do it for you for less than that, and you can afford it, *then you should hire someone to do it for you.*

My beloved and I were taking turns cleaning the house every week, and at some point she used this logic on me. She asked how much a neighbor would have to pay me to clean their house. The amount was far higher than a cleaning person, so we hired somebody. This is one of those times when it's good to think like an economist (it's not always). There's a saying in positive psychology: "If money doesn't make you happier, you're not spending it right." A study by business researcher Ashley Whillans tracked happiness and what people spent money on. Turns out you can be happier spending money to have more free time. She had experimental participants spend $40 on either something time-saving or on a material purchase, and people were happier when they bought themselves more time.[87]

I started with the cleaning example because it's common to hire a cleaning person. What I want you to do is to think about all of the other things in your life you don't like doing in just this way.

I'll give you another example from my own life. I'm an academic scientist, so part of my job (and calling) is to publish scientific papers in journals. One of the silliest, most annoying things about the process is that each journal requires its authors to format the paper according to that journal's specifications. Some want APA citation style, some want footnotes, and on and on. So when you want to submit to journal A, you format the paper one way. If it's rejected, you find journal B, and you might have to reformat the whole paper—not when it gets accepted, but *just to submit it*. Estimates suggest that formatting had a median cost of $477 U.S. per paper, which works out to $1,908 per researcher every year.[88]

Recently I submitted a paper and it was sent back because it wasn't in their preferred style. I went to some effort to reformat it, only to have a "desk rejection." It was rejected by the editor without being sent out for review, because the nature of the research didn't fit the journal's vision. Did I really need to change the formatting for them to make that decision? I won't mention the name of the journal to protect the anonymity of *The Journal of Creative Behavior*.

Rant over. Anyway, reformatting papers is a waste of time and I don't like doing it. So I now hire it out. I went online and posted a job advertisement for a personal and administrative assistant. I ended up hiring someone. She's university educated, and can do a lot. I pay her by the hour. Now, when I encounter a task that I don't want to do, or there are other things I would rather or should be doing instead, I'll ask if she can do it. If she can, I send it to her, because her hourly rate is lower than what mine would be for the same task.

If she can't do it, I try to find someone else. I had to have a paper reformatted in LaTeX, a complex computer science formatting system, and my assistant doesn't know how to use it. So I went on Fiverr.com and found a LaTeX expert who will reformat my paper for $35. Fiverr.com and websites like it allow you to connect with people to do gigs, often for a very low price.

This requires a change in mindset that takes a little time to get used to. You have to practice thinking about whether the task you have can be delegated.

You can apply these delegation strategies to your non-job endeavors, like child care, lawn maintenance, your taxes, and so on. You can also apply them to art projects. I've written a few plays, and even had a few produced. But the process of trying to get plays produced is very tedious. You have to submit them, get rejected, put them in formats, include cover letters, etc. So I just hired my assistant to do it for me.

Chris Bailey of *The Productivity Project* has these tips for finding an assistant: You should pay more for a quality person, because incompetent people will take more time than they give you. Always check references. A different time zone can be a perk, because they can work on things while you sleep, and you don't have to wait for it.[89]

This is great if you have more money than time.

Can you do this with your job? This is tricky, and you need to be really careful about it. You're getting paid some amount of money to do your job, and it might be that there are parts of your job that you could hire someone else to do for you for less money. This might not be allowed by your employer, so be careful. There are matters of data privacy, and so on, but your work conditions actually might allow you to do this. It's easier if you work freelance and can sub out personal tasks so you can take on more quality freelance work.

I have a friend who used to change jobs every few years. What he would do is negotiate his salary, and when they came to an agreement, he would basically say, "Okay, now how about this. You take 20 percent off of that salary and I don't work on Fridays." A surprising number of his employers over the years said yes. This was a friend with lots of other things in his life, and he wanted time for them. What actually happened, though, was even better than that. The company had projects that often needed intense work. So they'd ask him to work on Fridays, and he usually would. But after five such weeks, he'd have earned a week off of work, during which he could take a real trip, or intensely work on his other stuff for a full week.

What my friend was sacrificing was income, and what he gained was time to do many of the other things he found rewarding in life. For him, it was worth it. Whether such a situation would be good for your life is, of course, something you would need to decide for yourself. But a 20 percent salary cut might just be something you habituate to, and the time off might allow for a noticeably happier life of experiences and time nurturing social relationships. Just because our culture thinks you should work five days a week doesn't mean that's the optimal schedule for you.

WHAT SHOULD YOU DO EVERY DAY?

There's a real benefit to doing something every day. Daily commitments to things reliably build up over time, and when they become habits, they are unquestioned parts of your day that don't require as much willpower as non-habit activities. The problem is that there are many things that you are told to do every single day. It's one thing to say that you should meditate every day, or that you should exercise every day, or that you should read every day, but the bigger question is how many things are you going to spend your time doing every single day?

For example, in the past few years I have had nagging desires to do the following things every day: my physical therapy stretches, exercise, meditation, journaling, writing, napping, doing gratitude exercises, reading, clearing out my inbox, practicing my memory palaces, practicing guitar, reviewing flashcards, drawing, writing a poem, practicing programming, studying math, and playing with the dog. But if I gave each one of these things a half an hour every single day, it would take me over eight hours a day! Even if I was independently wealthy and didn't have to work, this is too much time to spend, particularly since most of it is sharpening the saw. If you're working eight hours a day on maintenance tasks, when are you ever going to get your real projects done?

Of course this is an even more ridiculous prospect if you have a full-time job, have to commute to it, and have a family to attend to.

Unfortunately, the only way out of this is to prioritize. Just because there's evidence that you should do something every day doesn't mean that you need to find a place for it in your schedule. There's just not enough time to do all of the things that evidence suggests you should do daily.

One way to do it is to figure out how much time you have every day to do things that you should be doing every day. Maybe you only have a half an hour, maybe you have an hour and a half. Make a list of the things that you wish to do every day, and then order them in terms of priority. Are there any things on this list that can be done every other day? Or perhaps once per week? Maybe instead of meditating and exercising every day, you exercise Monday, Wednesday, and Friday and meditate Tuesday, Thursday, and Sunday. Be careful though, because some things really should be done fairly frequently, either because they really need to become a habit, or because they require the daily frequency to be effective. For example, if you only work on a novel a half an hour every week, you will probably lose momentum and have a very difficult time making progress. If you have default half-hours for every day of the week, you can put in the things that you prioritize to do every day or every week.

The Half-Hours Method: Master Class

For years I rewrote my half hours on paper every day. Then I leveled up to using a spreadsheet on Google Docs. I like it because it's in the cloud, so I don't have to remember to carry a piece of paper with me when I go from home, to work, and to the bubble tea place to get things done.

The first column are the day's half hours, from 4:30 A.M. (sometimes I get up that early) until about 10:30 P.M. The second column is where I put what I'll be doing every half hour; this column changes every day. The next seven columns are for each day of the week, and they hold my regularly scheduled events for each day. For example, I eat breakfast, walk the dog, go to bed, etc., at the same times every day. I teach classes and have regularly occurring meetings on the same days every week. So on Monday, I copy and paste the Monday default column into the "today" column, and then work from there. It saves me a bit of time.

Competing Methods

Mine is not the only time-management advice out there. I'll address a few others.

One method is to work only on the most important, urgent thing in your life until it's done, and then move on to the next most important, urgent thing. This is the method recommended in the book *Eat That Frog!*, which explicitly states that "The ability to concentrate single-mindedly on your most important task, to do it well and to finish it completely, is the key to great success, achievement, respect, status, and happiness in life."[90] What's good about this advice, and why it helps some people, is it helps one focus on the most important thing and to actually get working on it. The problem with it is the advice to single-mindedly work on it until you finish it completely. In practice, this can be a disaster.

NON-PRODUCTIVITY SCENARIO: ADEN

Aden works on only the most urgent and important project he has until it is finished. When he finally gets done, he finds he has several projects on fire, that is, with imminent deadlines. When he turns to the next most urgent one, he realizes that he needs something to get started from someone else, who happens to be bad at responding in a timely manner. Aden is screwed.

In the sidebar "Non-Productivity Scenario: Aden" we can see how this can go wrong. Suppose you have a really important project at work. Say you have to do software development to meet a looming deadline. Is that more important than submitting your travel receipts, or making an appointment with the dentist for a checkup? Sure is. But you'll always have a looming deadline, if you're a software developer, which means that if you're following this method you don't call your dentist until your teeth are in such pain that it becomes finally more important than the current software deadline. Of course you've left it far too long.

Let's take writing this book as an example. Book writing is the most important project in my life. I estimate that it took me about 300 hours to write. At forty hours per week, the book would take me seven and a half weeks if I did nothing else. But this would have been seriously impractical. Most people cannot write for eight hours a day, every single day. The other problem is that nothing else would have gotten done in those seven weeks. No bills paid, no papers graded, no exercise. And at the end of it all, the most important thing would be my *next* book, so those other tasks would *still* never get done.

The half-hours method has you working on many projects every day, so the small things in your life that seem unimportant in the short term, but are important in the long term, like exercising and getting medical checkups, get done.

Further, very few people can actually work on only one thing all day, every day.

Another competing method is the "pomodoro" technique. For this, you break your day up into half hours. In each half hour, you work for twenty-five minutes (that's one "pomodoro") and then take a five-minute break. You use a timer to keep on track. After about four of them, you take a longer break of fifteen to twenty minutes.[91]

Like the half-hours method, having only twenty-five minutes to work on something gives you a valuable sense of urgency. There are a few things I don't like about it, though. First is that sometimes you are asked to estimate how many pomodoros it will take you to complete a task. Or you are supposed to assign a completable task to one pomodoro. Because of the planning bias, I think this is a recipe for frustration. In the half-hours method, you don't plan to finish anything in a given half hour, you just work on it.

The second thing I don't like about it is that it deviates from the wall clock. As long as you're doing twenty-five-minute bursts with five-minute breaks, you can stick to the natural half hours of the clock. But once you put in a fifteen- or twenty-minute break, you're starting your next pomodoro at 11:15 or something, and that gets over when? Okay, 15 + 25 is 11:40, then a five minute break . . . You need to do more math to know when things happen, and you need a day planner that breaks things in five-minute granularities. Even planning when you can eat lunch becomes a bit of an onerous task. This is probably why they use a timer rather than the wall clock. But timers are loud, and annoying to set every twenty-five minutes. With the half-hours method, you just do something different every half hour, as indicated on the clock, no timer needed.

BURNOUT

"Burnout" is a feeling you get about work or some kind of work-like activity. Burnout resembles depression in lots of ways: fatigue, loss of passion, emotional exhaustion, feelings of professional ineffectiveness, feelings of lack of accomplishment, and cynicism.[92]

In 2019 the World Health Organization updated the International Classification of Diseases (to take effect in 2022) to include burnout as an occupational phenomenon.

Long hours can contribute to burnout, but only if some other things are in place: a feeling of a lack of control, a mismatch between effort and on-the-job rewards, emotional energy expenditure, and high demands. Without these other problems, long hours are not so bad. People don't mind long hours so much if they feel they are being effective and being appropriately rewarded for it.

Caregivers seem to be particularly in danger of burnout. Teachers, social workers, and other care workers can experience a big drain on their empathy, and have the highest rates of burnout. But it can happen in any job. Forty-four percent of Americans say they feel burned out sometimes.

People who are burned out have more trouble regulating their negative emotions, and have more problems with memory and attention, but I couldn't see the causal relationship in the science, just a correlation. Possibly because of the high stress, people who are burned out have a higher risk of heart disease.[93]

Fixing burnout for yourself includes all of the things that you'd do to reduce stress in general, like exercise and socializing, but another

CHAPTER 2: HACKING YOUR BRAIN, HACKING YOUR LIFE

important thing is to reframe how you think about your work. Some people work for the money, some work for a career, and for others it's a calling. This kind of reframing can give your mind a different approach to your work.

Although burnout is related to stress, it's a bit different—it's work-specific. You can feel burned out at work but perfectly energized and engaged in activities outside of work. Sometimes the trick to curing burnout is as simple as changing jobs, for those who can easily do it.

There are things that individuals can do for themselves, but employers can also change things in the workplace to reduce burnout in their employees. Lots of places, like call centers, give people breaks in a crappy breakroom, all alone. They sit there on their phones for the whole half hour or whatever. But employers can try to encourage friendship and community by, for example, encouraging employees to take breaks together, with a comfortable place to go, and having breaks at the same time. When people take breaks together, they get more out of it. They can socialize, which attenuates burnout. But they also end up talking about work, and helping each other out with problems they're having.[94]

Giving workers some control and autonomy over their lives helps, too, as does simply removing work that goes nowhere—like streamlining the amount of paperwork that needs to be done.

A hack on the pomodoro technique and half-hours method would be to replace the fifteen- or twenty-minute break with a full half-hour one, and try not to commit to finishing your tasks within the twenty-five-minute pomodoro. You can keep the breaks, just remember to use the last five minutes of the half hour for a microbreak. Although I don't do this, you might want to, because breaks are very important.

I should add one caveat to my discussion of the half-hours method. Although it's true I've used it consistently since about 2004, I deviated from it for the writing of the first draft of this book: I decided I wanted to write 1,000 words a day, and during the summer of 2019 I did so, and wrote it in a few months (this book is about 150,000 words). It was very demanding, and I wasn't able to do much else that was intellectually productive during that time.

The reason I was able to pull this off was because I really loved writing this book. And when that first draft was done, I went right back to the half-hours method.

Taking Breaks

Time-management advice can sound like it is trying to put you on an endless treadmill of work. But breaks are absolutely necessary for your happiness and even your productivity. A half-hour break before a test will improve the test score, for example.[95] Breaks at work result in more positive emotions, decreased burnout, and better attitudes toward your job.[96]

It could be that most people are so great in the morning because a night's sleep is basically one huge break.

During the day, breaks should be short and frequent. Even a microbreak, like taking a walk around the office for two minutes, is helpful for getting back to clear thinking. Physical activity is more effective than sedentary activity—so the walk is better than sitting and playing *Bejeweled*.[97] If, at the start of your half hour, you feel sluggish, do fifteen squats, or go up and down a flight of stairs.

Social activity is better than solo activity. Completely removing your mind from work activities is better than trying to take a break while working—for example, listening to music or eating lunch while you work isn't as effective.[98] It really needs to be a break.

Lunch turns out to be particularly good for restoration of your cognitive abilities, particularly if you choose what you do for lunch[99] and are able to detach from work. Eating at your desk does your cognitive abilities no favors.

How to Read More Effectively

Reading is mostly sharpening the saw, but as far as saw-sharpening methods go, it's very effective, and it's worth spending some effort trying to optimize how you do it.

Let's start with the problem of what to read. There is so much good stuff out there, you'll only get to a tiny fraction of it in your lifetime. Not only that, reading is competing for your time with everything else you do in your life, including cutting with the saw. So picking what you read should not be done in the haphazard way that most of us do it.

It's important to acknowledge that people read for many reasons. The same person might read scientific papers to get some facts about the world, instructions for better use of software at work, stories to their children to help their education and to bond with them, a book they don't even like in their book club, a potboiler novel on the beach, and erotica to be titillated. This means that, from a time-management perspective, "reading" isn't a category that makes a lot of sense, because it's *what* you're reading and *why* that determines what goal it serves, not the fact that you're reading at all.

Reading for Knowledge

Reading is one of the best ways to learn about the world we live in. When you read, you can easily skip sections you think might be unimportant to you (thanks for not skipping a section called "how to read"). You can slow down or reread something difficult to understand. You can stop and talk about a passage or idea, perhaps even digest it over the course of the day before you return to it. You can highlight or take notes. All of these things are much harder to do with video, for example, and our brains process it differently, too.

When reading for knowledge, you want to be getting information that's true and important. Reading a phone book from the 1980s has historically true information about people's phone numbers, but reading the phone book is a waste of time because that information is outdated and wasn't even particularly important to remember even in the 1980s. It unmatters. Someone's phone number is information, or data, but knowing a phone number doesn't give you knowledge, in the sense that you don't better understand how the world works as a result of knowing a phone number. This is an extreme example, but there are lots of things out there you can read that have true but unimportant information. Reading useless stuff might be fun, but it doesn't sharpen the saw.

Some information is very *timely*. Timely information can become quickly outdated. Getting information that will be outdated soon is less valuable, in general, than information that will continue to be true. Thus, you should limit the timely knowledge you take in to only what you really need to know. If you're a stock broker, you might want to know about the current state of the stock market, but for most people, this information is too quickly outdated to be worth attending to. Of more importance would be the long-term trends in markets. This is more enduring information that might inform decisions down the road.

I'm going to argue for one of my more unpopular opinions here, and that is that most people should not read, watch, or listen to things about current events, or what is usually thought of as "news." Knowing about flooding in a town across the country, or knowing that somebody far away was murdered is irrelevant to most people's lives, and doesn't inform any decisions they have to make. Another way to put this is that it is information that is not "actionable." More specifically, it is less likely to be actionable than other, more promising kinds of knowledge. I've heard people argue that you need to consume news to know how to vote, but most people agree that they consume far more news than is required to make an informed decision about who to vote for. Wouldn't two hours of research before an election tell you everything you need to know to make an informed vote?

What your mind is fighting against, however, is the very strong feeling, or conviction, that current events are important. It *feels* important to consume news, and to get up-to-the-minute updates on what's happening. Some people describe the desire to consume news as compulsive.[100] But if you think about what you've done with your knowledge of current events over the years, you'll be unlikely to come up with much beyond knowing what everybody else is talking about on break at work. News unmatters. I try not to consume timely news for this reason. If the world changes in a way that isn't enough to warrant a change to a Wikipedia page, for the vast majority of people it's probably not worth knowing. I don't read or watch any news, and I still get far too much of it from people constantly trying to talk about it with me.

But it gets worse: not only does news unmatter, in many cases it antimatters. The news skews your perception of the world to the negative, stimulating your amygdala with fear, and giving you a strong conviction that the world is full of evil people. Bad things are more likely to happen suddenly, and progress often happens gradually, making progress less timely or newsworthy.

CHAPTER 2: HACKING YOUR BRAIN, HACKING YOUR LIFE

Part of the problem is that our minds are wired to remember rare, unlikely events.[101] The news caters to our minds, and what we find interesting. To do this it reports on unusual things, which are exactly what you *don't* need to hear about if you're trying to understand how the world works! You need to learn the rules, not the exceptions. The exceptions are misleading; your mind naturally ties exceptions together to form bogus rules.

As an example, people think that violence is increasing, when in fact it is decreasing. Violence is decreasing over time, but the progress is not smooth. This means that at any given time, violence might be going up or down, but over long time periods, the trend is down. But because the news salivates at violence, it breathlessly reports individual episodes of violence, and the minor, temporary increases in violence over time, ignoring the overall trend toward peace. As Steven Pinker notes, the public is exposed to only to the upticks, and is unaware of the larger downward trend. This leads people to think that violence is getting worse, when in fact the opposite is true.[102]

Although news slants strongly to the negative, studies of which stories are shared on social media reveal that people prefer to share positive stories. Perhaps people actually want their news to be a bit less scary and depressing than the news media thinks.[103]

Some people think that they can minimize the selectivity of news media by reading from multiple sources. This ignores the fact that for the most part all news outlets have similar ideas of newsworthiness. They might give different political slants on things, or differ in their focus on certain groups or geographical areas, but they all suffer from many of the same problems: priority for individual events and not trends, negative stories over positive ones, and recent changes to the world. Since these are baked into the very idea of what "news" is, you can't avoid these problems by reading different "news" sources. Thinking you can understand the world by reading multiple news sources is like saying you can get a nutritious diet if you eat several varieties of wood. To me, arguing about the relative quality of news sources is like arguing about which children's breakfast cereals are the most nutritious. Some might be better than others, but in general they are all bad for you.[104]

So what should you do if you want to optimize your reading for what matters, and thereby improve your understanding of the world? Put very roughly, "scholarly" stuff. That is, you want to read books and articles by people who have looked at a situation very closely, and analyzed it carefully, and are expressing this informed opinion in writing. Good scholarship tends to be or cite peer-reviewed journal articles. Good scholarship talks about the research that contradicts the thesis as well as research that supports it. If a book is described as a "polemic," even in praise, avoid it. News sources with a particular agenda tend not to report on findings that do not support their ideology. A pro-choice website is very unlikely to report on how traumatic an abortion can be, and a pro-life website is just as unlikely to report on how devastating it can be to have a baby you don't want or were not prepared to have. Don't read either of them.

The good news is that finding scholarly stuff has never been easier. Anyone with an Internet connection can go to scholar.google.com and for the most part only get search results that are peer-reviewed scholarly works, as opposed to the opinion pieces, news, and blog entries that contaminate vanilla web search results. Any time you hear a "fact" that seems important to you, type it into Google Scholar and see what the science says.

It's actually much easier to know if a journal article is peer-reviewed than even a book. Although books are quite prestigious in the eyes of the public, most publishers will publish any book they think they can make money on, and the truthfulness of the content is of lesser importance. You might be surprised to hear that almost all nonfiction books written for the general public are not peer-reviewed. This includes the book you're currently reading. When a big publisher publishes a "science" book, there is an excellent chance that *no experts* in the field read the book to check its accuracy. I have my colleagues read my manuscripts before publication to check the facts and reasoning, but that's something I do because it's important to me. My book publishers don't require it, because I write popular science.

You might pick up a book by a journalist on the science of sleep. It's very hard to know how much to trust such a book. The journalist isn't trained as a scientist, and it's likely no scientist read the book before it was published. Can you trust it? I would only do so if respected scientists endorsed the book in the cover blurbs. It's not that journalists can't write good stuff, but they're not subject matter experts.

The situation is different for academic publishers. If a scientist sends a book to MIT Press's academic publishing unit, they will send it out for anonymous review. If the reviewers think it's garbage, MIT Press will not publish it. Unfortunately, it's not so simple as this, because often academic publishers also have an arm of trade publishing (trade books are books for the general public).

So what can you do? There's no quick way to know if a book is legitimate or not, but here are some clues: It's written by an acknowledged expert in the field, a scholar of some kind who would have a reputation to lose by publishing a sloppy or inaccurate book. It's published by an academic press. It has lots of citations in the back. The last one is important because if you're skeptical of something, you can track down where the information came from and see if it actually supports what's in the book. Failing that, the next best thing is that the author vaguely refers to "studies" that have been done, even if they are not cited. The worst books just make claims without even a perfunctory reference to any evidence.

These criteria don't narrow it down enough, of course. There are still thousands of books to read that are well-written. How should you choose which of those to read?

One bit of good advice I got from Tim Ferriss is to try to read books "just in time" and not "just in case." This means that you should target the books you read to your current projects, and focus less on reading for knowledge that *maybe* you might use *someday*. For example, as I'm writing this book, I'm reading a lot about the science of productivity, happiness, and morality. I have books on my shelf that I'll need to read for my next book, but, interesting as they look to me right now, I'm waiting until I'm done writing this book to read them.

There's a balance, here, of course. It's good to read broadly; a little "just in case" exploratory reading can have unexpected benefits. And sometimes this "just in case" reading also has benefits of being pleasure reading, which I'll talk more about below. I just want to emphasize that your choice of books should be deliberate. Books require a lot of time, and you should have the balance of "just in time" and "just in case" that is right for your life.

Reading for Wisdom

By "wisdom" I mean applying knowledge and reasoning to make better choices in your life: knowledge is knowing that a tomato is a fruit; wisdom is knowing not to put tomatoes in the fruit salad.

Wisdom books are self-help and how-to books. They contain knowledge, but also advice on how to apply it in life. Advice, like knowledge, is subject to evidence, and you should use the same criteria for determining good and bad books as you do for knowledge books.

Reading for Pleasure

Sometimes reading is just plain fun, and reading for pleasure is a legitimate activity, even without any other benefits. This might sound obvious, but this knowledge is not always applied.

Decades ago, I didn't really appreciate this. I would find myself in a situation where I had something really unpleasant to read, such as literature written by dead people for an English class. I would read for maybe a half an hour, and then my mind would rebel. I'd consider reading the fun novel I had on the side. But then I'd talk myself out of it: "If I'm going to be reading, I might as well read this hard thing I have to read." Then I'd pick up the English literature, but I'd be stuck again, bored and out of energy. So I'd end up doing something completely different, like watching television. See the sidebar "Non-Productivity Scenario: Erika." If I'd had an appreciation that reading for pleasure is a fundamentally different activity than reading for, say, school or work, I would not have had this problem.

Reading for Mind-Expansion

Some books just give your mind a good stretch. I find science fiction quite good or this. Reading *Rainbows End* gives you a better idea of how society might be transformed by augmented reality than scientific papers on the subject; *Ready Player One* does this well for virtual reality and *The Diamond Age* does this for nanotechnology. Some novels and nonfiction books give you a fundamentally different way to look at the world that doesn't neatly fit into the previously mentioned knowledge and wisdom categories.

Reading fiction might very well improve prosocial behavior and empathy. In one study, white children read stories with African American characters, and this improved their attitudes toward the race—even more than children who shared a task with *real* African American children![105]

When to Stop Reading

A vitally important part of reading is knowing when to abandon a book. There is an ethic that people should finish what they start, and that abandoning a project means you're a quitter. This ethic needs to be jettisoned in general, but I'll just talk about reading books for now. If you finish every book you start, there are two bad things that will happen.

First, you'll read far fewer books. What happens is this: you will start reading a book, and it won't be good. A book can be bad for several reasons, corresponding to the different types of reading mentioned above. If you require yourself to finish it, you'll spend a lot of time with a bad book. This is unjustifiable. But further, everyone has a limit to how many books they will be simultaneously reading. If you're "reading" a bad book, chances are you'll spend less time reading *anything at all*, and over the course of your lifetime you'll read far fewer books.

The second reason is that it will make you overly risk-averse when it comes to choosing which books to start. If you have it in your head that you will have to finish any book you start, then when you pick up a big, thick, dense book, you might not give it a chance at all, in anticipation of a real slog. As such, you'll only dip your feet into books you're quite sure will be good.

If, however, you change your mindset, then you can freely take risks with books, and find some real gems. Rather than being proud of always finishing books, make it a point of pride that you *give up* on lots of books that aren't working for you. For me, I abandon about two thirds of the novels I start, and probably a quarter of the nonfiction books I start.

Active Reading

Although reading is an important part of life, keep in mind that reading is basically only sharpening the saw. Knowledge isn't an ultimate good, it's instrumental. This means that knowledge is only good when it's used for some greater purpose. Knowledge can be useless. For example, knowing whether there was an odd or even number of dogs named "Riley" in Boston on December 12, 1972, is useless information. With this in mind, we can strategize about what we can do to make reading as productive as possible, to read with the goal of using that information for something productive. You read so you can do something with the knowledge you're getting. I call this "active reading."

The first step is to optimize what you actually do while you are in the process of working through a book. You can discuss what you're reading with other people. This helps you make sense of it, and it also helps you remember it, because it is more deeply processed. Sometimes you can actually read a book too fast. Talking, thinking, and even sleeping helps your mind contextualize and remember things, so whipping through a book at lightning speed can be counterproductive.

WRITING IN BOOKS

Some of you might think that writing in books violates the divine plan of the universe. One of my mentors actually had to put in effort to even write on journal articles that came out of the printer! I think this comes from some idea of the preservation of purity. You should be aware that this ethic is holding you back from making reading as effective as it could be. The words of books might be precious, but the printed book is not; it's just a representation of those words.

But even if you can't get yourself to put ink in your books, you can use the plastic flag method, and then later skim the two-page spread and hope you can recall what you found useful there. It's better than nothing, and if you ever want the book pristine again, you can simply remove the flags.

For note-taking, if you can't get yourself to write in the book, or if you find even putting plastic flags in a book distasteful, then you can use an index card as a bookmark. When you want to take a note, write the note and the relevant page numbers on the index card. When you're done with the book, keep the index card in the book on your shelf for easy reference.

It is very important that you make notes while you read nonfiction. These notes might be mostly highlighting passages, but the idea is to mark which parts of the book you found interesting, or useful, and tag them so that you can more easily find and review them later.

Before the e-book revolution, I kept little stickies all over the house, and in my car. The paper ones don't last as long, so I used plastic ones. When I found a passage that was important, I would mark it with a pen, and then put a plastic flag on the upper right part of the right page of the two-page spread. This way, I could quickly flip to all the spreads that had any notes or highlights in them. I did this for years, but eventually dropped the flags, because I found that it was easy enough to flip through the book and see my pen marks. I'll still sometimes put in flags when there are only one or two things in the book I need to find in the future, or for things I need to refer to again and again.

After a book is read, and your notes are taken, it's time to *process* the book. This means going through your notes and highlights and somehow dealing with each one. What does it mean to "deal" with them?

You want to make it less likely that you'll have to read that book again in the future. So you go through each of your notes and highlights and think about each one. Why did you mark it for future attention? For me, it's often because it's relevant to some project that I want to work on someday.

I have a series of Google Docs, one for each subject area I'm interested in. I have one for imagination, one for feminism, and one for every book I'm writing. I call these literature reviews "litrevs," so I might have docs called "creativity litrev" or "productivity litrev," and each one holds references to everything I've ever read on that subject. So, for example, as I'm processing Eric Barker's *Barking Up the Wrong Tree*, I will put interesting things from that book in several different litrev documents, depending on their subject. This way, if I ever want to see everything I've ever read on the subject of, say, analogy, I can go to the analogy litrev and it's all there.

Each litrev consists of a series of entries that include what I found interesting and the book title and page number.

Because they are all in Google Docs, I can search for keywords across all documents (or just all litrevs, which are in their own folder).

You might not be in the business of writing books, but you still might want to reference these things in the future for writing opinion articles to the newspaper, or maybe even just to settle some argument you get into with somebody. It's very frustrating to know you've read something, but are unable to recall the details or even where you read it. For the most part, I don't have this problem anymore.[106]

In addition to adding to litrevs, I often also make a whole doc just for that book. I put in page numbers and what I found interesting. Often I copy and paste entries from here into the individual litrevs. With Kindle, it's joyfully easy! I just go to "my notes and highlights" on the web, then copy and paste the whole thing into the Google Doc.

By processing a book after you've read it, you increase your chances of being able to use the information contained in it for some future action. And isn't the ability to use what you've learned in books the reason reading is good?

Idea Capture

One of the most important parts of any creative endeavor, including all of the arts and sciences, is not letting the good ideas you have be forgotten. And believe me, no matter how good your ideas are, and how much they captivate you when you think of them, only a small portion of them will come back to you at a time when you can effectively use them. I know this because when I look over the ideas I've written down over the years, I often don't recognize them. They are completely unfamiliar, and I know they're mine because they're on my files or written in my handwriting. And some of them are good.

Sometimes you can simply remember the things you think of, but this has a cost. These ideas crowd your working memory, interrupting your train of thought so they can be rehearsed, and interfering with whatever you're doing. Writing down things as they come to mind allows you to forget them and get back to the task at hand, confident that you will read about them later.

You might talk yourself into thinking that what you need to remember isn't important enough to justify not writing it down. This is bad, so try to write it down within six seconds of thinking of it, and trust that you'll process that note later. When that time comes, it's harder to dismiss because it's written down as opposed to an easy-to-ignore idea in your head.

Preventing forgetting requires systems that capture ideas so that you can exploit them later. I'm going to talk about the many ways I use idea capture, for you to utilize as a starting point to develop your own systems.

The most important aspect of any idea capture method is that it is easy to do. Think of all the steps that you have to undertake to write down an idea as friction. You want to have a frictionless way to get ideas recorded.

When most people "try" to remember something, they mentally focus on it, or perhaps repeat the idea in their minds over and over again. Then they

CHAPTER 2: HACKING YOUR BRAIN, HACKING YOUR LIFE

get on with their business and hope that later they'll be able to 1) remember that they needed to remember something and 2) actually remember what they were supposed to remember. If either one of these fails, and at least one often does, then the idea is lost.

Given that our memories are so bad, it's shocking how much we trust them. Often people are told about a good book, or something that they should do, and they don't write it down. Although a reminder to see a particular movie might not be that important, if you're trying to do things that are productive, you really should have a method for idea capture.

Memory Palaces

The first strategy I recommend is to create for yourself what is called a "memory palace." A memory palace is a kind of mnemonic device: a cognitive strategy that functions as a memory aid. The basic idea of a memory palace is this: you choose a physical location, usually a building, that you know very well. You decide on a specific way to walk through it, and you specify locations in the building along the route. So every time you (virtually) walk through the house, you do it in the same order, attending to the specified locations in the same order every time.

For example, I have a memory palace of my childhood home. I start in my bedroom. I look at the turntable (location 1), then on top of my desk (location 2), then I look out the window and see what's under the tree (location 3). Eventually I work my way through all the rooms in the house, and around the yard. I have over seventy distinct locations in there.

This is something you mentally practice: using your imagination to walk through your memory palace, taking note of the locations, in the same order every time.

It should be fairly easy to memorize, because it is a location you know well, and you try to specify a route through it that makes some kind of sense. For example, in mine I follow the left wall to go from room to room. In a given room, the floor of the room is always first, then the next locations are around the perimeter of the room, from left to right.

When you have a memory palace in place (even one with only ten locations is incredibly useful), you can start using it to remember things. The way you do this is to create vivid images at the locations that represent what you want to remember.

Let's say I wanted to remember to call the dentist. I might imagine my dentist, spinning on my turntable (location 1). I might want to also remember to buy my sister a birthday present. So I'd imagine my sister sitting on my desk (location 2), happy to be getting a birthday present. Images of concrete objects are easier to remember than abstract concepts (buckets are easier than "justice" for example), and emotional, sexual, and violent images also stick in memory better. (For a great book on what kinds of images are most memorable, I recommend reading *Moonwalking with Einstein*.)

Later, when you have an opportunity to write down your ideas, you simply virtually walk through your memory palace and look at each location. Then you interpret the picture and write down the idea you had: call my dentist and buy a present for my sister.

In practice, I use this several times a week. When is it most useful? When you can't write anything down. I use it when driving or riding my bike, when I'm in a movie theater, having an intense conversation with someone, and any time when it's difficult (or rude) to pull out some paper and write something down. When I get home I just go through my palace and write down all the ideas. Then I "clean out" the palace by imagining removing those images from their locations and imagining them empty again. This actually works.[107]

As well as memory palaces work, you can't keep all of your ideas there and you need to get them into some nonbiological medium. Pen and paper is a great way to do this.

Index Cards

Personally, I use index cards. I keep a stack of them in my back pocket and replenish them frequently. I keep a pen in my front pocket, so I can always jot something down. I've used little pads in the past, but I'll explain why I now prefer the cards.

Index cards are the right size: large enough to get a complex idea onto, but small enough for your pocket. Carrying around a big notebook is nice, but there are times when you can't or just don't want to haul a big notebook around.

Often I'll write down book or movie recommendations on an index card, and I can immediately hand it to someone. When I'm ordering a bunch of food at a restaurant, I write everything down on an index card and simply hand it to the server, rather than trying to remember everything.

You can quickly file away your index cards into physical folders. If you use a folder system for your projects, you can just put each card into its appropriate folder, to be retrieved later when needed. I have a box for "writing fragments." When come up with some clever idea for fiction, I put it on an index card, and when I get home I throw it in a box. When need writing ideas, I have a box full of them.

Important cards can be placed somewhere special. Suppose you think of something very important. Do you want to be carrying that note on your person everywhere you go? What if you lose your notebook? With index cards, I can place that card in my desk drawer or inbox for processing, and don't have to worry about losing it as I run around town waiting for the notebook to fill up before I take it permanently out of my pocket.

You can also throw away cards when you don't need them. If you wrote down a reminder to call the dentist, and you called her, you can discard the index card afterward, rather than taking up space in your notebook.

Some of these are only possible because the cards have the benefit of being unattached to each other. That is, I can do something with an index card without having to rip it out of a notebook. I find that if I have a notebook, there is something in me that wants to preserve the integrity of the notebook, making me not want to rip out pages. But ripping out pages is necessary for discarding pages, filing them, leaving them in your suitcase so you'll see them the next time you pack, putting them in your work bag so you'll pull them out at work, or giving them to someone else. A notebook with half the pages ripped out of it is not aesthetically pleasing. You want to *like* the tools you work with, and I don't like mangled pads.

One thing that's nice about a notebook is that you have an automatic chronological record of the notes you've taken. You'll often have a lot of crossed-out things, but those are sometimes useful, too.

In my culture, many pieces of clothing designed for women lack pockets—or, even more baffling, have false pockets. The things men carry in their pockets women often carry in purses. If you don't always have pockets, you can keep index cards in a purse, perhaps fastened together with a rubber band or with a small binder clip.

Another thing I do with index cards is that I will write at the top of a card some problem I need to think about. I might need to create a backstory for a character in a novel I'm writing, or come up with interview questions, or brainstorm reviewers for an article. I keep these with the blank ones in my pocket. I can pull out my index cards at any time and see some small task that can be accomplished by thinking and writing solution ideas on the card itself. It's an easy reminder of some things I can do in my downtime. For example, the other day I had about five free minutes while waiting outside a restaurant for my guest. I pulled out my stack of index cards to see what I needed to do. At the top of one of them was written "interview questions about *Imagination*." So I used those few minutes to write down some interview questions I could think of just then, and processed that card later when I got home. Time that might have been spent in annoyed waiting was harnessed for problem-solving.

Idea Capture with Your Phone

Why not just use your phone? Well, you should also use your phone. Sometimes I do, too. But for me this only supplements, not replaces, a pen and index cards. As my calligraphy teachers taught me, keep your pens close. *And your penemies closer.*

The phone does have some drawbacks that index cards do not. It's not as easy to make diagrams. It's harder to give someone information with your phone. If you wanted to recommend a book to someone at a party, you'd have to get their email address or phone number (which they might not want to give you) and then email or message them the recommendation. The index card feels a lot more homey and less intrusive. You just write it down and hand it to them.

It's also socially awkward to pull out your phone in many situations: in the theater, at a dinner party, during conversations, even watching TV with a friend or lover. Pulling out your phone sends a message to everyone that you're bored. If you pull out your phone to take notes at a talk, people assume you're texting your friends or on Twitter or something. Taking notes on an index card is more socially acceptable in many situations.

That said, there are times when the phone is great, particularly when you're alone and won't break social taboos. There are several ways to use the phone for idea capture.

First, you can use a "notes" feature. This is only useful if you remember to process these notes later. Also, make sure your notes application is backed up in the cloud. It's easy to remember to process your index cards, because they fill up your pocket, and you see them every time you change clothing, or take a note. Another quick way to record an idea, if you're home, is to use a voice-activated device, like Google Home or Amazon Echo. Create a list called "ideas" and say something like "Alexa, add clipping your nails before you vacuum to the ideas list."[108] A weekly reminder in your calendar to process your phone notes and online ideas lists can keep you on top of them.

Second, you can use a cloud service, such as Evernote or Google Docs. You just open the right file, and put your idea in the right place. This is great, and sometimes I do this, but there is a cost associated with it that is so great that I can't rely on it. Opening the application, navigating to the right file, and finally typing takes a long time, and if you're say, walking with someone on the street, it might be too much time, and you might find yourself thinking "maybe this idea is not that important. Maybe I don't need to write it down." That's too much friction. If you find yourself having such conversations with yourself, you need to get easier modes of idea capture available to you.

What I prefer is using an app called zipnote, which admittedly sounds like a ridiculous application when you first hear about it, because all it does is send an email to yourself. You can just use your mail program to do that, right? Well, yeah, but zipnote takes away several button presses.

I have to click 1) the zipnote app icon, then 2) into the body of the email message. Then I start typing. That's two clicks.

If I use my email program, I have to 1) Click the email program app icon, 2) Click compose message (hopefully without getting distracted by items in my inbox!), 3) Type the first letter of my name, 4) Click my name, 5) Click in the body of the email. Then I can start typing.

It only saves you a couple of seconds, but it's shocking how the friction caused by a few seconds of time will make you say to yourself "nah, I don't need to record that idea." It gets too close to the six seconds you have to get an idea down before you start coming up with excuses not to.

Furthermore, sometimes when I send myself a message from my phone's mail application, it ends up only in my "sent mail" Gmail folder, not my inbox, so I don't even see it. Infuriating.

And when I say "start typing" I mean typing or using voice recognition. I am using voice recognition more and more, and it saves me a lot of time.

Anyway, the next time I sit at my desk to process email, I have a bunch of ideas, from me, in my inbox. Thanks, universe!

While I'm on the subject, the location of the zipnote icon (or whatever icon you use for idea capture) should be in the very lower-right corner of your home screen (lower-left if you're a lefty). The point is that you need to make it as easy as possible to capture an idea.

Zipnote, index cards, and little pads are great for when you're on the go. But when you know that you'll be taking notes, like if you are at a meeting or a talk, you'll often have a full-sized notebook with you to use. Notebooks have the benefit of being chronologically organized. I recommend putting the date before any set of notes you take on a given day. I put my email address on the first page, along with a "please return" plea, in case I lose it. On the first page I also put the date of the first note, so I can chronologically organize my notebooks by starting dates on my shelf for easy retrieval.

Notebooks are great, but one thing people forget to do is to process them later. If you write in your notebook a note to yourself to go to the dentist, if you just bury it in later pages of notes and never go back to it, you might as well have never written it down at all.

What I do is this: when I've filled a notebook, I put an item on my project list to "process" that notebook. I'll go over what that means in the next section.

For the last several years I've become enamored with the Livescribe pen and paper system. It's an electronic pen, and special notebooks with minute dot patterns on them. When you write with the pen on the paper, a little camera in the pen records your pen strokes. Then you can upload all of your notes onto your computer. So I have several notebooks full on my shelf, and on my computer—and with optical character recognition (OCR) I can search my notebooks to see, for example, all my meetings with "Sterling." You can keyword search your handwritten notes! Very handy.

But it gets better. You can also record sound with the pen, and then, later, click some text you've written and hear what was being said when that text was put to paper. So if somebody's talking a mile a minute, and you can't write everything down in time, or you can't understand your notes later and need to hear what the context was to understand it, or you need to remember something said for which you didn't take a note, it's awesome. I think all students should have this for their classes (ask your instructor if it's okay to record the lectures, though). I've filled over seventeen notebooks with this method at the time of this writing, so it's not some new tech that I'm still in the honeymoon phase about. It's worked for me for a years.

Processing Ideas

Once your ideas are out of your head and into some physical (or electronic) medium, then you need to make sure you put it in a place where it can be useful to you later. I call this step "processing" your ideas, and it's similar to how you process books, as I described above.

Whether it's emails from zipnote in your inbox, entries in your notebooks, or items on index cards, you need to do *something* with them. Processing is a pretty broad concept, so I'll just give some examples.

If it's a fragment of poetry, I will put it in a Google Doc called "poetry fragments." I will return to this doc when I need inspiration for poetry.

If it's a reminder to call the dentist, I will either immediately call the dentist, or put a note in my project list to do so.

If it's something I wanted to put in a book I'm writing, I might open the document for that book and write it in then and there, or I might make a project list entry for it. Things that take two minutes to do should be done immediately, things that take between two and fifteen minutes go on the Taking Care Of Business (TCOB) list, and longer things go on the project list.

If it's a phone number, I put it in my contacts.

If it's an idea for a future project, I put it in a list of "Someday/Maybe" projects.

And so on.

When I write down something I have to do in a notebook, I put a little square before it. When I do the thing, I put a checkmark in the box. This way, after I take a lot of notes, I can scan the left margin for empty boxes and quickly do all of the things I marked.

After an idea has been processed, then I cross it off (if it's in a notebook), archive the message (if it was an email), or discard (if it's on an index card).

With an idea capture system using some combination of index cards, emails, notebooks, voice-assisted technology, memory palaces, and a solid processing routine, you can ensure that good ideas don't get lost.

CAFFEINE

There are two forces at work in your brain that determine when you're awake and when you sleep. The first is your circadian rhythm, which is an internal clock roughly attuned to the twenty-four-hour day. (It's actually twenty-four hours and fifteen minutes, and nobody knows why.) This rhythm is kept in check, and corrected, by exposure to light. The other thing is sleep pressure, which is the buildup of adenosine over the course of waking hours. Usually these two components converge at bedtime to put you to sleep, during which your brain washes out the adenosine, a process that eventuates in your waking back up.

Caffeine works by blocking adenosine receptors in your brain. That is, the adenosine is still building up in the spaces between your neurons (the synapses), but caffeine keeps the neurons from knowing that it's there. These little particles of sleepiness are knocking on the doors, but caffeine locks them.

Caffeine has a half-life of seven hours, meaning that every seven hours half of those blocked receptors will be freed up for adenosine consumption, making you sleepier, sometimes very rapidly. This is why it's not great to have caffeine late in the afternoon, and certainly not before bed. What many interpret as insomnia is actually over-caffeination.[109]

If you feel you can't function in the morning without coffee (or some other caffeine), it can be a sign of two things: you are sleep deprived and are self-medicating every day with coffee, or you have grown dependent on coffee, or both. (You also might be a night person; see the next section.) This can happen because regular use of caffeine in the morning blocks your body's natural waking-up chemical routine. You can build tolerance to caffeine, as you can with almost every other drug, requiring you to consume more of it over time to get the same perky effects. Here's how it works: Because caffeine blocks adenosine receptors, long-term use of caffeine causes your brain to compensate by building more of them. Then, when you suddenly go without caffeine, the adenosine flows through your brain very, very easily, making you feel lethargic, depressed, and irritable.[110]

How to Sleep

You'd think people would not need pointers on how to sleep, but the modern age has many pitfalls that interfere with sleep and its many benefits. Nearly one out of five people in the world today are sleep deprived, and people often don't even know it when they are.[111] So yes, most people need pointers.

Before we get into good sleep practice, let's look at what science says about sleep's importance. Cognitively, sleep is terribly important because that's when your brain clears out all of the waste material[112] and turns short-term memories into long-term memories. Getting enough sleep contributes to mental and physical health and happiness.

Most people need between seven and eight hours of sleep at night. For the average person, getting fewer than six hours of sleep a night increases your chance of dying.[113] Of course, health and mortality are related to productivity, because you are less productive when you're feeling sick and are decidedly unproductive during every year that you're dead.

How much sleep do *you* need? One quick way to tell if you are sleep-deprived is whether your alarm actually wakes you up. If this happens every morning, you probably aren't going to bed early enough. Try going to bed earlier and earlier, in half hour increments, until you find yourself waking naturally before your alarm goes off. Other signs of sleep deprivation are feeling that you could fall back asleep around 10:00 or 11:00 A.M., or if you feel you can't function before noon without coffee. When you've been on vacation for a few days, and have caught up with sleep, how long do you sleep? Whatever the answer is, that's what you should be aiming for in your everyday life.[114]

Sleep Hygiene

The things you can do to make your sleep better or worse is called your "sleep hygiene." The research can be summarized pretty succinctly: don't have caffeine after 2:00 P.M., don't nap after 3:00 P.M., don't look at screens within an hour of bedtime (e-paper screens are fine; it's the blue light that's bad for sleep), try to stick to the same sleep schedule each night, don't drink alcohol at night, and turn off notifications on your devices.[115]

This is very important for people over forty, as sleep quality decreases with age. Older adults need as much sleep as younger people, they are just less and less able to get it.[116] If you have trouble sleeping, or are not getting enough sleep for some other reason, you should not blow it off. It's important, and you should do some reading about how to fix it. I recommend Matthew Walker's excellent book *Why We Sleep*. If you have serious problems, talk to your doctor.

Napping

Sleeping doesn't have to happen at night. Humans naturally have a biphasic sleep pattern of a long night sleep and a shorter afternoon sleep.[117] Although napping doesn't work for everybody, naps make you more alert, have a faster reaction time, learn better, and improve performance on just about everything. It also increases your health—one study found that the health increase caused by naps was equivalent to exercising every day. It can reduce your blood pressure and strengthen your immune system.[118] A ten-minute nap can help as much as an extra hour and a half of sleep at night.[119]

But naps need to be done right. Five minutes is too little, ten to twenty minutes is ideal, and more than that and you'll feel the costs: the sluggish, mentally foggy "sleep inertia." Naps of more than an hour can really impair you after waking.[120]

Some people claim that coffee puts them to sleep. What I suspect is happening for many of these people is that they are really tired, drink a cup of coffee, and fall asleep. But what a lot of people don't realize is that caffeine takes about twenty-five minutes to have its physiological effects. Coffee isn't putting them to sleep, they're sleeping because they're tired, and the coffee just hasn't taken effect yet.

This can be put to use with the coffee nap, or the nappuccino. When you're sleepy and want a quick nap, rapidly drink a cup of coffee and fall asleep (this often takes about seven minutes). Just as you should be waking up, around the ideal time of twenty minutes, the coffee will kick in.[121] You'll wake up with the combined benefit of the coffee and having napped. A study showed that this coffee-nap combination was better for performance than drinking coffee alone.[122]

Some people I talk to say they can't nap, because when they do they sleep for an hour or two. The usefulness of naps actually varies a lot from person to person, but I caution against thinking they don't work for you prematurely, because you might just not be doing it right.[123] It could be that many of these people would actually nap for the right amount of time if they were getting enough sleep at night. But if you're one of those people who get enough sleep at night, but also nap too long, set an alarm for twenty-five minutes (five to get to sleep, twenty for sleep). If you're not using this alarm, put your phone in airplane mode.[124]

In recognition of the importance of naps on productivity, some workplaces and universities have set up nap pods. They used to have napping rooms, but stopped that because people were using them to have sex.[125]

When should you take your nap? The best time to take a nap is usually seven hours after waking. Unless you're a night owl, this will mean between 2:00 and 3:00 P.M. In my default half hours spreadsheet, I have a nap as a regularly occurring event.

Your Chronotype

You've probably heard that some people are night owls and others are morning people. This is true, and you can use your "chronotype" to optimize your day. Science puts people into three categories: morning people, night people, and "third birds," who are somewhere in the middle.[126]

You can calculate what your chronotype is by finding your sleep midpoint—on days when you don't have much scheduled (weekends for most people). When do you typically go to bed and when do you typically wake up? Find the halfway point.

If your midpoint is before 3:00 A.M., you're a morning person. If your midpoint is after 6:00 A.M., you're a night person. If it's between 3:00 A.M. and 6:00 A.M., you're like most people, and are somewhere in the middle. These categories are sometimes called morning larks, third birds, and night owls. I tend to go to sleep around 9:00 P.M. and wake up around 5:00 A.M. So my midpoint is about 12:30 A.M. This makes me a somewhat extreme morning person.

For the most part, your chronotype seems to be genetic. Your chronotype is also likely to change as you age. Kids are morning people, become night people as teenagers, then start to become morning people again after age twenty. This continues for the rest of their lives. People over sixty tend to be even more extreme morning people than kids.[127]

Your chronotype determines at what points in the day you are better and worse at certain kinds of activities. There are three stages of waking hours: vigilance, the slump, and recovery.[128] This is the order for morning people and third birds. For night people, it's recovery, slump, and vigilance.

The vigilance period happens in the morning for morning people, ending at about noon. For night people, it's late afternoon and early evening. During vigilance, people are better at thinking logically, and even act more morally. A study by economist Nolan Pope found that students' test scores were so much higher in the morning that he recommended that one simple way to raise student test scores is to simply test them in the morning.[129]

For morning people, the slump happens in the early afternoon, or about seven hours after waking. For night people it's the middle of the day. During the slump, people are less happy, phone calls during the slump are more negative, people resort to stereotypes more often, and have less alertness and energy levels. Doctors and nurses make more mistakes.[130] This is why people try to nap or drink caffeine around 2:00 P.M.

Recovery is around 5:00 P.M. for morning larks, and it's in the morning for night people. During recovery, people are better at solving problems that require a flash of insight than they are in the afternoon, possibly due to lowered inhibition.

If you can plan your day according to your chronotype, you are likely to enjoy greater job satisfaction and productivity and have less stress. This is what happened when a company rearranged its work schedule according to people's chronotypes.[131]

Don't be a slave to what I've said about what is supposed to be your chronotype. It's important to be mindful of yourself, and figure out which hours during the day are *your* vigilance, slump, and recovery. When are you most alert? When does your brain feel like it's in a fog? Write down the half hours where these mind states tend to happen.

So how can you use your chronotype to schedule your day better? This is easier to do the more control you have over your schedule, but to the extent you can, you should try to break up your tasks into categories according to how much creativity and concentration they require.

Tasks that require concentration, logical thinking, and the catching of errors should be done during the vigilance phase. Tasks that require insight, creativity, and thinking outside the box should be done during recovery.

And what should you do during the slump? Tasks that don't require either of these things—running errands, doing administrative tasks, filling out travel forms, updating your project list, chores, phone calls to friends, napping, or taking a break.

I've estimated that my peak vigilance is between 5:00 A.M. and 10:30 A.M. I have this part of my spreadsheet colored in bright blue, a color I associate with vigilance. My slump feels like it's between 12:30 P.M. and 3:00 P.M., so I have that colored in a dull orange. My recovery seems to be between 3:30 and 6:30, so I color that green. My main goal in filling my half-hours with tasks is to make sure I get the more important things done first, but I temper that with the phase of the day. If I have a boring journal article to review, I just cannot do it during the slump. Luckily for me, the most important thing in my career requires vigilance, so I always do my writing in the morning, when I'm most vigilant.

HELPING YOURSELF BY HELPING OTHERS

So when optimizing your life, where does helping others fit in? It might be obvious, to many people, anyway, that if you want to optimize your life, then you should focus all of your energy on getting good things for yourself. When it comes to helping others, people can be categorized according to different "reciprocity styles." Broadly speaking, there are givers, matchers, and takers. Takers are always looking out for themselves, and when they do things for others, they do it strategically so that helping others helps themselves. Matchers are always using a sense of fairness. They are good to people who are good to them, and expect kindness returned in proportion to kindness given. Givers seem genuinely interested in others' welfare, and they do nice things for people to help them, without needing motivation from expected side benefits to themselves. Most people help those they are in close relationships with, without particular concern for whether their kindness is reciprocated to the same amount. People who are givers are like this with many strangers, too.[132]

So how do these different people turn out? Givers, in general, earn 14 percent less money, are twice as likely to be victims of crime, and are viewed as being 22 percent less dominant and powerful.[133]

However, there's a great deal of research to suggest that, in the long run, helping the people around you is one of the best ways to improve your lot in life. When a giver gives *everything*, with no thought to their own welfare, they burn out, have more health concerns, and get walked on.[134] But the givers who *also* care about themselves rise to the top.[135] Givers of different types are at the top *and* bottom of success rankings.

This is because we live in a world where your success in life is often dependent on cooperation with and endorsement by other people. If people think you're great, kind, and generous, or if they feel indebted to you, they are likely to help you out—introduce you to important people, notify you of opportunities, write you good recommendations, and so on. In our world, this kind of social capital is important for finances, happiness, and health.

If people think you're someone who's always looking out for only themselves, then they won't help you out. In fact, they are likely to try to thwart your efforts at success by sabotaging you. They might warn a potential employer about you. Needless to say, this can be really bad for your career.[136]

These categories hide nuance. Most people are a bit of a giver, a bit of a matcher, and a bit of a taker. There's an online survey that will show what you are.[137] Additionally, a person's reciprocity style might change over time. Sometimes givers act like matchers or takers in certain situations. For example, most people act like givers in their very close relationships.[138]

It seems that the optimal behavior is to help people by default, giving them the benefit of the doubt, but when you identify someone as a "taker," you stop going out of your way to help them, and take care that they don't take advantage of you.

Adam Grant pioneered this research, and describes it in his superb book *Give and Take*.[139] One of the reasons it feels so good to learn about this work is that it dovetails nicely what seems to be moral behavior with self-interested behavior. That is, the thesis of Grant's book is that being good to the people around you serves your own interests as well as theirs. You can help others to get ahead.

Volunteering

This is evidenced by a study of thousands of people over many years, which found that volunteering causes happiness. Volunteering is a way to get an interesting experience on the cheap that makes you happier,[140]

CHAPTER 2: HACKING YOUR BRAIN, HACKING YOUR LIFE

but only if the work is meaningful or enjoyable. Doing it out of a sense of duty doesn't seem to provide a happiness boost.[141] The ideal amount of volunteering a person should do to maximize their own happiness seems to be 100–800 yearly hours.

And the more socially isolated a person is, the more the volunteering helps.[142] Being a good person seems to make you happier. It also makes you live longer.[143] Even if you feel that you are a matcher or a taker, striving to be more of a giver makes everybody's life better, including yours.

Just as your productivity in general benefits from deep, focused work, it turns out that your giving behavior also should be concentrated. Doing about 100 hours every year seems to be the sweet spot for the amount of time spent specifically giving to others (for example, volunteering), but studies also show that doing these activities in chunks is more rewarding them sprinkling them throughout your week. That is, it's better to volunteer for two hours per week all at once than it is to do a little bit all over the place.[144]

Giving is also contagious. When you perceive those around you as being generous, it makes you more generous, too. This giving contagion spreads three people out before effects stop.[145]

3
HAPPINESS

Why So Productive?

What you've read so far covers most of the state-of-the-art science on personal effectiveness. Armed with this knowledge, I hope you're feeling really jazzed about how you can be superproductive. Now we're going to ask why on earth we might want to be like this. If you are more productive, does that make you any happier? Does it make you a better person?

Happiness is something that most of us strive for. We find it intrinsically valuable. This means that we don't need to justify our drive to be happy with anything else. But being happy (that is, experiencing positive emotions) has lots of side benefits, too. When people feel good, their thinking becomes more integrative, flexible, creative, open to information, understanding of others, and more attuned to the big picture. They feel that life is more meaningful, and that there is more unity between themselves and others.[1] Happiness, particularly from a diverse set of positive emotions, makes you more productive and healthier, too.[2]

You might wonder how science can study happiness. Can you even measure it?

Something I try to impress upon readers of all of my books is the idea that things should be scientifically tested. And if you think that something cannot be scientifically tested, don't be too sure. There is a good chance that somebody has done just that. I regularly hear people assert that measuring happiness is impossible, without looking into whether anyone has tried it or even testing this assertion against even a few minutes' of Internet investigation.

Happiness can be measured, and the measure you use depends on what you think happiness consists of. Some think of happiness (or its scientific, more inclusive term, subjective well-being) as preference satisfaction, which is how people's goals and desires are fulfilled. This is problematic, though; it only takes a moment's reflection to see that people often pursue goals that don't, in the end, make them happy once they get fulfilled. People underestimate habituation. They expect stronger emotions than they actually end up feeling. Furthermore, it's possible that some entities have goals or desires but no consciousness with which to enjoy their fulfillment. Although some scholars think that preferences are important even without consciousness,[3] I find it very implausible. As an AI researcher, I can code up an AI that has preferences in an afternoon, but without any conscious feelings about whether or not those preferences are achieved, it seems very implausible to conclude that this AI is happy or unhappy because of preference fulfillment or frustration.

Some scholars suggest looking at a basic list of good things, such as health, a good social life, and a decent income. But most people think that these things tend to *cause* happiness, and are not happiness itself. For example, a person can have all those things and still be miserable.[4] You might know people like that, who have everything one could want but are still unhappy. Most people, no matter what language they speak, have no trouble answering questions about how happy they are using a single dimension.

In this book I'm going to use the idea of happiness as a feeling—either moment to moment feelings, or a feeling you get when you evaluate your life in the big picture.

Simply asking people how happy they are is an imperfect but surprisingly good measure of how happy they actually are. Happiness researchers believe this because these reports correlate with lots of things we would expect would correlate with actual happiness, like how often people smile, heart rate, blood pressure, chance of suicide, evaluation by others, and activation in brain regions that kick into high gear when we look at cute babies.

It also predicts future behavior in ways we would expect it to.[5] Although there is some nuance to how it should be done, observing or placing people in different conditions, and asking them how happy they are can give us insights into the causes of happiness.

When people think about spending time in a way that makes them happy, they often think of relaxing. But careful research into happiness shows that this view is problematic. If you're really stressed out, relaxing can be great, for a short time, but in general, free time is unstructured and lacking in the goals and feedback that makes for a really rewarding experience. Ironically, for some people it can be easier to enjoy work than free time.

This gives us reason to think that productivity would make us happier. We have a basic need for mastery, and the development of competence when interacting with our environment. Even apes will solve puzzles just for the fun of it.[6] This drive for mastery, sometimes called the "effectance motive," seems to be evolution's way of getting us to sharpen the saw.

Flow

Mihaly Csikzentmihalyi ("mee-high cheek-zent-mee-high") did some remarkable studies showing that people were very happy in a "flow state." The flow state is characterized by complete absorption in the task, loss of a sense of time passing, and positive emotion. The more time people spend in flow states, the higher the quality of their overall reported experience. In one study, flow rates were 54 percent when people are actually doing their jobs (rather than taking breaks, gossiping, etc.). So people are in flow at work about half the time.

But during leisure time, the flow rate was only 18 percent. These were the people having dinner with friends, watching TV, reading, and so on. More than half of the leisure time was characterized as being a state of apathy. It didn't matter whether they were managers, clerical workers, blue-collar workers, or even those on an assembly line. They all enjoyed work more than leisure, and busier people are happier than non-busy people, even if they're forced to be busy.[7]

Of course, flow states *can* be achieved during leisure activities, such as playing squash, rock climbing, or playing the guitar. It's just that, in the real world, people are usually doing it wrong.

WHAT MAKES FOR A HAPPY JOB?

Although people on average don't like working, the kind of job you have can affect how happy you are doing it. On average, these are the things that affect how happy one can be expected to be on the job, in order of importance: doesn't interfere with family life, the job is not dangerous, you don't worry about work when at home, you have supportive coworkers, your work has variety and opportunities for promotion, and you have autonomy and some security.

You might be surprised to hear that wages come in dead last in importance.[8] So if you're making enough money to pay for your needs and have a little left over, be careful about taking a higher-paying job that has other bad qualities. It will probably make you less happy.

All this is surprising to me, because studies of how happy people are when doing various things suggest that people really don't like work (although being unemployed is worse). Working is ranked close to last no matter how you measure it, and this has been replicated many times.[9]

Although pure relaxation can feel good if you're stressed, too much of it makes us bored. The flow state happens when what we're doing isn't too easy, which causes boredom, nor too hard, which causes frustration, anxiety, and a loss of knowing what to do—which eventually leads to boredom, too. The sweet spot in the middle is where flow can happen.

Different people experience flow from different activities, but Csikzentmihalyi focused on things like sports and the creation of art. This gives just about everybody a good feeling, because sports require exercise and sometimes gets you outdoors, and art has an association of high culture. But you don't need to be improving your health or contributing to world culture to experience flow.

The flow research is in contrast with some other studies, which show that people are happier on the weekends and on holidays than they are on weekdays, primarily because they spend more time with the people they care about (this is true of good feelings, not life satisfaction, which the weekend doesn't seem to affect).[10] One of the top regrets of people on their deathbed is having worked too hard.[11] Working too hard leads to stress, fatigue, work-family interference, and less happiness.[12]

So are people happier working or not working? My reading of the science is that there are studies that support both sides, but it's leaning toward non-work time making you happier, as those studies are more recent and look at more people.[13] What we can say is that both work and leisure time can involve things that make us happy (challenges and the company of people we like). It's what you're doing with your work or leisure time that makes the difference.

Let's take a deep look at a leisure activity that people are spending more and more time on: video games.

Video Games

People can feel flow playing complex computer games, like *Skyrim* or *World of Warcraft*, but also from playing rather mindless games, like *Candy Crush* or *Tetris*. Computer games are so popular because they are engineered by very skilled people who design and test them to maximize flow states. With a multibillion-dollar industry full of professionals trying to create flow-inducing experiences, it's no surprise that for many people, computer games can be more enjoyable than actually facing real-world challenges. As a result, people often play video games *instead* of doing productive things. By 2014 just about all teenagers were playing video games (94 percent of girls, 99 percent of boys).[14]

A big part of the lure of games is the fact that they're challenging. This is very different from most relaxing activities, including most other forms of media entertainment. Gamers spend about 80 percent of their time failing at whatever they are trying to do in the game. You might think that an 80 percent failure rate would be demoralizing, but gamers find this exciting and optimistic. The study that found this compared game players to people merely watching games being played—the watchers were not nearly as aroused.[15] Only 20 percent of actions gamers take are "successful," in the context of the game, but they feel *really* good. Doing well in a game makes you feel productive. What's interesting about this is that feeling productive by *actually being productive* feels exactly the same as feeling productive by doing something unproductive, like playing computer games.

So if the work you're doing doesn't give you the same feeling of productive progress, the game will be a temptation. A 2009 survey found that 46.6 percent of workers play video games at work every day.[16] This isn't surprising, because real work is not optimized for being compelling, and video games are.

DOES RETIREMENT MAKE YOU HAPPY?

If people aren't happy at work, how about stopping work? Research on this is decidedly mixed. Some studies find that people are happier after retirement, others find that people are less happy. Being retired for a long time results in lowered feelings of personal control and, having control in your life is generally a good predictor of happiness. This is true for both men and women.

The *transition* into retirement is often very happy, especially if one is leaving a job with a lot of stress and strain. There appears to be a honeymoon period for these people, where they feel renewed energy and freedom for a time. Over time, though, if someone has depressive tendencies to begin with, retirement can exacerbate them.

There are some gender differences. A bad marriage seems to make retirement worse for women than it does for men, which is consistent with other evidence that women are more sensitive to relationship strife in general. But men's happiness in retirement is more dependent on finances than women's, which is also consistent with findings that men care more about money than women.[17]

So should you retire? It's hard to say in general. But when you're old, and you can afford it, you might want to if you are sick of your job. But it is worth asking if what you really need is a year off, or even a good vacation—this allows you to enjoy the honeymoon period, but not the longer-term negative effects of having nothing to do. Another alternative to retirement is changing jobs to something that's more rewarding, or at least new and stimulating. There is also the option of finding a long-term volunteer position or hobby—something to "wake up for."

Like video games, many office environments are also "artificial worlds." For many modern jobs, you're doing symbolic work that would have very little meaning to a hunter-gatherer: filling out forms, going to meetings, answering email, moving money around. None of these things have a visceral survival feel to them. If work is set up correctly, it will be rewarding. You'll be moderately challenged and get frequent if inconsistent gratifications. If you are lucky enough to work in a world engineered like that, you probably love working. But often dealing with these intangible entities all day makes it hard to see your effectiveness in a meaningful way. Setting up everybody's work environment to have the same perfect blend of drive and gratification that a video game has is hard. This is why so many people would rather feel productive by playing games than to feel productive by actually being productive.

So does this mean that you'd be just as happy playing video games your whole life as spending the same amount of time doing productive flow-state-inducing activities, such as painting or working?

No, because "happiness" is a word we use to describe an underlying complex of psychological states. The first is based on moment to moment emotions, pains, and pleasures. Emotions tend to have a "valence," meaning they are good or bad. Disgust feels bad, pride feels good, and contempt has a little of each. So one vision of a life of happiness means optimizing good feelings over time. If you spend more and more of your life feeling more and more positive emotions and pleasures (as a shorthand I'll call these "good feelings"), and minimize negative emotions and pain ("bad feelings"), then you're as happy as you can be.

But another important aspect of happiness is what psychologists call "life satisfaction." This is what we mean when we think to ourselves how happy we are with our lives. The interesting thing about life satisfaction is that sometimes you can feel more satisfied with your life by working on hard things, which, in the moment, don't give you a lot of good feelings. For example, you might struggle to create a business, and doing it might be stressful, scary, and full of sleepless nights, but when the business is up and running, you're very proud. In contrast, someone who smokes pot and plays Nintendo all day might have more day-to-day good feelings, but after a few years, have a low life satisfaction, and wonder what they have been doing with their lives.

Some things we do because we enjoy them, and other things we do because we just can't stop. Let's take picking at scabs. Lots of people do it, and just about all of them believe they *shouldn't* do it, and not only that, but they don't even enjoy doing it. Why do they do it, then? Why do some people chew on their nails until they are bloody? Many psychologists believe that pleasure and motivation are associated with different brain physiology. And unlike most things in the brain, this one's relatively simple: the difference between opioids and dopamine.

MEANINGFULNESS

Psychologists tend to agree on the importance of the two aspects of happiness mentioned: good feelings and life satisfaction. But some advocate a third aspect of happiness: meaningfulness.[18]

For example, Steven Pinker suggests that having children doesn't make your life more pleasurable, and often it can make it *less* pleasurable, but having children is very meaningful for some people, and that makes it worth it.[19]

The inclusion of meaningfulness as a component of happiness is debated in psychology.

If you artificially increase the levels of dopamine in someone's brain, they are likely to engage in compulsive behaviors. Some even get addicted to things. But this is not pleasure-seeking. I can feel the difference when I play different kinds of computer games. I'll take two games as examples. *God of War* is a platform puzzle and action game that takes place in an Ancient Greek–inspired fantasy world. The graphics are beautiful, the characters are great, the story is interesting, and the puzzles are fascinating. I somethings think fondly about things that happened in the game, years later, the same way I reminisce about travel experiences. Pleasure.

Compare this to playing *Tetris*, another game I've spent many, many hours playing. A game like *Tetris* is like busy work. It's monotonous, but you feel like you're making continuous progress.[20] This can feel very satisfying in its own way. Just like cleaning off a desk, clearing a screen of debris feels like a minor accomplishment. Busy work in the real world, like waiting tables or cleaning countertops, has similar immediate feedback that also feels good. With each plate of eggs served, or each wipe of the counter, the world gets a little better, and you can see results immediately. This kind of activity, both in games and in real life, is often described as being "in the zone."

Unlike the time I've spent playing *God of War*, I don't have great memories of particular things I've done in *Tetris*. It's kind of all the same: you have to fit falling blocks into the structure you already have. It took me a while, and studying psychology, to realize that when I play games like *Tetris* (or *Chuzzle*, *Puzzle Bobble*, or *Bejeweled*, or . . .), I'm not really enjoying myself all that much. Games like this improve mood,[21] but for me, I feel like I keep playing not because I'm enjoying myself as much as because I simply can't stop. Some games encourage compulsive behavior like this, but they don't create memories you can cherish later. It's just a bit more brainy than popping bubble wrap. If it sounds like addiction, it kind of is. About 8 percent of young American gamers have some pathological symptoms of addiction.[22]

Many people think that video games rot your brain, and some of those people play them anyway. The science is a bit more nuanced. There are some side benefits: it can increase hand-eye coordination and the ability to use computer interfaces, and they can even make you better at problem solving (the can-do attitude you need to master complex character-based games can give you a better attitude for real-world problems). But the problems you solve in the game world are not, generally speaking, real-world problems that need solving. Similarly, playing sports isn't productive, either, in this sense, it's just fun and good for your health.

Games can reduce fatigue, depression, and stress. The more kids play video games, the more creative they are (though the direction of causality is unknown).[23] Some neuroscientists think that video games make you a better learner. But you might be surprised to hear that the ones that are best at it aren't puzzle or role-playing video games, but violent first-person shooter games like *Half Life 2*, which can improve your visual abilities, visual and spatial reasoning, and problem solving; help you deal with pain; and even reason about scientific material.[24] There is even evidence that the most violent video games don't make people any less kind or helpful.[25]

I remember being in graduate school, getting my PhD at Georgia Tech. Part of what makes getting a PhD so difficult is that it takes years, and during those years there is very little external reward. At one point, I was frustrated with my own research. I didn't know what to do next. I was doing something completely new (which is the point of doing PhD research), which meant that nobody knew better than me what I should do, and I felt lost. At that time, my friend got the then-new Sega Dreamcast, and I started playing a fight game with him on it. One time, when I was playing it, I thought to myself "if only getting a PhD were as straightforward as getting better at *Soulcalibur*." When stuck in the real world, I turned to a game, where the missions are straightforward and the progress is steady and rewarding. Games like that are designed so that it is possible to do well at them. Even games you can't "win," like *Tetris*, at least make it possible to make progress.[26] PhD research programs have no such guarantees, which is part of why they often feel hopeless.

Complex games, in which you are in a virtual world solving complicated problems, seem to have an interesting effect on your mind: they engender a can-do problem-solving attitude toward the real world. The real world is full of problems that seem to have no solution, which can encourage a feeling of helplessness and sap one's motivation to get out of bad situations. But these games don't pose situations like this. In the game, you *know* there is a solution; you just need to find it. They can even encourage group coordination and problem-solving.

The reason so many people play video games is that they are fulfilling needs that the real world doesn't. And as games get better and better, more people will spend even more time playing them.[27] So if video games have positive benefits, what's the problem?[28] At worst, they are bad at delivering pleasure, compulsive, and don't make you better at anything. But even games that improve your skills are at best just sharpening the saw, and not cutting anything. Knowing what we know now, we can make a nobility list of video games.

At the bottom of the list, the worst games are twitchy puzzle games, which provide just a little compulsive reward, build few skills, generate no good memories, and offer nothing productive to the world. According to a theory of fun in video games, these might be described as "submission" fun, where you take comfort in passing the time by giving in to whatever the game throws at you, requiring little strategy, creativity, and offering little in the way of interesting ideas or fantasy.[29] These games unmatter and are the best candidates for being classified as a clear waste of time.

BRAIN TRAINING SOFTWARE

There are several "brain training" programs for sale. They are supposed make you smarter. Research done by the companies who make these games found evidence of this.

Independent research is less sanguine.[30] One study compared playing the video game *Portal 2* to people using *Lumosity*, a well-known brain training program. They found no effect of *Lumosity*, but the *Portal 2* players showed increases in problem-solving ability, spatial skills, and persistence on problems.[31] Another found only *Lumosity* benefits for cognitive flexibility.[32]

Video games seem to help you more than software made for brain training, and they're a hell of a lot more fun. Playing *Portal 2* was one of the best gaming experiences I've ever had, and it's so wildly creative that I felt that my mind was expanding as I played it.

But there's also the question of what you want to train your brain to be able to do. Video games and "games" like *Lumosity* might increase your visual attention and reaction time. These are very fast, perceptual skills. But when people talk about training their brains, what is really important? When you think of a really smart person, do you think of how rapidly they can identify a square among a bunch of diamonds? Or how fast they can react to a smiling face? You probably think of their ability to think about situations in multiple ways, or how they are creative, or find solutions quickly, or understand complex situations. You're not going to get better at that stuff by playing the simple "games" in *Lumosity*.

When I want to train my brain, I talk to people about ideas and read good books. The more I do this, the more ideas I have in my cognitive toolbox. I have more ways to think about things.

You can't get this by training your attention and memory.

Next up we have complex world games, such as role-playing computer games, that offer genuine aesthetic experiences (great set designs, stories, and so on) that generate memories that one can happily reflect on later. They also might help people feel more competent in the real world.

Near the top are collaborative, violent, first-person shooters, which are great for cognition and your social life. Seventy percent of contemporary gamers often play together, either in the same room or connected via audio and video, completing in-game missions requiring coordinated effort.[33] Though they are not productive, they are at least sharpening the saw.

There are some games that actually do make the world a better place. *Foldit*, for example, turns the problem of figuring out how molecules fold in 3D space into a puzzle video game. As people solve puzzles in the game, it provides information scientists can use to better understand biology. These games, sometimes called "serious games," are at the top of the video game nobility list.

At their worst, video games are distractions from real life, barely fun time-wasters. At their best, they contribute something to our world. But for the most part, games are just sharpening the saw, or recharging your batteries, akin to leisure time and other leisure activities, so you can actually do something productive. If you are a gamer, think carefully about what role games should play in your life.

Your Life Satisfaction

Optimizing your happiness means paying attention to two aspects: good versus bad feelings and your long-term satisfaction with how you've spent your life. Life satisfaction is typically measured by asking people how satisfied they are with their lives. As you might expect, scientific measurement of life satisfaction has some problems. See the sidebar "Can You Really Measure Life Satisfaction?"

Ideally, we'd fill our lives with activities that we love in the moment, but also do things we're proud of in the long run. The fact is, though, the activities that increase your life satisfaction are often difficult. The good feelings/life satisfaction thing is often a trade-off, at least in the short term.

Even when we know that doing something difficult would eventually make us happier, in the short term these things sometimes aren't much fun. There are lots of aspects of starting a business, raising a child, or getting a graduate degree that are really, really hard. You have to be able to stave off pleasure for a long time, with only an uncertain future reward to try to keep you going. Sometimes we just don't have the willpower to go through with it.

In the meantime, there are about a million fun things to do that you won't be proud of later. Let's take watching television as an example. Back in the 1980s, television was generally terrible. Even through the early 1990s, to say "I don't watch television" had an air of sophistication about it. But now, there are so many great television shows that many of us look at people who don't watch any shows as just missing out on some great stuff. We're really living in a golden age of TV entertainment. As the shows got better, so did the quality of the sensory experience. The sound systems are better, and screens are huge, flat, cheap, and gorgeous. Streaming technology allows us to have what is effectively unlimited choices when it comes to what to watch at any given moment. You could probably spend all of your time watching it and never get through all of the great TV and movies available at home. It seems like some people are trying to do just that. The average American watches over five hours of television every day. And that's average, which means that for every person watching one hour a day, there's probably somebody watching nine hours a day. Over the course of your life, that's over thirteen years.

CAN YOU REALLY MEASURE LIFE SATISFACTION?

When you report on your current state of happiness, or pleasure, you only have to evaluate how you're feeling *right now*. But when you judge life satisfaction, you have to evaluate your life so far. This depends on memory and all of its foibles.

One well-documented memory bias is that when you remember an extended event, such as a vacation, a trip to the dentist, or your whole life, you tend to recall the emotional impact of the very end of it, and the part of it that was the most emotional overall, and weigh those memories more than the other parts. You might have a great vacation, but if someone stole your rented bike, and you had a big delay on the flight going home, you might remember the vacation as being a disaster, even if you were quite happy for most of it.[34]

Another problem is that even though you're asking people to judge their lives in general, their responses are affected by their current mood or environment. One study found that they could increase someone's measured life satisfaction just by having a disabled person in the room at the time.[35] They are also affected by their own stereotypes. If you ask people how happy they are in the moment, you see small or no differences between genders and cultures. But when you ask people to remember how happy they were in the past, stereotypes come out: women remember being more emotional than men.[36] A study found that Asian Americans had greater moment to moment good feelings than those of European descent. However, the two groups remembered their lives differently, resulting in a greater retrospective sadness for the Asian Americans. They were happier, but remembered being sadder.[37]

Though the measure isn't perfect, there is reason to be sanguine about it. When psychologists want to be sure that a measure is doing what it's supposed to, they see if it correlates with things we think it should. This tests its "ecological validity." If our life satisfaction measures are any good, you would think that it would positively correlate with the things we might think are part of a satisfying life. They are: mental health, strong social ties, satisfying work and leisure time, strong social support, self-esteem, goal achievement, and frequency of sex and exercise. These correlations range from 0.48 to 0.68, which, by social science standards, are pretty high.[38]

The measure of life satisfaction was tested in another way. Psychologists asked people what the important domains in their lives were. People responded with things like family, career, and the performance of their favorite sports teams. Then they asked people how satisfied they were with these domains of their lives. The individual "domain satisfaction" correlated with general life satisfaction, suggesting that it's a meaningful measure.

The problem is, people who watch more television aren't any happier, according to scientific studies. In fact, in a study of 42,000 people from 22 countries, excessive TV viewers report *lower* life satisfaction, even after controlling for many other plausibly relevant variables. We don't know if watching TV *causes* unhappiness, but it's certainly linked in some way.[39] When you engage in a leisure activity that doesn't make you any happier, it's time for a change. Replace your leisure activities that aren't paying off with ones that really make you happy, build skills, make the world a better place, or do *something* good.

But if TV watching doesn't make you happier, why do people do so much of it? Part of the problem is that the act of *starting* to watch television is so darned *easy*. You just turn it on, flip around a few channels, or scan the recent additions to your streaming service, and start watching. In the moment, it's stimulating, and satisfies our craving for novelty.

Compare that with another activity that science has shown to be much more rewarding: getting together with friends. I'm an extravert, and I love hanging out with friends, but boy, is it a pain in the ass to arrange. Everybody's busy, you have to plan things weeks in advance, you have to go back and forth about when it's going to happen, and even then plans change and you have to go back to scheduling.

So you come home from work, tired, and you neglected to make plans two weeks ago to see a friend tonight. You could try to arrange something for the future, but that doesn't help you now. You're in the middle of watching a great TV series on Netflix, so what do you do? Click, the TV goes on, and another evening of engaging in a behavior that antimatters commences.

Like many day-to-day choices in our lives, the choice to watch television over getting enriching social interaction is the result of a conflict between our short-term and long-term goals, our good feelings in the moment versus our long-term happiness. Different parts of our minds lobby for one or the other, and we often pick the short-term benefit, be it eating junk food or watching television. This is especially true when we are low in willpower and impulse control—if the cognitive system is tired or distracted. The habit, reward, and pleasure systems take over, often holding you back from your long-term goals.

Life satisfaction is often attained by completing projects that took hard work. When you are on in your old age, will you be particularly proud of the many Instagram posts you looked at and liked, or all those emails you responded to really quickly, or those YouTube comment flame wars you contributed to?[40]

CHAPTER 3: HAPPINESS

Sabotaging our long-term happiness for a short-term gain is one way we frustrate our own lives. Why should it be this way? Why is it so difficult to stay happy?

Let's back up and look at humanity as a species—in fact, we can back up further to look at animals in general. Happiness, pain, pleasure, joy, sadness, depression: these are all valanced mental states. A state that is valanced has some amount of pleasant or unpleasant feeling about it. Some states, like thinking about the fact that peanut butter is brown, might have no or negligible valance to it. But some states, like hunger and joy, are valanced, and there are probably good evolutionary reasons we are capable of having them. Why would animals have evolved to have valanced states? The answer is obvious: to encourage them to do things that facilitate their survival and reproduction, and to discourage them to do things that jeopardize their survival and reproduction. That's why eating food feels good and getting wounded feels bad.

We can look at the set of valanced states, and the things that trigger them, as a system that pushes an organism toward survival and reproduction. When you think of the states in this way, it's easy to see that if the states were all positive or all negative, they would not be doing any good at all. If your level of happiness was always the same, then it would have no effect on favoring one action over another. It's not a regulatory system if there is no variance in the valance. Any creature blissfully happy all the time, or abjectly miserable all the time, would be unmotivated, and thus out-reproduced by others who strive more to increase good feelings and decrease bad ones.

It's not as simple as surviving and reproducing—it's surviving and reproducing *more* than other creatures who compete with you for resources, which are mostly members of your own species. Although we often think about evolutionary competition as being between predator and prey, most competition is between members of the same species. When you look at it this way, it's easy to see why evolution would make it difficult to be really happy all the time. Such people would have been out-reproduced by the unhappy people striving to make their lives better, because they evolved to find pleasurable those activities that encourage survival and reproduction.

All of this makes the search for happiness, for most of us, not a simple problem to be solved, but an ongoing issue that requires maintenance, changes in habits, and constant vigilance. For when you find something that makes you happy, your mind is made to get used to it, and to want to search for something more. This is called "adaptation," or "habituation."

Habituation

Even when you successfully choose things that actually make you happier, things get complicated. One of the fundamental truths of human psychology is that we adapt to things. Some new situation arises, and, over time, it becomes business as usual.

You can see this over the course of a few minutes every time you eat good food. The first bite of a delicious serving is explosively pleasurable, but as you keep eating, the amount of happiness increase declines, bite by bite. After you've been eating for ten minutes, sometimes you often barely notice the taste anymore. This is called habituation, and the reduction of pleasure associated with it is sometimes called the "hedonic treadmill."

The silver lining is that we also adapt to bad things. We get used to them, and they don't bother us as much as they once did. People recover from terrible events, though it might take years. Becoming widowed, divorced, or unemployed requires about five years for a full recovery.[41] Even if you lose the use of part of your body, a few months later, your mind usually springs back to its previous level of happiness. On average losing a spouse only seems to make you sadder for five to seven years. Getting divorced sometimes even leads to increased happiness, five years on (it might have been a bad marriage . . .).[42]

The devastating thing about the hedonic treadmill is that we are generally unaware, or outright disbelieving, of its reality. You probably know many people who believe that happiness is just out of reach—if they could just get that raise or lose five pounds, they would finally be happy. But when they lose the five pounds, and get that raise, they find that they are not any happier, and are looking to the next thing they want, as though happiness is forever just out of reach.

Part of why it's so hard to stay happy is that people adapt more fully and more quickly to positive experiences than to negative ones.[43] Further, long-term happiness seems to require a lot more positive experiences than negative ones. Happily married couples have about a five to one ratio of good to bad experiences with each other. These ratios are similar to productive and profitable business teams. Negativity seems to be about five times stronger in our psyche than positive things. The reason for this is probably evolutionary. When something happens to you that is good, it means you don't really *need* to do anything. On the other hand, negative experiences require you to do something different, so it makes sense that our minds would focus on these things. It's been good for the propagation of our species. But it doesn't bode well for individual happiness.

The world can be a mysterious place, and sometimes things happen to us that we don't understand. One thing that prevents habituation is when our minds return to an event again and again, trying to figure out why it happened. Positive things that happened in one's life are often easier to understand than negative events.[44] This means that we end up thinking about the negative events more often, because we have more difficulty figuring out why they happened, and how we can meaningfully fit them into the narratives of our lives.

Let's look at habituation to general wealth. A student might live in a crappy building, with so many roommates that nobody feels the responsibility to keep anything clean. When the student graduates and gets a decent-paying job, they move into a better place. They get married, combine incomes, and buy a house. Moving into a better place makes the this person happy, but often it doesn't last. She habituates to the new circumstance. It becomes the new normal, and her happiness returns to the set point. In part, this happens because we start hanging out with other people who have more money. She starts comparing her life to theirs, rather than the other students'. She can now take a vacation to somewhere warm once a year, but starts to envy her friends, who own a house in Bermuda.

Money and Happiness

This is not to say that money doesn't bring happiness. The issue is more subtle than that. First, more wealth predicts measures of life satisfaction more than emotional experience.[45] Second, more money makes you happier if you are poor—that is, if you don't have enough to meet your basic needs and have no spending money. But when these basic needs are met, additional increases in wealth and income provide only minor gains in happiness. How much money is enough? It depends on where you live, because some places are more expensive than others, and people have different attitudes about what "basic needs" are. As we'll get to later, making more money makes you happier the poorer you are.

Wealth seems to correlate with happiness even at a national level. Richer people in a given country are happier, and people in richer countries are happier than people in poorer countries. As countries increase their Gross Domestic Product (GDP) over time, happiness also increases.[46] This does not seem to still be the case anymore with rich countries. As they get richer, their average happiness doesn't continue to increase—an effect known as the Easterlin Paradox, though there is disagreement about whether this really happens.

But there is a bit of causation going the other direction: being happy helps you make more money. Happier people are more appealing to bosses and are easier to work with. Because they are optimistic, they don't think that working hard is fruitless, so they do it more often. They believe they have a future, so they save and invest more.[47] (But for some reason, the happiest people earn a bit less than the moderately happy.)[48]

HAPPINESS AND AGE

Most people are pretty happy.[49] But in addition to income, another factor is age. If you ask people of different ages about how well they feel, on a scale of 1 to 10, people's average happiness ranges between about 6.2 and 6.9. That's not a huge range, but it does change over the course of the lifespan. Between the ages of about eighteen to twenty-one, happiness is about 6.7. It drops until ages fifty to fifty-three, but only to about 6.2, and then it steadily rises into the eighties, where people are happier than they were as teens, and much happier than people in their twenties and thirties. This general trend holds across countries.[50]

In a 2018 study of 1.7 million people across the world, the yearly income at which an increase doesn't make you noticeably happier is about $105,000 per year in North America, $125,000 in New Zealand, and $35,000 in

Latin America (all of these numbers are in 2016 American dollars).[51] So if you live in Canada or the United States, and you make over one hundred grand a year, it's probably not worth putting much effort into making more money. Getting happier is going to require focus on other, nonmonetary parts of your life.

Further, if you're getting by okay, it might not make sense to pursue money and material gain too much. In a study of 100 American adults, Tim Kasser found that those who pursued extrinsic values, such as money, looking good, and social recognition, were less happy than those who pursued intrinsic goals, such as affiliation, self-acceptance, and health.[52] But Americans, anyway, seem to be falling for it. Three-quarters of Americans have garages too full to put their cars in. Just looking at this kind of clutter stresses people out, especially women.[53]

Money is so broadly useful, however, that these averages hide the fact that there are actually things you can spend money on that *will* make you happier. We'll get to what those things are later, but these numbers we've been discussing are for how people tend to *actually* spend their money in the real world. They don't always spend it in a way that optimizes their happiness.

The kicker here is that if you adapt to all things positive really quickly, then no matter what you do it's not going to make any difference in the long run. You're stuck on a hedonic treadmill and you never get happier. Luckily, because of the way your mind works, you can do things to make it work to your advantage.

Hacking Adaptation

What is adaptation? In this context, it means thinking about something less and less frequently. One study of people in luxury cars found that they weren't any happier than people taking drives in smaller cars, unless they were constantly focusing on the attributes of the cars while driving. This happens for negative things, too. Individuals who have lost loved ones experience sadness when their attention is drawn to the loss.[54]

The more you attend to something the slower your adaptation to it will be. This suggests a way that you can hack your attention to help you increase your happiness. You want to change yourself so that you adapt quickly to negative experiences and slowly to the positive ones.

If your positive experiences are more memorable, you can more easily attend to them more often, and adaptation will be slowed. One way to make your experiences more memorable is to do more dynamic activities that require effort and engagement. Studies show that when you engage in activities that are variable, they're more memorable. As a result you pay more attention to them, and attend to them more frequently, so you adapt to them more slowly. The lesson here is not to make your happy experiences routine.[55]

But this is also true for negative experiences. If your negative experiences are surprising, difficult to understand, and novel, then you'll be slower to adapt to them, too. Unpleasant dinners, dental procedures, and deadlines don't bother people so much if they're predictable and unvarying.

So to the extent that you are able, you will be happier by adapting more quickly to negative experiences, by making them less memorable and attending to them less. This can be done by making them routine. For example, if you have to get a shot every day, eventually you will adapt to it. But if you get one painful medical experience every couple of months, it is harder to adapt to.

You can make this work to your advantage with what you eat and with exercise. If you don't like eating raw broccoli, but would like to eat more of it, you need to make it a habit so you don't think about it so much. Sometimes while I'm working, I have raw cabbage cut into small chip-sized triangles sitting next to my desk for me to munch on. Because I'm focused on my work, I don't really notice the bitter taste of the cabbage, which I would not enjoy if it were the sole focus of my attention. I'd be thinking, "Huh, these sure aren't as good as Doritos." But they keep my mouth busy, they have a nice crunch, and keep my stomach from being empty, all of which inhibits me from snacking on things that are less healthful but taste better. By dividing my attention, I don't suffer from eating the cabbage as much.

You can make your eating of vegetables more of a habit. Can you eat healthful foods at the same time every day? Can you eat the same healthful foods every day? Doing so will make you stop noticing them, and how crappy they taste. In my culture, breakfast is the most routine meal of the day. It's rarely interrupted, it's often done at home, and you have a lot of control over it. As such, it's the easiest meal to hack and turn into a good habit. If you remove the sugar from your tea or coffee, or eat vegetables for breakfast (maybe in a frittata), over time it will be habit, you will stop noticing it, and you'll get those servings of vegetables you needed without suffering.

Likewise if there's a food that is not really good for you but you really enjoy, make it a special occasion, really savor it, and reflect on the experience often. This will make your adaptation slower. Don't have it every day.

I have a Vietnamese coffee every single morning, and this is not ideal for optimizing happiness because I take it for granted. I used to take my Vietnamese coffee-making supplies with me when I went on vacation, but recently stopped doing that. I find that having a break from my special drink when I travel allows me to appreciate it anew when I return home.

If you don't like exercise, doing it at the same time every day makes it more routine. You habituate to it, and stop noticing it so much.

Make bad things routine, and good things novel and memorable.

The Benefits and Drawbacks of Understanding Yourself

Popular psychology would have you believe that understanding yourself is always better. But it turns out that understanding why things make you happy can actually make you less happy. Here is how it works: just thinking about your emotions robs them of their power and how strongly they are felt. One of the reasons that journaling works so well for making you feel better about negative experiences is that it helps you understand those experiences. This is because your mind is able to let go of things that it feels that it understands. It is also why therapy is so helpful.

But the same is true for positive experiences. People are happier if they don't understand, and don't try to understand, why good things have happened to them. The mystery keeps the positive experience in their mind, and keeps them attending to it, and thus keeps them from adapting to it. Sometimes, simply identifying an emotion can eliminate it.[56] These results predict that journaling about good things reduces their positive emotional impact on you—though I do not know if this study has been done.

One study had people choose a poster to take home. One group was instructed to think logically about which one they wanted, and the other group was asked to make a gut decision. Later, the thinking group wasn't as happy with their choice. They chose different posters, and focused on attributes that were easy to talk about.[57]

Similarly, uncertainty about something that makes you happy will make the happiness stronger and last longer. One study told people that somebody liked them. The people who heard the name of who liked them were less happy about it than the people who didn't know. The very wondering about what happened seems to help keep the fact that someone likes you salient in your mind, and when you think about that, you get the associated happiness. Because the people in the mystery condition thought about it more often, they got a boost of happiness more often.[58] This discussion is also relevant to one of the recommendations of the Buddhists regarding focusing on the good things. To the extent that you can control your attention, you can deliberately make yourself attend to the positive aspects of things and not attend to the negative ones. This is essentially what gratitude exercises are all about, as well. One of the reasons to meditate is that it trains your mind to be able to attend to what you want to attend to, without being distracted by things that you don't want to think about. Because your mind naturally goes to negative things, being out of control of what you are able to think about, or having the "monkey mind," is detrimental to your happiness. By training your mind with meditation, you gain control over what your mind thinks about, making gratitude exercises easier. Meditation and willpower can be used to direct your attentional focus to reduce the habituation to positive experiences, and hasten the habituation of negative ones.

What You Don't Habituate To

Habituation is not evenly distributed across things in your life: some things you habituate to more than others. Let's take a common trade-off that people often face when deciding where to live. Living in town is convenient. You don't have to drive as far to get to most destinations, or you might not need to drive at all. But downtown homes are expensive and small. So people can live in town, and have a smaller place, or live outside of town, and have a bigger place. Many see this as a fair trade-off, and, given their budget, select a residence that is big enough to satisfy them while still being somewhat convenient. The big downside of living outside of town is that you have to drive around a lot, often including commuting into town to work. In many places people choose to live, there is *nothing* but other residences within a twenty-minute walking distance, which means that, effectively, every destination requires getting in your car and driving.

What people don't realize is that people habituate to house size fairly quickly. Move into a big house, and you fill it up with stuff and stop noticing it. Move into a downtown condo, you purge a lot of your books, feel cramped for a bit, but then get used to it.

However, people do not habituate to commuting. Commuting is stressful; the part of everyday life people have the worst time doing, and they never adapt to it. In fact, the amount one commutes has a direct correlation to social isolation. For every ten minutes one spends commuting, one has ten percent fewer social connections.[59] In general, the more people commute, the more miserable they are.[60] Note that this is true for driving, but if your commute is done by walking or cycling, you increase your happiness, presumably because you are exercising and because you're outside.[61]

Other bad things you don't habituate to are loneliness, lack of autonomy, and noise.

HOW MUCH DO PEOPLE COMMUTE?

Worldwide, people tend to commute an average of thirty-eight minutes per day. That's 300 hours per year, traveling between work and home, which is 10 percent of all work-related time.

In the United States people commute, on average, fifty-one minutes a day, and commute time is getting longer every year. It grew 5 percent between 2000 and 2012.[62]

But there are places where the commuting is far worse: nearly ninety minutes per day in Japan, and two hours, on average, in Bangkok![63]

Commuting is only one example of how people make choices that are not always in their best interest, and shows the difference between preference satisfaction and happiness. In the case of the commute, what people decide is a fair trade-off (in this case, a longer commute to have a bigger house) isn't fair at all, because they underestimate the more intangible costs of social isolation and stress, and also misunderstand that they habituate to some things more than others.

MODERATION

At this point some of you might be thinking that in general what we should be striving for is moderation. Too much of anything is bad, but a bit once in a while is fine, and we shouldn't worry over it.

Some version of this is right, but it's important to understand that what moderation actually means, for any given activity, is culture-specific. Let's take meat-eating as an example. In 2017 the average American ate an average of 98.6 kilograms of meat every year (that's 217 pounds, or about 50 chickens worth of meat). This is an average, which means that many people eat more, sometimes much more, and many people eat less. So an American, looking around at menus and how much people around them are eating, might think that eating 98 kilos of meat per year as moderate.

But people in many other countries eat a lot less meat. Like, a lot a lot. In fact, the average person in *every other country on Earth* eats less meat. In the People's Republic of China, for example, the average person ate only 50.2 kilograms of meat in 2017, about half that of America. The average Indian ate only 3.16 kilograms of meat in the same period![64] So what's a "moderate" amount of meat to eat?

Our ideas about moderation and excess are usually based on what we see around us day-to-day. But this method should not be endorsed thoughtlessly. When you think about moderation, keep in mind that your very idea of what is moderate is based on your experience in your culture, and your culture might be way off when it comes to optimizing happiness.

Do you think maybe, possibly, what Americans think of as a moderate amount of meat is much too much?

Achievement

One thing that people strive for is achievement. We make goals to achieve things, and when we finally achieve them, we get happy. Right?

There's a another bit of nuance to the relationship between achievement and happiness. Bring surprised by achievement gives you a boost of happiness, at least temporarily. But when you can see the achievement coming, your mind adapts to it before it even arrives, and when you finally actually do achieve it, you feel something more like relief than pleasure. That means that, rather than making you particularly happier, it brings you from a state of tension back to a baseline.[65]

So the relationship between productivity and happiness is subtle. You feel hedonic happiness when you're in a flow state, but you can be in a flow state for something that is ultimately really meaningless in your life, like beating the next level of the game *Diablo*.

For your life satisfaction, you need to do things that are meaningful to you. Part of what I want to get across in this book is that though we all might agree that choosing how to spend the resources in your life (time, money, etc.) is the most important thing there is, these choices are multi-factored and difficult to make. Right now we're focusing on only happiness, and we have two, and possibly three different kinds of happiness we need to consider.

American culture, and to a lesser extent Western culture in general, has cultural ideals about what we should be striving for, and these ideas can be wrong. By default, we adopt the values of our surrounding culture, but we should question them if we want to optimize our happiness, because every culture probably gets at least a few things wrong. To minimize this cultural bias, we turn to science, which strives for objective answers (with admittedly varying degrees of success).

So what does science say really makes people happy?

What Really Makes People Happy

There are three fundamental answers to this question. The first is that happiness is something you're born with—if you're lucky. The second is that happiness comes from within, how you direct your own thoughts and behaviors. That means that being happy depends on what's going on in your own head, and is independent of your external environment and the things that happen to you. This view is foundational to many philosophies, such as Buddhism. The third is that happiness comes from having a good life. The idea is that if you have a great romantic partner, meaningful work, and you are not in poverty, then you'll be happy.

Many of us start believing the happiness comes from without, but after a few university courses start believing that we're born with it or that it comes from inner attitudes.

Now it's time to see why all of them are important.

HAPPINESS IS IN-BORN

Although we often think of happiness as being a more or less direct reflection of how good your life is, a bit of reflection on the people we know shows that this isn't always true. Although in this book I'm focusing on the *change* in happiness that people can experience, studies show that happiness is remarkably consistent in an individual both across situations and times.[66]

Even though you can make a pretty good case that, objectively speaking, men enjoy a better standard of living in this world, it turns out that men aren't any happier than women (though women have more intense emotions in general). Good-looking people aren't any happier than everybody else.[67] What's important are your in-born *psychological* traits.

The Happiness Set Point

According to some, you have a happiness set point that you keep returning to. Like a happiness thermostat, life events and conditions can push your happiness down, but, for a lot of things, it's only temporary.

All of us know people who have objectively fabulous lives, but are continually negative and in a bad mood, perhaps even depressed and suicidal. You also might know someone who has a very hard life, but is unreasonably happy in spite of it. The fact that we find cases like this intriguingly reveals our underlying belief that happiness comes from having a good life.

There is a good scientific reason for these cases: some people are *lucky*. People who are naturally happy have won the cortical lottery.[68] The best predictors of someone's happiness are their personality traits, and these have a strong genetic component.[69] Your genes seem to be an important part of how sunny your outlook is on life. How important? About 50 or 60 percent.[70] What does this really mean? It means that when we look at large groups of people, and how happy they are, and lots of other factors, like quality of life, health, and so on, about 60 percent of the variance in happiness is explained by inherited factors. What exactly *that* means I'm apparently not smart enough to understand, in spite of talking to many people who understand statistics better than I do. What it *doesn't* mean is that the genetic *contribution* to our overall happiness is 60 percent, but since I can't get any better answer, this approximation of the truth is what we're going to go with.

People seem to have a happiness set point that is partly genetically determined.

Can you change your set point? It's hard. According to happiness researcher Jonathan Haidt, the scientifically most reliable ways to improve your happiness are meditation, cognitive therapy, and Prozac.[71] Talk therapy's great, and some studies find that it doesn't even matter much what kind of therapy you do.[72] Prozac helps, but only as long as you take it. Therapy has lasting effects.

Let's not dwell on the lucky and unlucky people of this world—what we want to know is how we can be happier by changing things under our control. We now turn to how happiness can come from within.

Happiness Comes from Within

Mindfulness

You might have heard about the importance of being "mindful." What does this mean? Although you will find differences in details, some of the commonly reported characteristics of being mindful are: paying attention to your immediate sensory environment and what's happening in your mind, and not directing your mind at particular memories, plans, or thoughts. Being mindful in your everyday life means paying attention to what you're doing without being distracted by fantasies, anxieties, or thinking about the past or the future.

Mindfulness *meditation* is when you sit there and do nothing but try to be mindful. It works like this: if something comes into your mind, you can note it's there, but you're not supposed to bring things to mind as an act of will. Obsessing about how well you did on last week's job interview, for example, is not being mindful.

You are also supposed to let those thoughts come and go. So if you think about your dog, you're not supposed to latch onto that thought. You should let it pass naturally, waiting for the next thought or sensation to take your attention. When you attend to things passing through your mind and sensory experience you are not supposed to be judgmental about them. For example, if you think about something you did wrong, you are supposed to resist judging yourself as being (or having been) bad. Counterintuitively, you are also not supposed to judge happy things. You should practice having a detachment from them, with a kind of a cool, detached "isn't that interesting" mindset.

So, for example, if you're walking, you can be mindful of your steps. You might focus your attention on the feeling of the ground beneath your feet. You might suddenly remember how your best friend in sixth grade screwed you over. Personally, I have trouble with this. I might be out on a perfectly nice walk with my dog, when my brain decides it's time to get angry over some slight from twenty years ago. If I don't redirect my attention to something else, my mind will keep playing that videotape over and over in my head. Watching it, I'll feel self-righteous anger, possibly at someone I care deeply about and currently have a great relationship with. This is clearly not good for me.

What's harder to accept is that you're also not supposed to latch onto memories or future thinking, even if doing so makes you happy. Why wouldn't you want to let yourself think of things that make you happy? One reason given is that a focus on the past is a focus on something that's not real. If you're thinking about a better time, it can make you feel worse about your present surroundings: the "here and now." Likewise, thinking about a possible good (or bad) future is problematic, because these possible futures are also not real. If you're thinking about a bad future, it could make you anxious, and if you're thinking about a good future, you might be setting yourself up for disappointment, as well as feeling worse about your present situation.

Happiness, according to the mindfulness philosophy, involves an appreciation of the here and now. One of the great insights of this philosophy

CHAPTER 3: HAPPINESS

is that no matter how good or bad your life is, there are always good and bad *aspects* to it. It's not hard to think about how a great life could have minor bad things about it, but seeing the good in a terrible life can be a little harder.

Let's try to think of what most of us would consider a pretty awful life, just as an example: Pat has a chronic illness that keeps her in constant pain. She was abused as a child, and is currently in an abusive marriage with a selfish partner. She is unemployed and has no prospects for meaningful work. She lives in a poor, rural area and she and her partner's bad financial decisions have put them in terrible debt. They cannot pay rent and have to start living in shelters or on the street.

Okay, so that sounds pretty bad. But the idea that happiness comes from within is, in part, a recognition that even in this life, there are probably great things happening all around. Walking down the street, Pat might see a cute bird, or the smile of a child. She might feel a cool breeze on a sweltering day. She might be able to read the greatest works of literature, free, at the library. If she has a mobile phone, she can access an incredible amount of quality content, free, at many places with free Internet access.

The point of all of this is that, to some extent, whether you are happy or not depends on what parts of your world get your attention. If you can focus your attention on the good stuff, you'll be happier. The extreme version of this philosophy suggests that changing anything about your circumstances is futile—happiness and misery are solely the result of what you focus on and your attitudes about them.

An example even more extreme than Pat's comes from this Zen story:

> A man traveling across a field encountered a tiger. He fled, the tiger after him. Coming to a precipice, he caught hold of the root of a wild vine and swung himself down over the edge. The tiger sniffed at him from above. Trembling, the man looked down to where, far below, another tiger was waiting to eat him. Only the vine sustained him. Two mice, one white and one black, little by little started to gnaw away at the vine. The man saw a luscious strawberry near him. Grasping the vine with one hand, he plucked the strawberry with the other. How sweet it tasted![73]

Even when you're about to be devoured by a tiger, you can still appreciate the taste of a sweet strawberry.

So, happiness solved? Not quite. To say that this is easier said than done might be the understatement of the century. And even if, in theory, focusing on the good would make you perfectly happy in any conceivable circumstance, even advocates of this method agree that being able to do so takes enormous amounts of practice—years of daily meditation and training your focus every day.[74]

THE SECRET TO HAPPINESS, OR THE PRACTICE OF HAPPINESS?

If there is a secret, or set of secrets, to happiness, it's not going to be the kind secret that, once you know it, you're happy for the rest of your life. It's more like one of the "secrets" to health: exercise. Just knowing that exercise will make you healthier doesn't by itself make you healthier. You actually have to, you know, exercise.

Jonathan Haidt put it well in his excellent book *The Happiness Hypothesis*: that we might have encountered some important secret to happiness already, but couldn't or wouldn't take the time to deeply process it, really understand it, and, most importantly, to *practice* living your life right so that happiness is easier to get every day. There are habits we all have that interfere with our happiness, and changing those habits is hard, and takes time, often months.[75]

You need training to be happy, not just a piece of missing information.

What about glorious experiences, epiphanies, that change people forever? Are those quick fixes? First of all, it's not easy to get them. But it's also true that most of the time the effects of epiphanies last for only a few weeks. Lasting change of the mind requires changing one's habits and behaviors, and this takes time. This also goes for self-help things you read, like this book. If you don't follow up with practice to change your behaviors, long-term change is unlikely.[76] Simply knowing what's in a book is unlikely to make you any happier, because being happier requires changes to your habits and skills, and that takes work.

If you're already a Buddha, no problem. But what about the rest of us?

The evidence for mindfulness is mixed. In general, people are happier if they are thinking about what they're doing, even if that thing isn't fun.[77] But one of the specific recommendations of mindfulness is to be in the present moment, and not focus on the past or the future. There is evidence, however, that thinking about happy things from the past makes you happy in the present. Nostalgia increases positive mood, optimism, pain tolerance, empathy, and creativity; reduces anxiety, boredom, and stress.[78] It also helps us refocus on who we feel we really are—to get in touch with our genuine selves.[79] Looking forward to things can also make you happy.[80]

Mindfulness asks you to pay close attention to what you're doing. Should you be mindful all the time? Some people recommend that you strive to.

But sometimes you have to do things that are really unpleasant. Suppose you have to take some bitter medicine, or eat some food that you don't really enjoy eating, but you should. Watching TV, or listening to a funny stand-up act, while you do it, makes you less mindful. And it might actually help you enjoy the experience more. You can take advantage of *un*mindfulness to get done the things that are really unpleasant to do. If you should be eating more kale, but you don't like it, munch on it while you watch a good movie, and you won't notice it as much.

When I'm walking my dog in the freezing cold (Ottawa is the second coldest capital city after Moscow) it can be a miserable experience. I just wish the dog would hurry up so we can get home. But the dog needs to do his business and get some exercise. I like to listen to audiobooks while I do this. It makes the time pass more pleasantly, because I'm not attending to how dark, cold, or rainy it is. I'm in the world of my book. I'm less mindful and happier.

But when it's a beautiful, sunny day, and I listen to an audiobook, sometimes I'll regret it afterward. I'd be absorbed in the book, but paying no attention to the beautiful park, or the dog that I love. At those times, it's better for me to be mindful. Of course, this doesn't always work. Sometimes I'll leave my headphones at home so I can be mindful but still find myself lost in my own thoughts for the entire walk. Mindfulness, even when you want to do it, is difficult and takes practice.

Happiness Comes from Without

Doing some things make us very happy in the moment. Studies of what people are doing when they are feeling the greatest pleasure have some unsurprising conclusions: people are happy when they are eating and having sex. Of course the secret to happiness isn't to eat and screw all day. Much as I love Little Debbie Nutty Bars, after two of them I've reached a saturation point. After indulging in sexual and gustatory pleasures, at some point people have had enough and they don't bring pleasure anymore.[81]

One way to look at it is that we have drives for things like sex, food, and water that slowly climb over time (at different rates for different people). This gives us motivations to satisfy these cravings, and when we indulge it gives us pleasure and simultaneously reduces those drives. But indulging when the drive is already satisfied isn't any fun.

Even when you are hungry or horny, but unable to satisfy those needs, there are things to be happy about in the world around you. There always are, and that's part of the wisdom of Buddhism. You just have to focus more on those good things.

Although your internal mental state affects how happy you are, and it's possible that you can make any situation seem good or bad, depending on how you think about it, in practice people don't have a lot of control over their minds. It takes enormous training to achieve those abilities. In the meantime, some life circumstances make it a hell of a lot easier to be happy, and others make it easier to be miserable.

Here's a metaphor I like to use. Suppose that happiness is like maintaining a certain speed while running. If you run too slowly, you are less happy, and if you're sprinting, you're ecstatically happy. Training your mind to be happy is like training for running—you get better with practice, and as you practice it takes less effort to maintain the desired speed.

But life experiences are like the steepness of the slope you're running on. Some things, like prolonged social isolation, make you run on an uphill slope, requiring more energy from you to maintain that speed. In these situations, being happy is harder, no matter what your internal skills are. Other things, like being well rested, or spending time with people you like, make it more of a downhill slope. It's easier to be happy in some life circumstances than others.

I like this metaphor because it does not deny that happiness *can* come completely from within, but the ease with which you can pull it off depends on what your situation. So while you're perfecting your attention, focus, mindfulness, and meditation practice, you should acknowledge that the different external situations you find or put yourself in will make it harder or easier for you, no matter how much composure you have inside yourself.

What other environmental effects are there on a person's happiness? Healthy people are happier, but people's subjective opinion of their health (as opposed to what a doctor would say about them) correlates even better. Happiness also seems to cause good health. Education has a small positive effect on happiness, especially for poorer people. Intelligence is not related to happiness at all.[82]

The big question, then, is this: what actually are the things that make happiness easier? We need to look to science to know, because we get this wrong surprisingly often when we follow our gut. That is, there are things out there that we think make us happy, but when we measure it scientifically, find out they unmatter or even antimatter. And there are also things that take a lot of effort in the short term, but in the long term really pay off. They might even make us unhappy in the short term, but overall they are worth it.

What Makes Happiness Harder

Some bad things you adapt to, like having a small living space, and other things you don't. It behooves you to focus on removing from your life those things that make you less happy, and that you won't get used to and eventually stop noticing.

I've already discussed commuting, and how a long commute makes you miserable, and you never adapt to it. Another thing is loud noise. If you live near an airport, or a busy highway, the constant noise takes its toll.[83]

Having no control in your life, a lack of autonomy, makes you unhappy. How can you change this? You might be in a job where you have no control over what you do when, or how you decorate your environment, or anything else. If you do, and it bothers you, you might want to try to find a different job.

Interpersonal conflict is another real downer in your life. If you really hate your coworker, boss, or spouse, your happiness is reduced and you won't adapt to it.[84] Women are particularly affected by bad relationships. If you have a really bad boss, it's time to transition. If you have a bad marriage, it's very likely making you less happy. (The moral question of leaving a marriage that makes you unhappy is a more complex issue.)

Stress

Many of the things I'm mentioning here are bad because they cause stress. There are two meanings of stress, depending on whether you are approaching it biologically or psychologically, but the meanings are related. In biology, stress is any response to a challenging environment. So being really cold can put stress on an organism, even if that organism isn't even conscious and has no psychology to speak of, like a birch tree.

In animals, stress results in readying the body and mind for danger. You might have heard of the "fight or flight" response, but it's more accurate to describe it as the "fight, flight, or freeze" response. In the presence of danger, if an animal thinks it might not yet have been noticed by a predator, it might freeze, because running away or fighting would reveal its position to the predator.

The stress response is very primal and is something we share with many other species. An in-the-moment stressor causes acute stress, but if the stress lasts a long time it is chronic stress. Most animals don't get this often, but a species that can have social ostracism, like baboons, can have chronic stress, too.[85] But humans have the dubious position of being able to experience more chronic stress than any creature that's ever appeared on the planet, due to ongoing problems in our social lives.

Chronic stress is a particularly problematic part of industrialized life. In a state of chronic stress, your mind and body are perpetually in a vigilant mode, ready for danger. It makes it hard to concentrate and learn, and it's bad for you, biologically, too. When your body is primed for dealing with danger, all the other, less important systems for immediate survival get slowed down—reproduction, digestion, and your immune system. For young people, this isn't as much of a big deal, but for older adults, the stress can increase the chance of anxiety, depression, slowed healing, and even heart disease.

You can find all kinds of pop-science advice about how to reduce stress, and luckily this advice is usually supported by science: stuff like exercise, being in nature, deep breathing, getting enough sleep, meditation, things like that. But what's rarely talked about is reducing the stressor itself. (The American Psychological Association's website, for example, lists five tips to help manage stress, and not one of them involves removing yourself from stressful life situations).[86] For example, if there is someone in your life who is causing you intense stress—a boss, a friend, social media, or whatever—getting some distance from that stressor is probably better than any amount of deep breathing!

It's not like stress comes out of thin air. There are things in your life that stress you out! So for me, the number-one piece of advice I give for stressed people is to *try to make your life less stressful*. Hang out with people who calm you. Hire a cleaner for your house, if housework stresses your marriage. Get a new job if your work environment is toxic. That kind of thing. Studies show that the three leading causes of stress are issues with money, family responsibility, and work.[87]

But not all stress is bad. When you're lifting weights, for example, you're putting yourself in a stressful situation, but ultimately it's good for you. Also, we want challenge in our lives. We want to face difficult situations and eventually overcome them. You can't do that without stress.

If a traumatic event isn't strong enough to cause post-traumatic stress disorder, it can sometimes have a long-term, positive effect on people, helping them realize that they are stronger than they had thought they were, and are better able to face stressors in the future because they know they can cope. Who benefits from trauma in the long term? People who are optimistic, younger than thirty years of age, and people who struggle to *make sense* of what happened. It's this sense-making that causes a change in the person's outlook.

Even a cancer diagnosis is often described as a turning point in one's life, or a wake-up call to what's really important: appreciating the present, and other people, and not materialism and prestige.[88] In the wake of a trauma, people will sometimes put their lives back together with a focus on some higher ideal—religion, love of the family, something like that—which gives them a deeper meaning for the rest of their lives.[89]

Interestingly, the experience of war, particularly between the ages of seven and twenty, leads to more cooperative behavior for one's in-group for at least a decade, but has no effect on how strangers are treated. This appears to be because disaster and uncertainty makes people focus on their community's culture and moral norms.[90]

Living in the City, Suburbs, and Country

People have strong preferences about where they like to live, so I don't expect people to move after they hear about data on how happy people are living in different environments. But if you're on the fence, or considering a move, there are some scientific findings you might find useful.

Who's happiest? People in the city, the suburbs, or the country? Being around nature makes you happier and has mental health benefits.[91] People who live in rural areas tend to be happier and have better mental and physical health.[92] Even though there are fewer people around, social connections tend to be better in smaller communities.[93] However, rural residents have a shorter longevity. They're more likely to have several health problems, including high blood pressure, obesity, and diabetes.[94]

Living in nature has some downsides, too. Although camping and living in a cottage in the wilderness gives you the feeling of living in harmony with the wild, it turns out that the very presence of humans terrifies just about every wild animal. Ecologist Justin Suraci played recordings of people reading poetry in the wilderness and tracked the behavior of the animals. They avoided the area. The mere presence of a person made bobcats more nocturnal and generally made animals flee—even top predators. The only animals who were attracted to the area were rodents, who were thriving because of the lack of predators. Being in nature makes you more happy and relaxed, but it scares the crap out of the other animals.[95] The fact that these animals are afraid of humans does not suggest, to me, they are just afraid of everything or afraid of anything that makes a lot of noise. In places where there have never been humans, like the Galápagos Islands, the animals have absolutely no fear of people. Having never been exposed to the dangers of humanity, the creatures have not evolved over the course of the history of the species, or learned over the course of an individual organism's lifetime, to be afraid of us. The animals that live near humans know better.

Living in the city has its own benefits. You might think that living in the country is more environmentally friendly, because you're living around more nature, but the opposite is true. Living in the city is much more environmentally friendly than living in the suburbs or the country. The concentration of people reduces the cost of moving goods from where they are grown, processed, and manufactured to where the people are. Smaller homes use less energy than the bigger ones outside the city, and densely packing homes (as in an apartment building) makes them cheaper to heat and cool, because shared walls mean less loss to the outside world. Another large factor is that living in the city reduces the need to drive everywhere. In the country, you can go for a recreational walk from your home, but for everything else you need a motor vehicle, and there's less public transportation. (We'll discuss ethics later in this book, but it's also more difficult to buy ethical meat in a small town. With a big enough city, you can have stores that cater to unusual preferences. In Ottawa, where I live, there are a few stores that sell meat raised in ethical conditions. In my hometown, I don't think there is a single place you can buy chicken that was not raised in a factory farm.)

Cities are also much more productive. The efficiency of a place is negatively correlated with the number of gas stations per capita. The fewer gas stations, the more efficient the society, and dense cities have fewer gas stations per person.[96] Large numbers of people support technological advancement. The largest land masses support the largest populations, and they experience the fastest innovation. Important factors are having a university around, lots of immigrants, and a dense population. And because teams that are close together are more impactful, dense populations of lots of people generate the most ideas, innovations, and scientific advances.[97] People born in New York City, for example, are 80 percent more likely to be in Wikipedia than someone born in Bergen County, New Jersey.[98]

Although people are happier in places with lower population density, the causal relationship is not completely clear. More politically conservative people, for example, tend to be happier *and* tend to choose rural homes. So are happy people choosing to live outside of cities, or is living outside the city making people happier? It's a complex issue that science hasn't figured out yet.

What Makes Happiness Easier

Gratitude
The positive emotion of gratitude appears to be one that has multiple benefits. Gratitude is a positive feeling you get as a result of a personal benefit that was not intentionally sought after, deserved, or earned, and because of some other well-intentioned person's actions.[99]

CYCLING IS SAFER THAN LIVING

In a world where so many things seem to be a trade-off—broccoli is great for you, but doesn't taste as good as peanut butter cups, environmental restrictions are good for the planet but often bad for business—bicycles are one of the few shining gems of this world that have no bad effects.

The first thing that comes to many people's minds is that riding a bike isn't safe. It turns out that this is one of those cultural myths that just won't go away. When you look at the science, it's very safe. In terms of accidents, it's no more or less dangerous than walking outside or driving.[100] If you wanted to die riding a bike in the United States (which is not even known as a particularly cycle-friendly country), you'd have to ride an estimated fourteen billion miles, far more than you could ever ride in several lifetimes.[101] Your legs would be *ripped*.

But given that biking involves exercise, it actually *increases* your overall life expectancy, even when you take the probability and severity of potential accidents into account. Studies show that for each mile you ride a bike, you are expected to gain thirteen minutes to your lifespan. In contrast, you *lose* eighteen seconds of life per mile driven. Exercise helps prevent many of the top causes of death. So rather than asking if it's safe to cycle, the better question is whether it's safe *not* to cycle. This is one of those times when common sense lets you down. If *feels* safer in a car than on a bike, but it's just the opposite.

So what do I mean by the title of this sidebar, about cycling being safer than living? When we have lots of data on what people spend their time doing and when they die, we can crunch all those numbers together and find out how risky or safe it is to do *any* particular activity, even if we have no idea of any causal relationship between the activity and mortality. One way to look at it is in terms of fatalities per million exposure hours. So if, say, 100,000 people cumulatively do an activity for a million hours, how many would die? For skydiving, the number is 128.71. For bicycling, it's .26. For doing *all things they looked at put together*, it's 1.53. That means that an

CHAPTER 3: HAPPINESS

hour spent cycling is better for you than living in general (including cycling!).[102] This should not be a surprise, since any kind of exercise, or doing anything good for you that people don't do enough of, will result in a lower number than 1.53.

Cycling is better than not cycling by a huge margin. Estimates range from 13 to 1 all the way to 413 to 1. Keep in mind that if it was just 2 to 1, cycling would be twice as safe as not cycling.[103]

It's also better for the environment. It takes much less material to make a bike than a car, and the bike uses less energy to move.

For the same reasons it's more environmentally friendly, biking is cheaper than driving a car. Cars cost about fifty cents a mile to drive, all told, while bikes cost only a nickel.

Biking also alleviates traffic congestion. When you're biking in a city, for example, traffic doesn't really affect you very much, because you just ride past all the traffic jams. I actually like biking when there are a lot of cars, because they all drive more slowly. It's when they have the road to themselves that cars really speed. But biking is also better for road congestion, because the more people bike, the less congestion there is on the road, because bikes are so small.

Finally, biking is just fun. If you haven't really been on a bike since you were a kid, try it again. It's exhilarating.

Taking the bus is also way better than driving in almost all the same ways as biking is, except for exercise and fun.

"Gratitude journaling" is a scientifically tested way to get more feelings of gratitude, although the effects are generally small, and results inconsistent.[104] It's pretty easy to do: every night you write down five things you're grateful for. In a study by Robert Emmons, some participants did gratitude journaling and others wrote about hassles they were dealing with or compared themselves socially to others. The gratitude journalers had more reports of enthusiasm, alertness, attentiveness, determination, and energy. They were more likely to help others and offer emotional support. They enjoyed longer and better sleep, and spent more time exercising. Beyond self-report, their family and friends also reported that they appeared to be happier and more pleasant to be around.

Like many things, though, you might habituate to journaling, and see fewer effects. I try to do this exercise a few times a week. It helped to motivate me to put them on Facebook, but I got some feedback that it was making some people compare their lives unfavorably to mine, so I made it a special Facebook group. I'm pretty sure my mother is subscribed.

One theory for why it doesn't always work is that by the time people sit down to record a good thing in the gratitude journal, they've already thought about it and incorporated its meaning into their lives. As such, writing it down doesn't have much additional impact. A way around this is to think about what you might have done so that you *didn't* have that good thing in your life. This *It's a Wonderful Life* exercise gives people a fresh way to appreciate the good things in their life and looks to be a promising method.[105] If gratitude journaling leaves you cold, you might try this.

Exercise

Exercise is great for you. It reduces your chances of getting cancer, heart disease, diabetes, colds and flus, osteoporosis, depression, Alzheimer's disease, and anxiety. It's better for your cognition in general, and, most relevant to this book, makes you happier.[106] The problem is, we're just not getting enough exercise. Eighty percent of people don't get enough exercise to prevent chronic diseases.[107]

It doesn't need to be intense. Moderate exercise, even walking for twenty minutes, helps enormously. There are so many ways you can exercise, it's worth exploring and experimenting to find something that works for you. I really don't enjoy exercise for exercise's sake. I liked dancing, and did it almost every night, but then started going to bed early, and the clubs I liked started getting going later. I live in a city that's a frozen wasteland six months of the year (nobody should have firsthand experience that 40 degrees below is the same in Fahrenheit and Celsius . . .), so outdoor exercise is challenging. I ended up playing squash, which I can do year-round, so long as there's not a pandemic happening. I find it fun, because it's a game, and I do it with other people. But some people enjoy running on a treadmill while listening to books. Recall my friend who only lets herself watch TV if she's using her stationary bicycle. There are many ways to exercise, and you'll like some of them more than others. So there is room for you to be creative and find something that works for you.

In terms of when to exercise, there are different benefits for different times of the day. Exercise in the morning increases your metabolism for the day, exercise at midday helps stave off feeling low during the afternoon slump, and working out in the evening reduces stress you've built up during the day. What's most important is that you do it, not when.

Experiences, Vacations

When you spend money on yourself, it can be broadly classified as paying for possessions or experiences. Although shopping itself is an experience that might be pleasurable, in general the possessions we buy don't make us as happy as we expect them to. We habituate to possessions, and often we forget we even own them.

Experiences are more rewarding, and dollar-for-dollar buy you more happiness, in general, than buying possessions. Part of what makes experiences rewarding is that you not only enjoy yourself during them, but you might be able to enjoy the memories afterward, too. Whenever my beloved and I go on a big trip, we use the photos we took and create a hardbound photo book. We periodically page through these books, relive the experience, feel pleasure, and refresh our memories of the trip.

Some people think that a great life is one in which you enjoy extraordinary experiences. What you consider extraordinary might be different from what somebody else does, but in every case the experience is rare and special. Although these experiences are usually very enjoyable, there are two problems with having them. First, if your experiences are too different from those of the other people in your social group, and they can't relate, your happiness will actually suffer, in the long run, from having extraordinary experiences.[108] The other downside is that memories of these experiences can make your normal life seem worse. I have a friend who lived in Lebanon, and the shawarma there was so good that she can't eat the shawarma in Ottawa, which I love to eat. Who's better off?

On the other hand, routines make your life feel more meaningful, and this finding found an effect after controlling for other factors, such as religiosity, positivity, and mindfulness.[109] Being happy involves striking a balance between what you know makes you happy, and provoking curiosity with new, challenging experiences.[110] The former more reliably brings pleasure, the second increases memories and opens us to new opportunities for discovery of pleasure-generating activities. Everyone needs to find their own balance.

DOES RELIGION MAKE YOU HAPPIER?

Religious people are happier than nonreligious people—and it doesn't seem to matter what the religion is.[111] This is because of two unrelated reasons. First, religious people tend to be a part of a religious community that gives them social ties. In fact, even if you're not particularly religious, it might be good for your health and happiness to go to church or some other social religious gathering anyway. Religious hermits don't benefit from the social benefits of religion.[112]

The other reason has more to do with feeling a part of something greater than the self. Being religious or spiritual causes some positive emotions that lead to feelings of well-being, such as love, awe, elevation, gratitude, and peace, but not amusement or pride.[113] Religion seems to cause a small but real increase in happiness, and seems to help the most for people in difficult circumstances, such as illness, poverty, or dealing with the death of a loved one.[114] Religion also seems to change people's standard about what it even means to have a good life—atheists tend to judge how good their life is based on a summation of their day-to-day happiness, where religious people judge the quality of their life more by the standards set by their religions.[115]

This might be a part of the explanation for why religious countries tend to be less happy. As Steven Pinker notes, "many of the world's most religious societies are hellholes."[116] There are those who think that more religion would be better for the world. It should give us pause that the least religious nations on earth, such as New Zealand and Canada, are the best places to live (by almost every common-sense measure), and the most religious nations often have a terrible quality of life. Even in America, the states that are most religious are the most dysfunctional.[117] Gross Domestic Product is inversely correlated with national religiosity. Each of the ten most religious countries has a GDP below $5,000.[118] How can religion be good for the world when more religious countries are worse off in almost every measurable way? Is religion causing poverty and national unhappiness, or the other way around?

Studies suggest that religion affects happiness very little, in general, but can act as a stress-buffer. If your life is fine, religion doesn't increase happiness very much. This might help explain why religion is more prominent in poor, difficult countries: it's the same reason there are more sick people in the hospital.[119] It's not the hospital that's making them sick.

Vacations aren't just fun. They *increase* productivity at work, predict better performance reviews, lower absenteeism, and make people more likely to stay with their jobs.[120] However, vacations are not a panacea. In a study by Jeroen Nawijn, people rated their vacations on a scale of how relaxing they were. After the vacation, those who said their trip was "relaxed" had no more reported happiness than people who did not go on vacation. It was only people who reported their trip as being "very relaxed" who enjoyed increased happiness upon return.

I've been talking about vacations here, and I want to distinguish them from "travel." Just as you read for different reasons, you get out of your home for a while for different reasons, too. Sometimes you want to experience another culture, and get out of your comfort zone, expand your horizons, and see amazing sights you've never experienced before, eat foods you've never had, and meet people from other cultures. This is travel. It broadens your mind and produces amazing memories. However, it's also stressful. You tend to move all over the place, worrying about getting ripped off, navigating train schedules, struggling with languages, finding places to sleep, getting exhausted, feeling guilty for spending the day in bed because you're exhausted . . . Sometimes you get "decision fatigue" from having to decide even where to eat all the time. I think we all have an idea of what this kind of travel is, and it's different from a beach vacation or a cruise, where the primary function is relaxation. If you are really stressed out at work, and really need a break, perhaps what you need is a vacation, and not stressful travel. Don't go backpacking across Africa and then wonder why you weren't relaxed. Figure out what you need before you plan your trip, and decide how much of a vacation versus travel it should feel.

But even this change in happiness lasted a maximum of two weeks, after which there is a return to pre-trip happiness levels. The length of the vacation had no effect.[121] Another study found that trips shorter than eight days risks your not being able to forget about work and completely relax, but after day eleven, people start to miss home and their routines.[122] This suggests that a relaxing vacation, for most people, should be about nine or ten days.

But vacations don't budge your happiness much.

Of all the happiness-boosting and misery-alleviating tips I've put in here, I've saved the best for last. If there's one thing science shows makes you happy, if we had to pick one thing to be the "secret to happiness," here it is: strong social connections.

Hang Out with Friends and Loved Ones

Social species, be they fruit flies, squirrel monkeys, or human beings, do not thrive when alone.[123] Being lonely for too long can cause depression, suicidal thoughts, depression, violence, and mental breakdown.[124] But being around people you like can make you happier even if you're not particularly lonely. If you're looking for the secret to happiness, you can't do much better than this: the single most important thing you can do to be happier is to spend quality social time with people you like.[125]

Recall our discussion about the relationship between money and happiness. Part of the reason that money has a limited ability to make you happier is that people tend to spend it on the wrong kinds of things. When you're poor, you use your money to pay your bills and buy food—that's great, because spending money on necessities is a very efficient way to reduce misery. But once your basics are covered, your discretionary income, above and beyond the necessities, can matter or unmatter, depending on how it's spent.

Spending money on material goods does not seem to have a positive effect on your happiness, but spending money so that you have more time to do what you want, and especially spend time with people you care about, does. So if you find yourself in a position of trying to decide between say, a new car you don't really need and an unnecessary trip to see friends, take the trip.[126] Money makes you a bit happier, and one study found that having a good social life is worth about £85,000 of extra income per year, or about $166,700 USD at that time.[127]

Introverts

But what about introverts? Lots of people identify with this personality type. What I'm saying is all well and good for extraverts, but what about people who value their own solitude and don't want to talk to everybody all the time? Some people like to keep to themselves, like solipsists, who never actually talk to anybody.

First of all, talk about introverts and extraverts suggests that everybody is one or the other, but that's not true. Most people are somewhere in the middle. As with almost every psychological measure we have, people fall on the introversion/extraversion continuum in a normal distribution, or a bell curve. That means that the extreme introverts and extraverts are actually pretty rare, and most people are somewhere in between—ambiverts. So to classify each person into one group or the other is a little misleading, making one think that it's a bimodal distribution rather than a normal one. The issue is further complicated by the fact that even extreme extraverts often act in introverted ways and vice versa.[128]

That said, some people actually are really introverted or extraverted. If you want to test yourself, take an online personality test by doing a search for "personality test OCEAN." (Ignore the Myers-Briggs; it's not scientifically validated.) Extraverts tend to be happier (have more positive affect), at least in cultures that value lots of outward behavior, like America, Canada, and England. Being extraverted is better for your happiness and health than even exercise and a good diet.[129] Interestingly, when introverts act like extraverts, they're happier, too. It's just harder for them to do it.[130]

But introverts have their own advantages. Interestingly, introversion is different from shyness, which is a fear of social interaction.[131]

Should you sacrifice your social life to work harder? For most people, working really hard doesn't pay off in terms of happiness. But when we look at some of the heroes of history, some of them worked their asses off. The problem, though, is that many of them sacrificed their relationships to do it. Einstein, for instance, made a draconian contract with his wife, including a requirement that she leave the room without question whenever he asked, and a release from any expectations of intimacy.[132] Mozart was composing until he died, even during the birth of his first child, and only after his life ended did he start decomposing.[133] Super-high attainers tend to have more depression and anxiety because they sacrifice their relationships.[134] Keep this in mind if you do any hero worship—there's more to their lives than that one thing they're best at, and sometimes something has to give.

Being world-class has a cost.

WHAT ABOUT ONLINE SOCIAL NETWORKS?

Unfortunately, the kind of social interaction you get with online social networks does not have the same kind of benefits that face-to-face networks do.[135] One simple reason is opportunity cost: time spent online isn't quality time talking to friends.

One study had young girls engage in a stressful test, and had their mothers console them afterward. The girls who got phone calls felt better, but text messages had no effect.[136] The problem is online and text interactions *feel* like they're doing good, when generally they are not. It's the illusion of being social.[137] Some have attributed the decline in empathy in young people to their decreased social interaction as a result of online social networking.[138]

There's a difference between being really good, and even mastering something, and being one of the very best in the world. The Earth holds over seven billion people at the time of this writing, so which one is going to be the best, say, stock broker or drummer? With that many people in the competition, there will be at least a few of them with enormous in-born talent and a *willingness to sacrifice everything else in their lives* so they can let their obsession blossom unimpeded. To be better than them, you'd need to sacrifice just as much. Even if you have the in-born talent to compete with these people, would you really *want* to?

So who should you socialize with? The short answer is "friends and family," but it's not that simple. It goes without saying that most people have at least a few people in their families they can't stand, and some people don't even really like their friends all that much.

To go one level deeper, you need to be with people who make you happy and you have lasting attachment with. This is in contrast with advice from many great thinkers from a long time ago, like the Buddha, Saint Augustine, the stoics, and Yoda, who tell you to avoid attachments.[139]

Ultimately, it's not the objective number of and quality of social ties you have that matters, but the psychological states that result from them. That is, what matters is how connected or lonely you feel. You can have a lot of social contacts and still feel lonely, and you could also have no meaningful interactions with anyone but your spouse and feel socially fulfilled. Surprisingly, loneliness is partially heritable. Feeling lonely makes you less happy, more easily stressed, find it more difficult to have fun, and even makes you die sooner. It increases the risk of death about as much as smoking, and is worse for your health than physical inactivity.[140]

One problem that I found is that my friends are so busy that it's hard to see them very frequently. If you see a friend only once every two months or so, the personal conversation can't really get much past catching up on each others' lives. You have to have more frequent interaction with someone for them to be intimate friends. Only when the major events in their lives are already well-known to everybody can you have the time to delve into how they're dealing with those things, and to put what happened recently into perspective. So I find myself in a position of having lots of relatively distant friends (this is more common for men than it is for women), which I don't like.

It made me stop and try to be more deliberate with how I go about socializing. I made a list of the people that I make an effort to socialize with and thought to myself, which of these people do I want to be closer to? I put them at the top of the list, and try to interact with them more often.

Who's fun and positive to be around? Who can I talk about science with? Who gets along well with my beloved, too? Who lives close by? Who can I have deep conversations with? Who doesn't make me feel like I've lost a pint of plasma afterward? Who reciprocates? The people who maximize these traits went to the top of the list. I'm not saying you need to use a list, but if you find yourself lacking in intimate friends, you might want to make your own list and prioritize some people in your life over others.

I also make lists of people at work I can have lunch with, people near my house who are spontaneous (not everybody likes to be invited to dinner at the last minute), and people who we only invite to parties, but never one-on-one (often because they cancel so frequently). With my friends categorized, I can target certain ones depending on what I need. Do I need a laugh? Do I need a deep conversation? Do I need to work out some science thing I've been thinking about? I recognize that there is something distasteful-feeling about ranking friends like this, but I felt I had no choice. Without some explicit prioritizing, I just wasn't getting what I needed out of my social connections.

Socializing events differ in terms of who you socialize with, but also what you do during the socializing. You can socialize by complaining about people, gossiping, talking about a book in a book group, talking about the wine when you should be talking about the book in the book group, playing board games, catching up on events in your life, having heart-rending, soul-bonding conversations, or discussing philosophy. All of these things generate different effects for you. Some people really enjoy just catching up with friends, but might not want *all* of their socializing to be like that.

SURVIVALISM

Typical "survivalists" concern themselves with how they will fare after the world goes to hell—usually some kind of imagined societal collapse. They learn skills of self-sufficiency. They stockpile supplies, try to live off the grid, and often get lots of guns and fortifications.

As my biologist friend Alexander Gill points out, though, these people don't realize is that survival is fundamentally about cooperation. Bullets have a shelf life. Canned food expires after a few years. So what's the long term plan, here, guys?

If those people really cared about surviving an apocalypse, instead of stocking up on guns and canned food, they should join a community that survives without the world's infrastructure, like the Amish, and learn how to farm, build houses, forge tools, and so on. Survival, and thriving, requires *some* kind of community, even if it's a relatively small one like an Amish town.

If society does collapse, I bet that after fifteen years the Amish will have outlasted the gun nuts.

A (Good) Marriage and (Maybe) Children

Marriage

Partnered (usually married) people tend to be happier. There are a few reasons for this.

First, when you hear something like "married people tend to be happier," you should immediately wonder what the causal direction is. There are three possibilities: 1) marriage makes people happier; 2) happier people are more likely to get married; and/or 3) there is some third factor that causes both happiness and marriage, but neither one causes the other.

The problem with survey data is that it is very hard to find causation. Doing so requires measurement over time, and, ideally, with experimental interventions. Of course, we cannot take two groups of people and tell some of them to get married and some of them to stay single, so experimental exploration is out. But what we can do is look at how happiness changes over time, and see how being married versus not being married affects the expected trajectory over thousands of people.

Are happier people more likely to get married? Yes. Happier people tend to make better marriage partners. If you were trying to decide between a happy and unhappy person, who were otherwise very similar, which would you marry? So when you look at married versus unmarried people, it's not a random sample—really miserable people don't get married as often.

But marriage seems to have a causal effect on happiness, too. There is no disagreement that marriage makes people happier in the short term, but there is debate about whether it has a lasting effect.[141] What makes it complicated is that people adapt to their situation—our old friend habituation. So when you start living together, you get happier, and when you get married, you get even happier. This makes sense, as people don't tend to get married unless it's something they really want.

But then, after about two years, the honeymoon period is over, and people have habituated to their new situation. Most scientists think that marriage makes you happier in the long term, but a few believe that you habituate completely, returning to your base level of happiness.[142]

Being married also has health benefits. Married people live longer, and being unmarried tends to lead to worse health-related behaviors, and this effect is stronger for men.[143]

IS THERE A DIVORCE CRISIS?

People are getting divorced more often than in the past, for sure. Most people interpret this as a sign that marriages are worse, or that people are unable to keep commitments. But I want to question whether or not this is so bad.

First of all, there is some evidence to suggest that although the divorce rate is increasing, the marriage *dissolution rate* is not. Basically, people's marriages used to end because somebody died, but people are living a lot longer now. If you are in bad marriage, but only have ten more years to live, you might not bother with divorce. But if you have another twenty years to live, divorce seems much more attractive, because you have time for a new life without your partner.[144]

The other thing is that people feel that they have more freedom than they used to get out of bad marriages. We can't assume that divorce has always been easy! In different cultures at different times, getting a divorce at all is difficult or impossible, and sometimes women didn't feel that they could even support themselves if they weren't married. Those things have changed in most of the world. So we can look at the increasing divorce rate as a good thing: marriages might be just as good or bad as they always have been, but now people are empowered to better their lives by opting out of the bad ones.

Divorces don't always bring misery—if the conflict is bad enough, it brings happiness. If the conflict is mild, the divorce makes things worse.[145] In any case, the divorce rate seems to have leveled off in the 1990s.[146] We should be very skeptical of thinking there is some kind of divorce crisis happening.

Children

Scientists generally agree that having kids does not make people happier, on average. Some argue that it makes your life more meaningful, but not

CHAPTER 3: HAPPINESS

happier. Some say that if you *want* kids, then having kids will make you happier, but having kids when you don't want them makes you unhappier, which brings the average down for parents in general.[147] And some studies see no effect of having kids. People with children are three times more likely to tell Google they regret having children than childless people are to tell Google they regret not having children.[148] So in terms of your own happiness, it might not matter much to your happiness. Having a child in the United States can expect to cost you about $300,000, your time included, but for many people it's worth it.[149]

Some studies show that children bring on a temporary boost in happiness. People are excited when they are about to have kids, and excited when they get them, but then level off back to normal levels after about a year.[150]

Some people don't want children, but then, at some point in their lives, suddenly do. This is popularly known as the biological clock, but scientists call it baby longing or baby fever. Interestingly, there are gender differences in baby longing. Women (on average) peak in their twenties and then their desire for children goes down from there. Men start out (in their teens) wanting children much less, but their baby longing gradually increases into their twenties.[151]

The lesson I take from this is that if you're a man who wants kids, you probably will continue to want them, but if you're a woman, your desire for kids probably will taper off. So women who are generally on the fence about it, but feeling a currently strong urge, might consider that they are contemplating a very long-term solution to what might be a short-term problem.[152]

Many people are very proud of their children. Unlike many long projects, by having a child you sort of commit yourself very hard to the responsibility, time, and cost. You can neglect to use birth control just once, and then be saddled with decades of work, the product of which you *might* be very proud of. Either way, you're going to do the work. In contrast, nobody accidentally starts writing a multivolume fantasy novel series in a moment of weakness and says, "Oh well, looks like I'll have to finish it now." Children are perhaps the ultimate precommitment: when you have kids you sign up for a project you *have to* follow through on.

As for the morality of having kids, I'll get to that later.

LOVE HAS MANY FLAVORS

Do you believe in "true love"? According to Jonathan Haidt, this modern myth has these tenets: True love is being madly in love. If you find yourself in true love, you need to marry that person. If love is true, it will never diminish. Thus, if you fall out of love, you married the wrong person.[153]

Love is not only one emotion or feeling. It's a combination of lust, infatuation, intense focus, friendship, attachment, and compatibility. We only call what's happening "love" when enough of these things are involved. And different combinations make the different kinds of love we talk about in our culture: puppy love, love for a friend, love for a baby, and finally, the stages of romantic love.

This might sound wrong, but even people who are true believers in love know that over the years love changes. Romantic love, of course, tends to have elements of lust and infatuation, and perhaps, strongly put, obsession.[154] Later, the love turns to a more friendship and attached phase.[155]

Different hormones and neurotransmitters make the different feelings happen. There are hormones that make us attach: oxytocin and vasopressin. Oxytocin is the same hormone that attaches mothers to children, which explains why some mothers feel like they are "in love" with their babies.

Norepinephrine surges through us when we start getting into somebody. This hormone makes us focus on a single person very strongly. That's why we can't stop thinking about our new love.

Dopamine causes the intense pleasure we feel around that person.

I imagine that in the friendship-style love stage, the dopamine is still there, the oxytocin/vasopressin is still there, and the norepinephrine is reduced. Thank goodness. Can you imagine being obsessed with your love for your whole life? In fact, that's the idea I explored with my cowriter in my play *Medea: The Fury*, based on the Greek myth. The idea is that when Medea falls in love with Jason as a result of getting struck with Cupid's arrow, she never leaves the initial rush stage of love. So eleven years later, when they have kids and Jason is thinking of his career and such, Medea is still absolutely obsessed, lustful, and infatuated with him. Drives him crazy.

Let's look at the idea of love at first sight. I think just about everyone has had the feeling, upon seeing or meeting someone for the first time, a very strong attraction. I think we have also all felt, on some occasions, that after talking to the person for five minutes that attraction is reduced considerably. These cases are not remembered as "love at first sight." Why? Because it did not end up developing through any of the other stages that we associate with love.

Now, when you feel that instant attraction and end up dating and getting very attached, then retrospectively you might want to call it love at first sight, even though it's probably the same exact feeling you had when meeting the other attractive people who ended up being jackasses.

"Love" is a problematic term, which is probably why some people don't know and have to really try to figure out whether they are in love or not.

Love for your baby or puppy, or your parents when you are young, is made up of feeling good, attachment, and focus.

Love for your parent (as an adult) is feeling good and attachment.

Love for a new romantic partner (infatuation) is feeling good, attachment, focus, and lust.

Love for a not-so-new romantic partner is the same with a reduction in lust and focus, with an increase in attachment.

Love for your favorite movie is just feeling good and focus.

Pets

What about pets? Are they a substitute for children? For me they certainly are.

Many people think that pets are good for their happiness. But the science on this is much more murky. Indeed, pet owners, especially dog owners, are happier than others. But what we don't know is whether having a pet makes you happier, or if being happier makes you more likely to have a pet. It could be that having a pet is good for your happiness if you want a pet, but isn't good for it if you don't want a pet.

Having a dog requires that you go on walks, which, as we've seen before, gives you social and exercise benefits. It usually gets you outdoors, engaging in physical activity, and is sometimes social. I always meet more people when I have a dog. Although many of these interactions are superficial, at least they are positive. Another good thing about owning a dog is that it forces you to take breaks in your day to walk around outside.

One study found that the longer one owned a pet, the greater their empathic concern, which in turn is linked to many benefits in relationships, such as quality relationship maintenance behaviors and commitment.[156] This suggests that having pets might cause people to be better in their romantic relationships.

So marriage makes you a bit happier, but it has to be a good marriage. How about divorce? What we know is that divorced people are less happy, but that's like saying that people in the hospital are less healthy. It doesn't mean divorce caused the unhappiness, but is likely rather a reaction to it. But in general people are lonelier after divorce, which reduces their happiness.

A study of male scientists and artists showed that marriage has a deleterious effect on scientific productivity. One reason for this might be that the (often subconscious) desire to be productive and to achieve might be to attract a mate, and after this happens, the drive reduces, possibly because of a drop in testosterone. (The same study found that marriage also reduces crime.)[157] It seems that there might be a productivity/happiness trade-off when it comes to marriage.

Happiness and Memory

When we reflect on our life satisfaction, what we're doing is a curious thing. We're looking at our lives, for as long as we can remember, and making a judgment about the whole enchilada. The curious thing is that when we do this, we do not, and cannot, review *all* of our memories—to do so would take years, even if we could recall everything. We make a judgment based on a vague feeling of satisfaction or dissatisfaction about our memories in general.

In mindfulness practice, you can train yourself to focus on the perceptions in your environment that bring you joy. Similarly, you can focus on memories that make you happy and satisfied, or you can focus on memories that make you feel like crap. (Note that a focus on any memories at all is *not* practicing mindfulness.)

It's not just the goodness or badness of memories that matters, it's how you appraise them. Let's take a day when you might feel excruciating pain. Not a good memory, right? It turns out that it makes a difference *why* you had that pain. Childbirth, which, by reputation, is one of the most painful things a person can experience, is remembered much more fondly than the equivalent amount of pain caused by getting attacked by an alligator. The latter feels much more meaningless, where having a child can feel transcendent. My mother says that the day she gave birth to me was the happiest of her life.

It was probably also one of the most painful.

It might feel like you don't have any control of this framing, but studies show that merely making up meaningful reasons can make you feel better about bad memories. A study found that a bothersome arm pain could be made much more tolerable by imagining that the injury was the result of enemy agent activity during an espionage mission.[158]

If you want to maximize the happiness you get from memories, then you need to do things to help keep those memories in place. Photos can do it, but often people never look at their photos again.

I've heard people say that travelers take too many pictures, and don't spend enough time savoring the experience in the moment. I agree with the second part but disagree with the first. Yes, you should savor your experience, and not simply take a picture for later, but on the other hand, yes, you should take pictures. Whenever my beloved and I get back from a trip, we collect the photos from the trip and put them in a photo album. She designs it online and orders a hardcover book for the trip. We keep these together and periodically review them. We've learned a few things from this exercise. First, pictures are vitally important. Over the years, they seem to be the only things we remember about the trip. Those other memories don't get rehearsed as much as the memories triggered by the photos, and eventually are forgotten, or merged with other, similar memories from other trips. So take lots of pictures, and choose the best for the album. Second, having the book in physical form helps remind us to review it. We have all of those pictures on our computer, but they're buried in folders and there is no visual reminder to look at them. If you don't want to print out your photos, then it might be good to set up repeating calendar reminders to review the experiences. Third, albums allow for captions, which help contextualize the pictures. A photo might cease to be meaningful to you if you can't remember what it's a picture of or what meaning you intended to communicate to your future self when you took it. This is particularly important for names of people! I have many pre-Internet travel photos of people I met on my trip with absolutely no idea who they are. We also find that looking at these albums helps us appreciate our lives and the bond we have with each other. Finally, looking at these albums has diminishing returns. Like nearly everything else, if you review them too often they lose their punch. This isn't a problem, because over the years we have more albums to review anyway, so we see each individual one more and more rarely.

Because memory works in some predictable ways, you can modify your life to make your memories improve your life satisfaction. For example, our minds are pretty good at detecting differences between things. Things that are similar tend to get all grouped together and classified as the same. When this happens, your memories of individual instances are basically forgotten.

When I look out my back window, I see drama unfolding among the squirrels. There is chasing, mating, fighting, chittering. I enjoy this, but there's a problem: I can't tell one squirrel from the others. Some are black and some are gray, but aside from that, I can't tell who's who, even though I try. When I see one squirrel with a pizza crust in his or her mouth being chased by another squirrel, that's kind of interesting. But the next day, I see a squirrel shivering in the cold, and I can't tell if it was the same individual who had the pizza or not. If I could, I think I'd get a lot more out of squirrel watching.

Let's say you have a favorite vacation spot, some resort in Cancún. You love it so much you go every year. There's a benefit to this: you have a very high confidence that you will love it, and you're pretty much guaranteed a lot of happiness (in the form of positive affect) when you're there. But a funny thing happens over the years—you'll have a general, vague memory of being in Cancún, but the individual years will be forgotten.

If you go to the same resort in Cancún every year, it's reliably a great experience, but over the years your memories of those trips will blend together. You won't be able to distinguish what happened and who you met one year versus another. It will be one generic memory.

Now, let's compare this to a person who goes to a different place for vacation every year. One year it's Cuba, the next it's Barbados, the next it's Portugal, and so on. This person is taking chances. Maybe they're not going to love where they are going. But as time goes on her life will feel like it's actually been longer than the person who went to the same resort every year for a guaranteed good time.

If you go to different places every time, it's harder for those memories to interfere with each other. Having a long life is good, and your life will *feel* longer to you if you have more distinct memories. One of the primary ways we perceive the passage of time is through the number of distinct memories we have of that time period. The more memories we have, the longer that span of time feels.

So you will have some objective lifespan: the number of years you live. But you will also have a subjective lifespan, based on how long your life feels like it has lasted. Your subjective lifespan is a function of, in part, the *variety* of experiences you have. Does this mean that you should try to do things differently every single day?

No, you should be selective about it. For one thing, changing up your routine every single day takes a great deal of mental effort. Choosing to have a different, memorable breakfast every morning takes a lot of planning, and planning has a cost in terms of mental energy that could be better spent elsewhere. Steve Jobs wore the same outfit every day for this very reason.

Also, if you're changing things up all the time, you will have a hard time building habits. Let's take breakfast as an example. Some people have cereal for breakfast. Cereal is delicious, but it's mostly empty carbohydrate calories. Cereal's not good for my health, makes me feel sluggish afterward, and I'm hungry again by 10:00 A.M. If I have eggs and steel-cut oatmeal, I'm filled up until lunch, it's healthful, and I like to eat it. So that's what I have most of the time.

Can I remember all of the egg breakfasts I've had over the last twenty years? Of course not. But I really appreciate that I got in a habit of eating well. I don't have to plan it, think about it, or force myself to do it, because it's a habit. It doesn't contribute to my subjective life span very well, but it's worth it for the other benefits. In the mornings, when I like to get writing done, I'm not distracted by breakfast decisions.

There is a bright side to your mind's ability to merge memories of similar things: You can try to merge memories of bad stuff so you *don't* remember them. Let's take exercise. There are ways to make exercise more fun, but let's suppose that for your own reasons the exercise that you have to do isn't fun, you just need to do it—or perhaps you have to do painful physical therapy. Doing the exercise as a routine, doing it in the same way every day at the same time, means that each individual instance of exercise will be less distinguishable in your memory from the other ones, and you'll remember it less. When you reflect on your life satisfaction, you won't remember all the time you spent doing boring exercise as much, and you'll feel better about your life in general. As we learned in the section on how to hack your habits, this is also the best way to keep doing it.

Another trick you can employ to help you forget bad experiences is to crowd your memory afterward. When something traumatic happens to you, you can impair your mind's ability to encode that memory by crowding your hippocampus with noise. Studies show that playing *Tetris* after something bad happens reduces your ability to remember that bad thing. Let's suppose you live in a situation where there is a lot of arguing and social tension. Reexperiencing memories of these fights can hold your mind hostage in the future, keeping you miserable and unable to focus on other things. Occupying your mind right afterward, with a video game or a strong visual experience, like watching *Spirited Away*, can help you forget.

Memory adds an interesting dimension to good and bad experiences. A good experience is worth more if it's remembered, and a bad experience has a lower cost to your happiness if it's forgotten (I'm only talking in terms of happiness here, not the things you might learn from a bad experience). Specifically, the more often you recall a memory, the greater impact of that experience on your happiness—for better or worse. So if you can't remember something at all, it ceases to continue to make your life better or worse. If you only recall it once or twice, it's not affecting you much more than it would having been forgotten.

Similarly, the longer you live, the more opportunities that memory has to affect you. A great, memorable experience early in your life can be recalled more often than one later in life, simply because you have more time to recall it, making it more valuable.

Memory is important, but is it necessary? Is there value to experiences that you can't remember at all? Suppose you could have an exquisite week-long vacation that you wouldn't remember at all, or a decent vacation that you would remember. Which would you pick? Part of the rationality of this decision depends on how long you have left to live. If you're probably going to die within ten years, maybe pick the first one.

But we don't want to deny any value to un-remembered good experiences, because ultimately all experiences will be as good as forgotten when we die. If a vacation you can't remember is valueless, then what is the point of giving a good life to a being that will eventually die and not remember anything? In a way, our entire lives are like trips we won't remember.

If we look at someone in the past who lived a miserable life, we can't say it doesn't matter, simply because that person cannot remember the miserable life anymore. We strive to give beings good lives *even though* they won't remember those lives when they're dead. To think otherwise means that *everything* unmatters, because everything, even if it takes the heat-death of the universe, will eventually die and not benefit from accumulated memories anymore.

There Is More to a Good Life than Being Happy

Our culture's obsession with the pursuit of happiness sometimes appears to assume that a happy life is a good life, or even that there isn't anything more to a good life than a happy one. But many have argued that there is more to well-being than being happy.

To help appreciate this, suppose someone were to offer you the opportunity to spend the rest of your life high on drugs. Suppose you were confident that with this offer you'd be healthy, live a long life, and be completely blissed out and in ecstasy the whole time. But your experience would not be rich—that is, it wouldn't feel like you were living a life, you'd just be high and incapable of cognitive function. You'd just be an entity that feels pleasure and does nothing else. If the idea of drugs turns you off, suppose it was non-drug brain stimulation machine instead. Would you do it?

If *pleasure* was all there were to life, there's no reason not to. Life satisfaction is valuable, as we discussed, but in the pleasure machine you wouldn't have low life satisfaction, either. You'd be too drugged up to have any opinion about your life at all.

Some people might get in the machine, but many would not, including me. We value something about experiences, not just the pleasure or displeasure they bring. Now imagine that the person says, "Oh, okay, how about, instead of the pleasure machine, we put you in the *experience* machine?"

The experience machine idea is from philosopher Robert Nozick.[159] In it, you have all of the experiences of a real life, but it's the best experience you could have. That is, rather than how real life is full of people being crappy to you, depression, and Internet flame wars, the "life" you'd have in the experience machine would be optimized for happiness, life satisfaction, meaningfulness, and every other thing you care about with respect to subjective well-being. You wouldn't even know you were in it—you'd take it to be real, as you do in dream experiences. Maybe you have a happy virtual childhood, marry a wonderful virtual person, and achieve great virtual things. The difference is that the entire thing is churned up by virtual reality software, the objects, people, and issues you encounter there are not real in the physical sense, and the beings you encounter there have no consciousness. That is, you might experience having fifty years of a wonderful marriage, but the spouse in the experience machine is a computer program subroutine that does not have its own subjective experiences.[160]

Most people would not enter the experience machine, either,[161] but not always for reasons that philosophers expect. The thought experiment is intended to show that people prefer reality to a happy simulation, and that there's more to a good life than happiness. But when you talk to people about it, you get different reasons. Philosopher Filipe de Brigard asked people if they'd get *back* into the experience machine: Suppose you woke up and a man told you that you'd been accidentally put in an experience machine (it was supposed to have been somebody else), and that the life you *thought* you'd been living was actually a simulation. Now, do you want to go back to it, or stay out and start experiencing some new, foreign real world you know nothing about?

This is a bit like the choice Mr. Anderson had to make in *The Matrix*. Note that in that movie, Mr. Anderson is not shown to have any social life at all—no friends, no wife, no family. Lots of people in de Brigard's experiments chose to go back into the simulation, even though it wasn't "real," because they didn't want to give up the life they knew. Reality wasn't as important to people as many philosophers thought.[162]

If we assume that there is more to a good life beyond happiness and pleasure, what are these other goods? Scholars have suggested many, including rationality, having true beliefs, aesthetic experiences, and achievement of valuable things (rather than achievements in games).[163] But the one I want to focus on in this book is moral goodness.

Is being happy a moral good all by itself? It's certainly good for you, but what about everybody else? People who are happy are nicer people,[164] and experiments have shown that making someone happy makes them more likely to help others, as long as it doesn't endanger their own mood.[165]

To start, there are even moral questions related to the pursuit of happiness, because some things we chase after in the name of happiness, like prestige and displays of wealth, are zero-sum games. What I mean by that is that we can't have a society where, for example, everybody has high prestige. Having it feels good, but the very concept of prestige being high is only in relation to people with low prestige. People compete with each other for this prestige.

The same goes for displays of wealth. As we will discuss soon, there's an excellent chance that, if you're reading this book at all, you are one of the richest people who has ever lived. But whether or not you *feel* rich depends on how much money you have relative to the people around you. If you are the only person in a wealthy neighborhood to have a private jet, you might feel good about yourself. It's the same feeling that the person in a poor town might feel if she were the only person with a tin roof, where everybody else has to put up with straw. Prestige is relative, and not everybody can have it.[166]

What this means is that *your gain in prestige is always at the cost of other people's prestige*. This makes the pursuit of prestige a moral consideration. As a community, such as a nation, gets richer, it just means that people need that much more to gain that prestige and ability to show that they're wealthier or more prestigious than the people around them. There is an increasingly expensive list of things you must display to qualify for the respect you're chasing. By playing the game, you're raising the bar, and making the problem worse. If you already have a lot of prestige, and keep trying to get more, you're kind of hogging it.

I'm as guilty of this as anyone. Although there are outward displays of class and wealth that I find distasteful, such as huge diamonds and luxury cars, there are things I won't give up. Some examples include a full set of teeth, and an accent that signals a higher class. We all have *some* community we are trying to impress; it's good to understand that, reflect on it, and decide how you want to live your life, rather than just going with the cultural flow unquestioningly. It's hard, though, because we instinctively tend to try to gain prestige and to look good compared to the people around us, according to whatever values the culture happens to have.

Luckily, as we saw in previous chapters, many of the things that really make you happy are not zero-sum, such as interesting activities and socializing. Getting happiness from these kinds of things doesn't hurt anybody. It's not like your playing with your kids means that other people can't play with theirs, or reading a book or listening to music takes that pleasure away from someone else.[167]

Does this mean that it is unethical to pursue zero-sum resources, such as prestige or grant money? Not necessarily, but it's important to realize that there is a cost to obtaining these resources that is paid by other people. Although you're taking from others to benefit yourself, we need to think about the greater moral impact of what you're doing with that resource. It's okay to take prestige from other people if you do more good with that prestige than they would have.

Let's take my writing of popular science books as an example. When I published a book, that means that there is one other manuscript out there that isn't getting published. This is because publishing houses can only publish at a certain rate, and they have to turn lots of projects down. I've already published several books, so in some sense it's greedy of me to publish more. Why not give somebody else a chance, maybe someone who has never published a book before, and whose career would benefit more from it?

Although this troubles me, I also have to think about the effect of these books on the world. That is, are my books improving the world more effectively than the book that *would have* been published had mine not been? It's impossible to know what that book would have been, so I have to think about the *expected* value of this theoretical book. If it's a general nonfiction book, it might not be a science-based one. As I believe that, in general, science-based books being out there improve the world more than non-science-based books (your values might differ from mine), the expected value to the world of my book is greater than that of the book that got passed over (which has a lower than 100 percent chance of being science-based). Now, if I knew my book pushed out another science book, then I have to take a hard look at whether *my* science books are better for the world than others.

The same goes for grant money, for fame, and all the other zero-sum resources: whether it's good or bad to pursue them depends on what you do with it.

When we're looking at morality, we often think of religion and philosophy. But can science help make you be the best person you can be, morally? In particular, how can science help you be the morally *best* person you can be? We will turn next to how to optimize your goodness.

PART II
IT'S NOT ABOUT YOU

4
WHAT WE THINK IS RIGHT AND WRONG

Happiness, Huh?

Earlier I talked about reciprocity styles—there are givers, matchers, and takers. Studies show that being a giver, and acting like a giver, tends to be good for you as well as the people around you. It's worth asking, though, if being a giver like this actually leads you to being the *best* person you can be. That is, being a giver is good for you, but is being a giver actually good for the world? If you're reading this book, for example, you're probably rich. By this I mean that if you are literate and have access to books, you're probably one of the richest people alive today, and almost certainly one of the richest people who have ever lived.

There are websites where you can put in a few pieces of information and find out where you rank in the world in terms of income. At the time of this writing, an American making $45,000 per year is in the top 0.41 percent of the richest people in the world.[1] Sometimes people talk about the "1 percent" richest people. If you're an American making more than $33,000 per year, *you are the 1 percent.* Even an American making only $6,000 per year is still in the top fifth of the world.

These numbers are adjusted for buying power, which means that even though the same amount of money goes further in Ethiopia than in

America, these numbers take that into account. If you hear that someone is living on $1.50 per day, that means they are (trying to) live on what a buck fifty buys you in America, not Ethiopia. The only caveat is that because even the poor in America are so rich, there are fewer extremely cheap, low-quality goods available for purchase. There just isn't a market for them, because people have so much money.[2]

Being a giver, as presented in books like *Give and Take*, seems to be about being generous to the people you meet. But unless you're living in a truly poverty-stricken place, it is very likely that the people around you are rich, too. When I look at the moral aspect of being a giver, I have to ask: how much help do the rich people around me really need? I'm not saying that the people around you don't matter, and that you should feel free to be a jerk to everybody. What I'm saying is that maybe you shouldn't judge how good a person you are solely based on how much you help other rich people—where by "rich people" I mean just about everybody you've ever met in the industrialized world.

In part 1, I discussed a variety of topics about self-improvement, focusing on productivity and happiness. If you're ready to level up, in this part we will delve into the next stage: being a good person. And because I'm an optimizer, I'll talk about what it takes to be a *really* good person. As good as you can be. And even if you're not particularly interested in optimizing your goodness, you might want to keep reading anyway, because thinking carefully about and reviewing scientific findings relating to this issue bring up some fascinating, mind-bending ideas.

Ethics is a philosophy topic, primarily concerned with what is actually right and wrong, and the reasons why. Before we get to that, it is helpful to have a grasp of what we *think* is right and wrong. That is, how real people on the ground think about moral situations, and how they act in ways that have ethical implications. This will give us a grounded understanding for the more highfalutin ideas to follow.

When We Think About Right and Wrong

Not all thinking is moral thinking. People think and do things all day long, and only some of it involves moral thinking. For example, when somebody's deciding which comedy movie to watch on Netflix, they're often not considering any moral implications of their choice. When somebody is sitting

CHAPTER 4: WHAT WE THINK IS RIGHT AND WRONG

with their legs crossed and feel the urge to cross their legs the other way, they don't consider whether or not it's ethical to do it.

Our minds seem to operate under the assumption that if you are not *spontaneously inspired* to think about the moral implications of an action, then that action is, by default, morally acceptable. When you confront someone about their actions as being morally problematic, and they never thought of the action in a moral way, their knee-jerk reaction is to disagree with you. If it were morally wrong, wouldn't their mind have detected it?

I find this more than a little disturbing. What we have is a situation where moral thinking only gets triggered once in a while. Arguably, every choice you ever make has moral implications, in that any other choice that could have been made might have a resulted in a world a little bit better or worse than it turned out. In this sense, every choice is an ethical one. But it is also clear that most choices are not ethical from a psychological perspective, in that most choices are made without our minds making any moral judgment, aside from the fact that our immorality detectors, which presumably are constantly monitoring what we do, don't alert us to anything. They don't weigh moral considerations for many actions. These mental functions behave like smoke detectors, which sound an alarm in the presence of smoke for the purposes of preventing fire. To suggest that all of our choices are morally considered because of the existence of these psychological immorality detectors is like saying that everything that happens in a household is done with consideration of fire risk, just because there are smoke detectors in the house.

Many of our moral instincts are emotional. The amygdala is a part of the brain closely associated with emotions, particularly the negative ones. It subtly monitors what's going on, looking out for moral infractions. When it detects one, it goes off, triggering a moral response—and, possibly, a more thorough consideration by your frontal brain areas, specifically the ventromedial prefrontal cortex, which works to consider every angle when making a judgment.[3]

But in general, we rely on a subconscious process to decide what things are morally relevant. Although we *can* think carefully and consciously about a moral decision, we don't even start to do so unless someone else brings it up, or, more commonly, our own subconscious process lets us know that it's an issue worthy of moral consideration in the first place. If we want to optimize our morality, we need to know if we can trust that process.

When you are forced to think about the moral implications of something over and over, it can *become* a moral issue for you—your subconscious process can get attuned to it. When I was a kid in the 1980s, it was common for the boys to tease each other by calling each other "faggot." At the time, I didn't think that calling someone a "faggot" was any worse than calling them any other mean name. It never even crossed my mind that it might be. Later, I learned that lots of people find it hurtful to hear the word, and my initial reaction was to blow off their concerns—because that was something that didn't *feel* wrong back then. But after hearing about these concerns many times, and thinking about it, I concluded that using that word was wrong in the same way that using racial slurs is wrong. Now, I get a primal, emotional moral response, in a way I hadn't in the 1980s, whenever somebody uses the word.

Similarly, our values influence when we consider something to be an opportunity for moral consideration. You might live most of your life oblivious to the moral implications of eating meat, but then, after you learn about the conditions of farm animals, or about the impact of what you eat on the environment, or the economic ramifications of where your food comes from, suddenly your choice of food becomes a moral issue, when it never was before.

The gradual moralizing of more and more things is a part of moral development, both for the individual as well as society. Much of the latter half of this book is going to be about thinking about more and more things in a moral way. But there is a downside to this.

A study by Fieke Harinck brought people into the laboratory and asked them to negotiate what penalties should be given in hypothetical criminal cases. One participant was made to be the defense, wanting to lessen penalties, and the other prosecution, trying to increase penalties. They negotiated four cases at once. Some of the cases were set up so that there could be a win-win outcome of the negotiation. Some pairs were told to act selfishly. This means that the defense negotiators were told to try to get the smallest penalties for the accused, and the prosecutors were to try to get the biggest. Other pairs were asked to think about the negotiation in moral terms. That is, the prosecutor was supposed to try to get higher penalties for the accused because it was more just, and the defense to try to get lesser penalties for the same reason. Contrary to what I would have expected, the pairs told to be selfish were more likely to reach win-win situations than the pairs told to be moral. It turned out that when people were thinking morally, in terms

CHAPTER 4: WHAT WE THINK IS RIGHT AND WRONG

of justice, they were less willing to negotiate. When our minds moralize, things get more absolute, and we can become uncompromising. The selfish pairs thought about which outcomes were less important to them, and were willing to give in a bit on those.[4] When people start thinking in a moral way, they start to think in absolutes.

Religion and Morality

Where does morality come from? For many religious people, they believe they get their morality from their religion—and often think that atheists have no morality at all. (This is a myth; studies suggest that atheists don't score any differently than religious people on moral dilemmas.)[5] It's clear that some very specific moral beliefs, like Jews not eating pork, certainly have a religious origin, but what about basic ideas about right and wrong?

If you ask people about what their religion says is right and wrong, it often lines up pretty well with what they believe is right and wrong. But there are reasons to suggest that the causal direction is the reverse of what you might think—to some extent, people's religious beliefs reflect their moral beliefs, not the other way around. That is, people don't agree with their religion as much as their religion agrees with them.

Let's take the Christian version of the Bible as an example. It's a long, complicated book, full of stories with ambiguous meanings and moral proclamations that appear, on the surface, to be contradictory, particularly between the Old and New Testaments. This means that when a person or a culture tries to figure out some ethical code from scripture, it requires interpretation: what things mean what and what parts are more important than others.

For example, Leviticus 20:27 could sensibly be interpreted as suggesting that we should kill people who can commune with the dead: "A man or a woman who is a medium or a spiritist must surely be put to death. They shall be stoned; their blood is upon them." What counts as a medium or a "spiritist"? Is it someone who has conversations with dead people? People who talk to the dead? People who *try* to talk to the dead? A church or religious leader might have answers to these questions, but whoever it is needs to do some interpretation.

I have heard Christians say that some event was a sign or message from a deceased loved one. Many Christians pray to get guidance from their dead parents or grandparents. It wouldn't be a brain-dead interpretation of the Bible to conclude that these people were, effectively, spiritists. At the same time, most contemporary Christians don't believe that someone who prays to her dead mother should be stoned to death.

On other occasions, though, Christians will recruit Old Testament passages to justify other moral beliefs. So what makes them endorse particular Old Testament passages? They often choose to accept the parts that feel right and ignore the others.[6] That's what modern religions do: they pick and choose the parts they feel are moral. And thank goodness! The Bible advocates some activities so brutal—like dashing infants against rocks—that you couldn't even depict them in a contemporary novel for tweens.

And then there are some Christian moral ideas that don't even come from the Bible, like eating fish on Friday.[7] The point is that there is more—a *lot* more—to a religion's moral beliefs than can be found in scripture alone (the vast majority of religions the world has ever had didn't even have any scripture).

In the modern industrialized world, people have a choice of religion. They can go to this church or that one, or a reform synagogue or a conservative synagogue, depending on how well that community matches up with the moral beliefs they already have. As we will see, these moral beliefs come from genes and the culture in general, of which religion is only a part.

Even experimental studies show that people *think* they get their moral beliefs from what they think God believes, but it's actually the other way around. Nicholas Epley surveyed religious people about their own moral beliefs, as well as what they thought God would say was right and wrong. Not surprisingly, these matched up pretty well—people agree with God. But then the people read persuasive essays that changed some of their minds on some moral stance. If they get their moral beliefs from what they think God believes, now the participants' beliefs should be in conflict with what they think God believes? Right?

Wrong. When people changed their own moral beliefs, they *changed their opinion of God's beliefs at the same time*. A brain scan revealed that when thinking about God's beliefs, your brain looks more like it does when thinking about your own moral code than when thinking about others'. All of this suggests that when people try to come up with God's beliefs, they are, unconsciously, just looking at their own.[8] It appears that many people are pulling fast ones on themselves—they choose their religion, or interpret their religion in a way that fits their morals, and then use the religion to justify those morals to themselves and other people.

To the extent that religion *does* affect our morality, though, note that the most popular religions around today were created over a thousand years ago, and are quite possibly out of touch with some of the nuance of our technological, hyperconnected world.

Morality vs. Selfishness

The most obvious moral dichotomy in morality is conflict with being selfish. Taking advantage of others to benefit yourself is generally viewed as immoral. You probably don't need to hear about data that shows that people are selfish, but here's an entertaining one: *U.S. News & World Report* asked readers, "If someone sues you and you win the case, should he pay your legal costs?" Eighty-five percent of people said yes. But for some people the question was phrased differently: "If you sue someone and lose the case, should you pay his costs?" Only 44 percent said yes to this one.[9] So some part of people's ideas of rightness is based on "what's good for me?" whether they are aware of it or not.

Notions of fairness tend to be biased by our own interests, too. The real world is messy, with lots of complicated factors to consider when we're trying to allocate scarce resources fairly. How much to weigh each of these factors isn't always clear. What people tend to do is weigh those factors that favor themselves. In negotiation, this distortion of fairness notions leads to bad effects.[10]

Selfish behavior can begin innocently enough. Part of it stems from the fact that harm done to us is judged to be of a greater magnitude than the same harm done to someone else. This is the trap of retribution: what we think is a measured response to being harmed is often more than what is called for, and is perceived by the harmed as being *even more* disproportionate. This is how people, and groups of people, can get into escalating cycles of revenge.[11]

When I talk about selfishness, I'm talking about a person's own well-being being given too much weight in a moral decision. Psychopathy is a mental illness that has many symptoms, but its most striking one is a complete lack of caring for others. Persons with psychopathy often know what other people are thinking and feeling, but just *don't care*. Having a drink of water is self-interested but not selfish, because (under most circumstances) you're not hurting anybody else by drinking water.

Do People (Act as Though They) Care About Themselves or Their Genes?

When people are acting selfishly, it seems obvious that they are acting with their own well-being as the only consideration. But there is another way to look at it—that in many instances when people behave selfishly they are acting not for the benefit of themselves, per se, but for the benefit of their genes.

According to the "selfish gene" theory, organisms behave in ways that increase the probability of better representation of their own genes in future generations. What does this mean? Well, each person has DNA strands in every cell of the body. Each person's DNA can be thought of as consisting of some number of genes, and these genes help make each of us who we are. Animals have a biological goal to reproduce, and one way to look at what this means is that organisms have a (nonconscious) goal to create more beings with their genes. But what this also means is that if some *other* being (like your sister) also has the same genes, we should care about their reproductive success, too.

The selfish gene theory helps explain a phenomena we see across species called "kin selection." A creature will care about another creature to the extent that there is genetic overlap. That is, the more genes in common it has with another creature, the more it cares about them. Care can be measured in the risk one is willing to take to help another, the resources one is willing to sacrifice for another, and so on. This works for wasps as well as humans. The moral implications are obvious and of no surprise to anyone: people treat family members better than strangers. Even psychopaths treat their relatives a little better.[12]

What's not obvious about it is that people care about *themselves* because they have 100 percent genetic overlap with themselves. In this light, personal selfishness is just one consequence of kin selection.

Each of your parents and children has half of your genes (relative to other people—each human has about a 85 percent genetic overlap with mice),[13] so the theory predicts that you would care about your parents and children about half as much as you'd care about yourself. Your grandchildren have half of that, and so on. This degrades pretty quickly. Second cousins only have about 1/32 of your genes.[14] This explains, in part, why people treat their relatives better than strangers, and close relatives better than distant ones.

You will also note that people will back up these behavioral tendencies with explicit moral pronouncements. That is, some people will come out and say that treating your relatives better than strangers is a legitimate moral position. This is interesting, because we perceive *personal* selfishness as being in conflict with morality, but many see *genetic* selfishness not only as morally acceptable, but as morally required! For example, if someone had the opportunity to prevent their own child's leg from breaking, or prevent the legs of two stranger children from breaking, most people would think badly of a parent who chose in favor of the strangers over their own children.

The Expanding Circle

Selfish behaviors benefit only the self, and kin selection expands good treatment to relatives. The idea of a widening class of entities that get moral consideration is sometimes called "the expanding circle." Some have argued that the circle has expanded over the course of history,[15] but we can also talk about it in terms of an individual's moral development.

What is the next layer of the circle? Your friends. These are people who help you, and you help them. Treating your friends better than strangers is called "reciprocal altruism." Whereas with strangers people keep track of debts and favors owed, with friends (as well as family), the bookkeeping is much looser.[16] Most people will only start keeping track of favors owed after one party or the other really starts to slack off in the giving department.

Reciprocity is a moral stance shared with some other animals. Being able to engage in reciprocal altruism requires certain kinds of computations of mind. It could be that these same computations make possible the detection of cheating, and the outrage it evokes. We'll talk about violations of fairness later. Even bats will punish other bats who don't help out, though reciprocity is generally rare for non-primates.[17]

We also engage in being good to acquaintances and even strangers—we will give strangers money, for example, in exchange for goods and services. But this is not friendship. One of the hallmarks of friendship is that you *don't* keep track of every little thing. If you do find yourself keeping track, it's a sign that the friendship is deteriorating, or you're not really friends at all.[18] Suppose you had a good friend over for lunch, and afterward she pulled out her wallet and asked how much money she owed you. You would probably scoff at such an idea, and maybe even be offended. This is because she is treating what you did more like a business transaction, and it can be interpreted as a signal that the friendship isn't strong.

The Social Group

The next layer of the expanding circle is your immediate social group.

There is no doubt that human beings are intensely social animals. Before civilization, humans lived in hunter-gatherer groups of about 150 people. This is a larger social group size than any other mammal, but it's still pretty limited. Throughout most of human evolutionary history, these social groups had a lot of relatives in them, but only about a quarter of them were close relatives.[19] Humans have a strong in-group loyalty that affects moral treatment of other people. Put another way, people tend to have less moral consideration for people in their out-group.[20]

For small groups, this could be the result of reciprocal altruism. Everyone in a community helps one another, to everyone's benefit. A problem with this is that somebody might try to be a freeloader. They might take advantage of the generosity of others for their own benefit, without giving their fair share in return. These individuals undermine trust in the community and this encourages others to be more selfish. It might be that because this is a strategy that some adopt, we have sophisticated, special-purpose "cheater detection" mechanisms in our heads. Think about how you feel when you've been waiting in a long line for something, and you see someone trying to surreptitiously cut in ahead of you (the scientific term for this person is "buttinski"). It might drive you absolutely bonkers. When we perceive others to be cheating, we often feel anger, and people will sacrifice their own resources to punish individuals like this.

One time I saw a hit-and-run. A car was trying to pull out of a parking spot, and it brushed up against another car and drove away. My cheater detection mechanism went berserk, and I wrote down the license plate of the offending vehicle. I felt particularly righteous about doing it because the car that was hit was owned by a *Star Wars* fan, as evidenced by a bumper sticker. I left a note, with the license plate number, and a message: "May the force be with you." I felt an immediate social bond with a fellow *Star Wars* fan—a person I have not and won't ever meet. Similarly, people can make quick in-group/out-group judgments about others if they share (or don't share) a nation, religion, or favorite sports team. Humans naturally form groups of about 150 members, but humans have the unique ability to have arbitrarily large social groups based on just about any conceivable criteria.

Our in-group bias shows clearly in news reporting. When there is a disaster, the amount of coverage it gets depends on its "cultural proximity" to the country doing the reporting: For example, American newspapers report more heavily on Italian disasters than Guatemalan ones, even if more people died in Guatemala.[21] The cultural proximity of one country to another depends on physical closeness and how many residents visit that country.

The in-group bias is very strong. We have a few reasons to believe this. First, the in-group bias seems to be a cultural universal. Anthropologists find it in every culture they examine. Second, a study of men found that giving them more oxytocin (the hormone that bonds mothers to infants) makes men more cooperative—but only with their in-group. Third, people tend to cooperate when forced to make fast decisions in experimental settings. Fourth, it makes sense evolutionary: giving out-group people the same benefits as the in-group would mean that other selfish groups would take advantage.[22] As I explored in my previous book *Riveted*, religion appears to be a force that evolved to facilitate being good to your in-group.[23] Fifth, the in-group bias shows up very early in development.[24]

THE OUT-GROUP AND ANIMALS

This doesn't mean that people can't feel anything for strangers. Sometimes this is actually caring for out-group members, and other times it's a reframing of what the in-group consists of. There is currently a debate in moral psychology literature concerning whether we cooperate with *everybody* instinctively, or only favor the in-group. That is, experiments where people are put under cognitive load sometimes do and sometimes don't show more out-group cooperation.[25]

Expansion beyond your social group means including out-group human beings. This is the basic view of humanism, a broad moral philosophy that puts the quality of human experience as the highest good.[26] The idea of "human" rights is that everybody has some rights, merely because they happen to be human beings. A humanist is likely to agree that, from a moral point of view, everybody on Earth should have the same rights when it comes to freedom from oppression, or free speech.

Here is where we start to see political differences. Left-leaning people more easily think of all of humanity deserving equal moral consideration, and right-leaning people are, on average, a bit more nationalistic.[27]

Beyond the human out-group, the circle expands to include animals. The distinction between humans and every other living thing is likely a very strong part of our moral psychology. When you look at thousands of languages, you see that if a language has only two words for living things, they make a distinction between human and nonhuman.[28] Although this severe distinction might be a part of human nature, should we use it as an important demarcation in ethics?

In general we tend to restrict moral considerations of animals to their feelings. For example, few people talk about the dignity, honor, or disgrace of animals. We care about their quality of life in terms of how they feel. So what's relevant here are what are technically called "valanced states," which simply means those states that can be some amount of conscious pleasantness or unpleasantness, happiness, sadness, pain, or pleasure. But because the term "valanced states" doesn't exactly roll off the tongue, I'm going to use the word "feelings" from here on out, even though that's not quite right, either. For animals it gets a little tricky, because it's very unclear how much happiness, misery, pain, and pleasure members of any particular species are capable of experiencing. Does a dog have feelings? A pig? A mouse? A mosquito? A cabbage?

Let's back up and think about what we *know for sure* about who can experience feelings. The answer is . . . you.

It's generally taken for granted that if you think you can feel things, you can. This goes back to Descartes's insight that you have to exist, because even if what you think of as the real world is just an illusion or a hallucination, there has to be some entity that is being fooled. That's what "I think, therefore I am" really means—it's not some assertion that thinking defines what's important about you or anything, it's just that the very fact that you're thinking is proof that you exist.

However sure each individual is about his or her own consciousness, our certainty about even other human beings drops off considerably. We instinctively believe and act like other people have feelings, and when we are trying to be rational about it, we conclude that the other people in our lives have feelings because they are so much like us, in looks, behavior, and brain physiology, that it makes sense to infer that their inner experience would be similar to ours. But this is an inference because we can't perceive someone else's experience, we can only reason that it's there.

And sometimes we are wrong. Sometimes when we are dreaming or hallucinating, we meet characters who we take, at that moment, to be actual people with feelings. It is only later that we realize that they weren't people at all, but our own imaginings. In other words, in dreaming, we all have had experiences where we felt pretty certain that the beings we are interacting with have conscious states, and were wrong about that.

So if we are wrong about our experience in dreams, how do we know that the people we meet in the real world have feelings? Could this experience, too, be a part of a larger dream? It's an easy idea to dismiss but a hard one to reason out of, because the certainty we feel about other people being conscious when we're awake we also have when we're dreaming. I'm not saying one should assume that one is the only being in the universe that has feelings, I just want to make it clear that our belief that other people have feelings is an inference we make, albeit usually subconsciously.

Animals

We circle the drain of uncertainty as we consider animals that are less and less like us. It's hard to deny that a chimpanzee has feelings. Similarly, raccoons and even rats have many of the same brain structures as human beings, particularly those involved with emotion. Importantly for morality, the parts of the brain that seem to be the most active during emotions are very old brain systems with parts we share with all mammals, birds, and even reptiles. They process what looks like fear and rage in the same brain circuitry. What we don't know is whether or not they are conscious of those emotions.

It's tempting to throw up your hands and make no decisions about it. *We have no idea.* But we really can't do this because of the vast moral implications. We have to treat animals a certain way, and our actions reveal our implicit ideas about whether they have feelings or not. So even though our current science can't know for sure what's conscious and what's not, we need to come up with informed opinions about it so we know how to act.

Let's take animal research as an example. At a university, when scientists want to do research on animals, including human beings, they have to apply for ethics approval. The ethics committee reviews the proposal and estimates the harm that would be or could be caused, weighs it against the estimated benefits, and decides whether to approve the proposal. Many universities have a policy that if the creature is a bug (an insect, spider, worm, or something like that), you don't even have to apply for ethics approval. That is, a biologist can run experiments on fruit flies or bumblebees in the laboratory without getting permission from anybody. At my university, if the animal in question is as or more complex than a mouse, then you need to apply for ethics. There's no great scientific reason for this arbitrary line, but for ethical reasons we need *some* arbitrary line, and the line between bugs and mammals is at least clear and intuitive.

Things get trickier with invertebrates. An octopus or squid is so alien-looking that we don't get the usual cues of emotion that we are used to with mammals: squealing, hair raising on the back of the neck, baring teeth, things like that. Outside of Pixar movies, a frightened octopus looks much like another octopus. So it's easy to dismiss the idea that they have feelings.

But this is changing. In the past decade, we've found that octopuses have remarkable cognitive abilities. You can watch Internet videos of them crawling out of enclosures, opening jars, getting food, and sneaking back in. They can learn through observation.[29] I have friends who stopped eating octopus because of these findings. The findings don't tell us that they have feelings, but many assume that higher cognition makes it more likely that they have some kind of consciousness. But we don't know.

So some ethics boards have made special rules for octopuses and squid. But you can see that we're pretty in the dark about this. As a result, our rules for what we're allowed to do to animals are embarrassingly inconsistent.

Our laws reflect our moral inconsistencies when it comes to animals. You can go to jail for doing things to a dog that are legally and regularly done to animals raised for food. Though extreme cases of violence are brought to the law's attention, ordinary violence to and suffering of animals are generally tolerated if humans have some benefit. We use animals for nutrition, or work, or for company. Unfortunately, it costs more to give animals bigger

CHAPTER 4: WHAT WE THINK IS RIGHT AND WRONG   193

pens and enriching things to do. We also are entertained by animals, with circuses, fishing, and zoos. In these contexts, harm to animals is thought to be justified and lawful.[30] Later we will revisit whether this is morally acceptable.

Anything That Can Suffer

Once we accept that some nonhumans have feelings, our moral considerations expand to them. As the conviction that nonhuman animals can suffer has grown in acceptance around the world, so have our moral considerations for them.

As a basis of comparison, it's instructive to take a look at how societies used to treat animals. A very long time ago, and even today in some nonindustrial societies, animals were treated respectfully. Humans killed things to live, like just about every other living thing does, but it was done in the context of animist religion, which, roughly speaking, attributes a soul and consciousness to all or most living things. Thus after killing a large mammal, there might be some religious ritual done to mollify the spirit of that animal. Living in hunter-gatherer societies, humans had little power to make animals suffer compared to what we have today. Still, over the course of generations, nonindustrial societies did manage to affect the environment in substantial ways, mostly in terms of causing extinctions through hunting (like the mammoth), deforestation, or by introducing invasive species.

This all changed with the agricultural revolution. More and more animals became domesticated, and with that change several new ways of looking at human/animal relations promptly took prominence in the world. New religions, with a distinctly different feel, sprang up around the same time as agriculture.

Keeping animals in pens removes the autonomy of animals that was so widespread in the past. Belief systems became more about how to keep them alive for human benefit, rather than trying to please their spirits. When we look at some practices around the world at this time, we see a callousness to animal suffering that's difficult to even read about. In medieval France, for example, people liked to watch burning cats. In this form of entertainment, these cats would be in wicker cages, and sometimes they'd be doused in a flammable liquid, set on fire, and chased around town, to the delight of onlookers.[31] Of course, slavery of human beings was also rampant at this time.

In modern days, we find this kind of thing reprehensible. Two out of every three Americans surveyed are uncomfortable with how food animals are raised, let alone cat-burning entertainment.[32] But at the same time our eating choices put vast numbers of animals through incredible suffering as they are raised for our food.

If the animals are put through more suffering than people would tolerate, then why do these conditions persist? The reason for this is simple: most people are completely unaware of the quality of life of the animals they eat. If they knew, they'd be really upset. But our experience of meat as food is sanitized. When we see chicken breasts on sale at the grocery store, those chickens might have been raised outside, walking around, picking bugs out of the grass, with a great life until they were killed, but, more often, they were raised indoors, packed together with other chickens for the entirety of their short, miserable lives. When people get an inkling of what factory farming is like, they often don't *want* to know where their food comes from. Factory-farmed meat is cheaper, so we have cognitive dissonance. Stores try to distract us from the horrors of factory farming by appealing to nationalism: I see lots of advertisements that say things like "We use 100 percent Canadian eggs!" Like I care *where* the chickens get abused.

It's ironic that though we have much more of a collective consciousness about animal suffering than medieval societies did, we actually cause much more animal suffering than they did, due to our incredible level of factory farming that only a fraction of people really know about. Our hearts are in the right place, even if our spending habits don't reflect it.

This increased psychological care for animals is a further expansion of the circle to include all things that have feelings. Hopefully, the factory-farming practices will catch up to the care we feel for animals, but it is likely that before that happens, lab-grown meat, or higher quality meat substitutes, will take over because they'll be so much cheaper. We also might genetically knock out the genes that allow farm animals to suffer from pain.[33] I predict that peoples' complicity with factory farming will be looked at by future generations similarly to how we look at the medieval French burning cats for fun.

Some groups are trying to get chimpanzees and other intelligent animals to be recognized by the law as legal persons. This is not the same thing as saying they are people—the law has a very specific definition of personhood that is bestowed on people and also corporations. (That's right, a corporation can be a person, but not a chimp.) In many places, even some human beings used to be considered nonpersons by law, such as women and slaves. Currently, a corporation enjoys the freedom of expression, but not a right to life. Nonhuman animals are a potential third category in the law.[34]

It might even be that nonhuman animals need to have three categories. Sue Donaldson and Will Kymlicka argue that, when it comes to animal welfare, there are three important categories: wild, domestic, and liminal. Domestic animals, like pets and livestock, are our dependents, and perhaps should be viewed as a part of our community. Liminal animals are not domesticated but have adapted to human civilization. These animals include crows, rats, raccoons, and pigeons. Wild animals are better off left alone. Their rights, perhaps, should be defined in terms of sovereignty. They have positive rights when we should restore the environments they live in.[35]

Ecosystems

The circle expands further for some people. Protecting the environment one species at a time is slow and difficult, and as hard as it is for some people to accept that animals suffer, it's even harder to get them to want to protect animal environments, which don't feel anything at all. But some are trying to do it, and trying even to get legal rights to, say, a watershed or a mountain. When these battles are fought in court, we see arguments about how personhood should be reserved for human beings only, and that things like watersheds are "artificial constructs." Interestingly, the same lawyers who argue this also benefit from the fact that the corporations that hired them are seen as persons in the eyes of the law![36]

These battles have resulted in a few cases where ecosystems were given rights under the law. In 2014, the Te Urewera National Park, in New Zealand, was turned into a legal entity that owned itself. It had to be managed in a way that respected its rights. In 2017, New Zealand also put into law that the Whanganui River had rights as a legal person. This issue is discussed in David R. Boyd's book *The Rights of Nature*.[37]

This might sound ridiculous to you, in part because an ecosystem has no opinions, feelings, or preferences. Perhaps even a company can have preferences, but a river? Although it might be hard to wrap your mind around, it's worth trying to do. The law, and life in general, has lots of instances where we have to protect entities that can't speak for themselves: children, the very elderly, and the very sick. If we can infer what's good for them with some confidence, then we can make things better for them. The same goes for an ecosystem or a river. It's not hard to agree that devastating a fish population due to overfishing is a bad thing for the ecosystem.

Still, this can feel easy to dismiss. I get it. But the expansion of rights has often felt unthinkable to the people at the time. In the past, suffragettes fought to get women equal rights. Abolitionists fought to make slavery illegal. The ongoing battle now is for animals and parts of nature.[38]

Morality, Personality, and Emotion

When people make moral choices or judgments, sometimes they are based on beliefs and thinking, and sometimes they are based on emotions. Doing something out of concern or empathy for someone else is different, psychologically, from doing the same thing because of a belief about moral obligation. However, brain studies suggest that emotions are just about always implicated in moral thinking to some degree.[39]

In general, the strength we feel of a moral judgment correlates with the strength of the emotion that comes along with it.[40] There are lots of emotions that trigger morality. Some are the prosocial emotions, such as empathy, sympathy, concern, and compassion. They are complex feelings directed at the states of others. There are emotions that are used when judging others, such as anger, jealousy, contempt, fear, and disgust. Negative feelings of the self are morally relevant, too, such as guilt and shame, but there are also positive emotions directed to oneself, such as self-righteousness, gratitude, pride, and elevation. Feelings like loyalty, affection, and love are not thought of as moral emotions per se, but when you feel them there are often moral ramifications.[41]

Anger

Although one can sometimes feel anger at oneself, typically anger is a feeling we direct toward others. It tends to be triggered by perceptions of intentional harm or unfairness. One feels anger most intensely and most often when the victim is someone we love, followed by those perceived to be in the in-group. Anger is a negatively valanced emotion, and as such people will usually try to do things to reduce it. Anger at moral infractions seems to trigger feelings of retribution. People are more satisfied with stories that end in people getting their comeuppance than with stories that end with forgiveness and redemption.[42] Imagine how much less exciting the *Kill Bill* movies would be if they ended with everybody making up and resolving to be better (that said, redemption stories have their own kind of power).

Sometimes anger is suggested to be a real scourge on the world, and that eliminating anger completely would make the world a better place. While it might be true that getting angry makes the individual less happy, anger seems to play a role in keeping people acting morally. That is, the expectation that certain actions will trigger anger in others keeps us in check. Many experiments have found that cooperation increases dramatically when there is the mere possibility of people being able to punish each other.[43] Perhaps punishment, and the anger that often triggers it, isn't the best way to make people behave, but it's one of the ways we currently have, and removing it without a better replacement might make the world worse—not that we're in any practical position to eliminate anger from the world!

Guilt

Of all the moral emotions, guilt is most clearly associated with morality. The amount of guilt felt is roughly proportional to how close one is to the victim of one's actions, and the severity of the harm. It's felt when you wrong someone, but can also be felt when you perceive yourself as having a benefit you don't deserve. Examples of this include survivor's guilt, sometimes felt by Holocaust or AIDS survivors.

Like anger, guilt is an unpleasant emotion that people will take steps to reduce. Popularly, guilt is feeling "I did something bad" and shame is "I am bad," but behaviorally they are distinguished by guilt's response being reparation, confession, self-criticism, and apology, while shame tends to cause social withdrawal and avoiding eye contact.

And as with anger, guilt affects how we behave without anyone even feeling it—often people act better to avoid a potential feeling of guilt in the future. One might, for example, have an opportunity to cheat on one's spouse, but refrain because they know they'd feel guilty about it later. This might be in addition to feelings of concern and compassion for your spouse, which also might encourage fidelity.

Although we can feel guilt about some things and not others, in some sense the emotion behaves like a single substance in the head. We know this because experiments show that reducing guilt for one reason has behavioral effects on unrelated activities. One experiment, for example, asked people going in and out of confession for a donation. The people going into the confession donated about 40 percent of the time, and people leaving donated at a rate of about 20 percent. This suggests that the confession, which presumably had to do with moral actions unrelated to donation, reduced feelings of guilt so much that the compensatory effect expected from the donation was of less value, so they were less likely to donate.[44] This is one of many examples of how our moral psychology responds to situations that don't make any sense when thought about rationally, and supports the idea that we should be circumspect about blindly trusting our moral emotions.

When you experience something bad happening, there are two different feelings that are easily confused. One is sympathy (or concern, compassion). When you're concerned you care about someone else, and this feeling is associated with a lower heart rate. But when you see someone else getting hurt, you also might feel distress. Distress is you feeling uncomfortable with what *you're* experiencing. It's associated with a *higher* heart rate. The behaviors these emotions facilitate are often different, too. Sometimes when viewing a moral infraction that makes you just distressed, you will try to escape the situation without actually trying to right the wrong. This might happen, for example, when you are watching a disturbing video about factory farming. You simply turn the channel rather than trying to do anything to help with the problem. This is more likely to happen when you're distressed than when you have concern. Of course, you can feel both at once.[45]

How easily you feel emotions like these, in general, differs from person to person. For example, some people are easily disgusted, and they seem to just be born this way.

That much isn't too surprising. What is more surprising are the moral ramifications of this: that your in-born propensity for this or that emotion affects your moral outlook on life.

Disgust

Disgust evolved to help us avoid things bad for us—what not to eat, touch, or have sex with. For food, the disgust emotion helps us come up with an expected value of consumption, combining with feelings of hunger and beliefs about food scarcity. So if you're starving, and you find some meat that's gone off a bit, your disgust system will try to get you not to eat it, while your hunger system will try to get you to eat it. Among other internal processes, these will inform the decision of whether you eat or not.[46]

You can have disgust directed at something in particular, say, someone throwing up. But feelings of disgust transfer over to other areas, as we saw with guilt. Being disgusted at a nonsexual stimulus, like rotting food, can make you less sexually aroused, even though the food has nothing to do with the potential sexual partner.[47]

When it comes to disgust for people, we have evolved to get an emotional reaction to things we can see and smell. So a person covered with sores might make us feel disgusted. The evolutionary reason for this is that someone covered with sores is more likely to have an infectious disease, so disgust helps us steer clear of such a person, which, on average, keeps us healthier. Disgust is typically triggered by bodily products, death, and certain aspects of sex, such as incest.

Being disgusted for those things isn't so bad. Many people are set to the proper calibration of disgust. People who are disgusted too easily, on the other hand, can be at risk of malnourishment, and people who are not disgusted enough are more at risk of food-borne pathogens. The empirical evidence for these findings are mixed.[48]

But the problem is that, like many things in our foundational psychology, disgust is a blunt instrument, and we feel disgust for ways people look that have nothing to do with infectious disease: being obese, having burns on the face, or anything else really out of the ordinary. At the same time, we *don't* get a disgust reaction to someone with lung cancer, because (for much of the course of the disease, anyway), they look perfectly healthy.[49]

People also use their disgust reactions to judge many (but not all) morally deviant behaviors.[50] For example, merely thinking about incest causes a strong disgust reaction in people across cultures. Jonathan Haidt ran studies showing that people's moral aversion to incest was an emotional response. He did this by creating moral situations that "dumbfounded" people. That is, he described scenarios of, say, a brother and sister having sex that did not have any of the rational reasons for thinking that it's wrong: they did it only once, used three kinds of birth control, were not upset by it, and so on. In experiments, people kept trying to rationally argue why it was wrong, but eventually said things like, "I can't tell you why it's wrong, I just know it is!" Haidt calls this "moral dumbfounding."

Haidt ran a study in which he associated a neutral word, like "basket," with a feeling of disgust. This can be done with priming or with posthypnotic suggestion. In both cases, people are unaware of the association put in their minds. Then, when asked to judge the morality of some action, those who saw the word "basket" in the scenario were more likely to judge the action harshly than if "basket" didn't appear.[51] This is one of many examples of how our emotional state, regardless of any thoughts of moral principles or reasoning, can affect our feelings of right and wrong.

What is unsettling about these bleed-over effects is that it feels so contingent on the evolutionary history we just happen to have. How secure should we be about our ideas about right and wrong when those feelings are based merely on what made our ancestors more likely to reproduce? The feelings are fickle and inconsistently applied.

A study of over eleven thousand people across thirty nations found that the more parasite risk there was in that country (usually this means hotter nations) the more conservative and traditional the residents' views were.[52] It's clear that moral and political positions are not always decided with evidence and reason. To a great extent, they are affected by emotional levels we are born with, and even the geographical conditions of where we live.

When we think somebody is disgusting based on how they look, we have harsher moral judgments on them when it comes to disgust-related moral infractions. For example, if an ugly person is accused of prostitution, people tend to have a harsher opinion of them than if they are beautiful. But we see no difference in the harshness we have for ugly or beautiful people when it comes to stealing materials from the office—a moral infraction, but one that violates our sense of fairness rather than triggering a disgust reaction.[53] Blaming an obese person for prostitution more than a thin person is clearly an indefensible moral position. This is yet another of many reasons why we need to be circumspect about how our emotions affect our moral judgments.

We evolved these moral emotional responses for good evolutionary reasons, but these reasons are not sufficient to justify contemporary *moral* reasons. The fact that our emotions cause inconsistent moral judgments is evidence of this. We simply can't trust our emotions and instincts to determine right and wrong, which is one reason why people have come up with ethical theories to help guide us.

The Good of the Many vs. the Rights of the Few

Many see morality as a struggle between selfishness and helping others. But even if we agree to help others, there can be disagreement on how to do it. Often this involves conflicts between the good of the many versus the rights of the few. In some circumstances, you can hurt one person to help out several others. This is more of a true moral dilemma, in that it pits two moral values that we all have to some extent: that you should not hurt people (at least, not people who don't deserve it in some way), and that you should do things that benefit the most people.

In some sense, taxing the rich to provide benefits for the poor falls into this category. The government takes resources from people with the good intention of helping out the less fortunate in society. Most people find this acceptable to some extent, but there are examples we find repugnant, too.

In philosophy, notions of rights and responsibilities to people fall under the umbrella of deontology. In this view, morality consists of a bunch of rules like "don't kill anybody" and "help out people in need." One of the important aspects of this way of thinking about morality is that it doesn't have any explicit notions of magnitude—that is, the number of people helped or hurt, or how much people are helped or hurt, aren't part of the rules.

So suppose we are going to put a medical clinic in a poverty-stricken area, and we have several potential locations in mind. Placing the clinic in *any* location means that some people will have better medical care, and others will not. People's lives might even be on the line. The difference in location just determines, in a fairly predictable way, which people will die or suffer and which people won't. Although different deontologists differ on what the rules exactly are, deontology generally doesn't have much to say about where the clinic should go. But if it seems clear to you that you should put the clinic where it will do the most good, that's because there's more to moral psychology than deontology-style rules.

Once we start talking about where the clinic will do the most good, or reduce the most harm, we introduce magnitude into the decision-making process. Although medical help is simpler than simply life-or-death, let's just look at lives saved as a shorthand for medical care in general. If we have two possible locations, and we have reason to think that placing the clinic in one would save twenty lives per year, and the other location would result in saving an expected thirty-five lives per year, then, assuming everything else about the locations is the same, it's kind of a no-brainer to put the clinic in the second location. We all have functions in our minds that process moral rules, and we also all have functions that can process magnitudes of goodness or badness. These two ways of thinking, rules versus magnitudes, roughly map onto large brain areas.

For a long time, scholars have argued about whether morality was a function of conscious, deliberative processing, or unconscious, emotional processing. Turns out they were both right, but they were talking about different parts of the brain. Roughly speaking, there are two systems in your head that sometimes have conflicting "opinions" about right and wrong. System 1 is evolutionarily old, fast, unconscious, emotional, automatic, shared with other animals, and mostly located in the back and lower part of the brain. System 2 is newer, more prominent in humans, slow, rational, conscious, uses reasoning, and is mostly located in the front and outer parts of the brain (the neocortex).[54]

We often have fast judgments about the morality of something, accompanied by a strong feeling, suggesting that System 1 has something to do with moral psychology. Jonathan Haidt tapped into these with his clever examples of moral dumbfounding, where people have *only* a feeling of wrongness, and no good reasons.

But this initial feeling can sometimes be undone and reversed by *reasoning* about it, suggesting that System 2 also plays a role. For moral thinking, System 2 seems to be implemented in the brain by the dorsolateral prefrontal cortex, which is just behind your forehead. It reasons about magnitudes, and often ends up thinking of right and wrong in terms of doing the most good, and avoiding the most harm. This corresponds to the other big idea in the philosophy of ethics, utilitarianism.

Let's look at a tough dilemma: it's wartime, and you are hiding with a bunch of people from the enemy. It's pretty certain that if any of you make any noise in the next hour, the enemy will find and kill everybody. You have a baby who regularly makes noises due to a cold. Is it morally justified, or even required, that you suffocate the baby?

When people are given this problem, they often take a long time to think about it. System 2, with its more utilitarian approach, is saying, "Well, if you don't quiet the baby, everybody, *including the baby*, will be killed by the enemy. So you must suffocate the baby." Meanwhile System 1 is like, "You don't kill your own baby, and I don't need to hear about anything else about the situation to know that!"

The brain areas for Systems 1 and 2 are very active while people are trying to come up with an answer, and the response is determined by which system "wins" and is the most active. When the different parts of the mind generate two different behavioral responses, there is a part of the mind that picks one. This job seems to be done by the anterior cingulate cortex.[55] If you occupy System 2 with another task (putting it under "cognitive load") people are more likely to say "no, it's not okay," because System 2 is too busy with other things to compete with System 1's opinion.[56]

Similarly, you can trick people's System 1 by putting them in an irrelevant emotional state. People shown funny videos before a moral dilemma are more permissive of moral transgressions, because the dilemma doesn't evoke the same emotional anger or disgust outrage as those who watched a neutral video. "I feel all right, it must be okay!"[57] People who have emotional deficits, due to some medical condition, tend to think about morality more like a utilitarian.[58]

How did human beings come to have these two different systems? When we look at some part of the brain doing one thing, and some other part of the brain doing some other thing, there is a tendency to jump to evolutionary arguments. The fact that we have found a brain area that does something, and the fact that our brain structure is influenced by our genes, allows some people to make the leap that the brain structure, its function, and the details of that functioning are all evolutionary adaptations. There's a right and a wrong way to think about this.

First of all, keep in mind that *all* of your psychology is a result of brain processing. So even things that you learn during your life, like where you parked your bike this morning when you got to work, is a memory because of some change in brain state or processing. Every word you read of this book is changing your brain in some physical way, even if we don't know how to measure it yet.

We're very ignorant about how exactly this happens, for most mental phenomena. So what we usually get are, at best, broad areas of the brain responsible for broad activities. We say that the old brain and the limbic system (System 1) are implicated in rights and duties morality. But that's like trying to understand the Canadian government by saying "government seems to happen in Ottawa." It tells you nothing about the parliament, the house of representatives, let alone the role of the Usher of the Black Rod.[59]

What people do, subconsciously, is kind of assume that the behaviors we've found brain areas for are evolved, and for the behaviors for which we have yet to identify a brain area, it must be learned. But of course that's silly. We should not use our ever-changing knowledge of cognitive neuroscience to determine what we think is nature versus nurture. We need better standards than that.

Let's start by asking what it even means to say that some behavior evolved.[60] First, it could mean that the brain and mental processes that do moral cognition evolved, but all of those processes (emotion, reasoning) might also be used for other things. This is almost certainly true. Second, it could mean that what evolved was a specific set of brain functions that do moral cognition, and perhaps little else, but moral reasoning was not an adaptation. Finally, it could mean that moral cognition is an adaptation, which would mean that not only did moral cognition evolve, but it evolved *because* it promoted our species' reproductive success. If you're confused about the difference between the last two, you're not alone. Basically, structures, metabolism, and behaviors (phenotypes) arise in species because they are adaptive, or because they are by-products of other adaptations, or because they just evolved by accident (for example, through genetic drift).

How could we get psychological characteristics without genetic evolution? There are a few ways. Let's take the belief that rain makes the ground outside wet. This belief is universally endorsed by every known world culture and every adult human, but it's not evolved. It's useful, but it didn't *need* to evolve because the general-purpose learning systems that humans have are pretty much guaranteed to arrive at this conclusion on their own, so the belief itself doesn't need to be hard-coded in genes.

Another way is through cultural learning and evolution. You don't need a culture to learn that rain makes the ground wet, you can learn it all by yourself, but you do learn from other people the grammar of your language, what to wear, that you shouldn't check your phone in the middle of a conversation, and so on. Cultures themselves learn things over time, too. A culture might have a taboo about eating a particular kind of fish when pregnant or lactating, and indeed this might be good for the health of the mother and baby, yet the individuals in the culture might be unaware of the causal relationship between their taboo and the health effects. It's just that, through cultural evolution, those who respected this taboo lived longer, and passed on that bit of culture to their kids.

In summary, every psychological tendency you find is the result of some combination of 1) an adaptation, 2) the by-product of other adaptations, 3) evolution due to genetic drift, 4) cultural learning, or 5) individual learning.

When it comes to moral cognition, we have an array of possibilities. It's further complicated by the fact that even if there is some adaptive reason for morality, there's no doubt that at least *some* of any individual person's morality is learned from their culture.

What kind of evidence would we want to support the idea that some moral tendency is an adaptation? All too often, the only evidence presented is some plausible rationale of what that adaptation might be. A particularly notorious example is zoologist Desmond Morris's *The Naked Ape*, which is chock full of speculation about everything from why humans got so smart to why men find cleavage attractive in a blouse. That book is almost completely lacking in anything resembling evidence.

But there is something completely alluring about adaptationist stories that give laypersons and scholars alike a very certain feeling of something understood. This feeling attenuates critical thought, because it can *feel* like there's nothing more to be explained. This all-too-common uncritical reaction to evolutionary claims drives many psychologists bonkers.

Such adaptationist stories are not sufficient for a well-grounded science claim of an adaptation having happened. For evolutionary psychology, there are some other kinds of evidence that can be brought to bear. Let's go over the kinds.

Cross-Cultural Tendencies

Evolution takes a long time, and most evolved characteristics that we have evolved before we were even human, let alone before humans spread around the world from their starting place in Africa. So if something in our behavior evolved, it *probably* evolved before humans spread across the world, which would mean that you should find the tendency in most or all human cultures. As such, if a trait is found across many cultures, it can be used as support for the theory that it evolved.

There are a couple of caveats here. First, some things evolved *after* humans split up, like skin color, epicanthic folds, and the ability to digest milk as an adult. We see variation in human beings in different parts of the world, and nobody thinks that epicanthic folds are learned by the culture.

Second, culture is an incredibly powerful thing that can override genetic tendencies with enough "cultural force." There's little doubt that humans evolved to want to have sex. This claim is not falsified by the existence of small groups of celibate monks. The idea is that if you look across cultures, the traits we evolved to have will be *mostly* present and *rarely* negated by culture.

Third, something like the belief that rain is wet is not evolved, even though it is cross-cultural, simply because it is an inevitable tendency of a learning being in a world where rain is obviously wet.

Presence of the Tendency in the Uncultured

Basically, if very young beings have the tendency, it's more likely to be evolved, because they have not had a chance to learn much of anything yet. Usually, this means running experiments on newborn babies.

Again, there are caveats. First, we're constantly discovering amazing things that human children learn in the womb. They can hear muffled sounds and react differently right away to their own language versus one they've never heard.

Second, it's not like your genes make a baby and then shut off. The idea that your genes are like a building's blueprint—used to create a body and then discarded—is false. Certain gene sequences are triggered by environmental stimuli later in life, such as those involved with puberty. The physiological changes at puberty are certainly evolved, even though they are not present in infants.

Cross-Species Tendencies

Something is more likely to be evolved if we can see it in many species. Let's take walking as an example. Humans, caterpillars, robins, and lizards all have their own kind of movement along the ground. These are animals from very different parts of the genetic tree. But the fact that "walking" in some form or another is so common suggests that it's a damn useful thing for a creature to have. The same goes for digestion, chewing, and so on. Evidence that something is very useful is supportive of it being evolved because it provides a good (metaphorical) "motive." More accurately, we evolved the ability and desire to learn to walk.

We can go further: if we see traits across species that are related to each other, then it might be that the trait evolved in a common ancestor to all of them. Let's take the existence of hair follicles (and in most cases, hair) on mammals. All mammals, even killer whales, have hair follicles. This is strong evidence that hair evolved before all of these mammals differentiated into their current species. The fact that birds and lizards don't have hair suggests the follicles evolved for mammals after they split from birds and lizards in evolutionary history.

The Tendency Is Heritable

If the trait cannot be passed from parent to child, then it is not something that resulted from genetic evolution. For example, if a woman breaks her arm, her children will not be born with broken arms. The problem with finding heritable genetic traits is that parents pass on culture as well as their genes to their offspring. If the parents like the Beastie Boys, they are likely to play it in the house, increasing the chance that their children will like them, too. This is a result of cultural inheritance, but not genetic inheritance. This makes it very challenging to determine what's genetic and what's cultural, because kids tend to inherit their parents' genes and culture. This is why studies of separated twins are so valuable in this regard.

Evidence of a Genetic Pathway

If we have found genes that are implicated in a trait, it adds support to the idea that it's evolved. This is a very challenging biological problem, in part because we often are inferring causation from correlation. Most traits are highly "polygenic," meaning that they are the complex result of many genes and environmental triggers that cause them to express.[61]

Selective Pressures from Long Ago

Evolution generally takes a great deal of time. Recent human technological progress means that the world we live in today is different in a great many ways from the kinds of environments we spent most of our evolution in. One might speculate that something evolved because people often live alone, or have to deal with money, but these hypothesized evolutionary pressures are characteristics of a modern, industrialized society, not the hunter-gatherer bands of the past. An evolutionary argument is stronger if the selective pressures are those that were likely to be present for a hunter-gatherer tribe in Africa thousands of years ago (the so-called "environment of evolutionary adaptation").

Moral Bias

The best evolutionary psychology doesn't stop at the theory-generation stage. The best science involves theory, but then tries to find ways to test it with hypotheses.

At least part of our morality is evolved, and much of it is learned through enculturation. For many people, moral thinking is something done without much reflection about any possible underlying theory. The structure of morality, in the average person, is kind of a hodgepodge of emotional responses, rules of thumb, and beliefs. Because it is a hodgepodge, and not a coherent theory, it is prone to inconsistencies.

Part of this hodgepodge are our moral biases. A bias, in this context, is any rule of thumb or mental shortcut that generally works in frequently encountered situations, but can lead to predictable errors in unusual ones.

Moral philosophers and psychologists try to discover what these biases are by examining our own moral intuitions and by running empirical

CHAPTER 4: WHAT WE THINK IS RIGHT AND WRONG

experiments. One is the principle of double effect, which holds that an action is morally worse if there was an intent to harm someone as a means to achieve some goal. This also means that if someone is hurt accidentally, or as a side effect of what they're trying to do, it is not judged as harshly.[62] Although people reliably behave in ways that show that they have the principle of double effect installed somewhere in their minds, they're generally unaware of it, and need to learn about it in a philosophy class. This is similar to the grammar of one's own language, which is intuitively known, but for it to be explicitly known needs to be discovered by linguists and then explicitly taught to people in school. They have an implicit understanding of morality and grammar, but have to figure it out or be taught an explicit theory. And the explicit theory is never a perfect fit.

Because these biases are unconscious, they are not only hard to discover, but very difficult to correct. That is, when you use biases you're not aware of it and have to use conscious reflection to infer that you might have been using them.

Let's look at the moral bias involving physical contact. Studies show that when one person is harming another by being in physical contact with them it is judged more harshly than if they're not in physical contact. Military drone operators, for example, are very distant from the people they kill. They are operating a robot by remote control, and the drone itself is high in the air, making the people affected by the drone's actions look small on a computer monitor. This distancing is likely to make people more callous about killing, because it superficially resembles playing a video game more than actually killing people. Sometimes drone operators refer to civilian casualties as "bug splats."[63]

Why might we have this bias? Even fifty thousand years ago it was very hard to kill someone from afar. So the mind's shortcut for detecting that one person is intentionally harming another is to look for physical proximity or contact between the agent and the victim—strangling them or perhaps throwing a spear. What's interesting is that when you ask people whether they think that physical contact *should* be a factor, they say no, that proximity should be irrelevant.[64] That is, when they use their considered judgment, they do not *endorse* the idea that harmful actions done through physical contact should be judged to be worse. People will reject this bias in the abstract, but when you ask them about particular situations, this bias rears its head. Most people's considered judgments are often not even in line with their own moral intuitions in many cases.

Another bias we have is our moral insensitivity to numbers. We're just not built to think very clearly about numbers greater than about five or six. When it comes to moral value, this number insensitivity has been shown again and again in psychology.

For example, everybody thinks that twenty polluted rivers is a worse situation than two polluted rivers, but when you ask people how much money should be spent on cleaning up two polluted rivers, they tend to suggest the same amount of money as people asked about cleaning up twenty rivers.[65]

A study by Elizabeth Dunn asked some people to predict how sad they would be to hear about five people dying in a fire, and others how they'd feel if they heard that 10,000 people died in a fire. The latter group predicted more sadness than the former. But when other people were given news reports about these deaths, the two groups were equally sad. Hearing about 10,000 people dying makes you no sadder than hearing about two people dying, even though it's much, much worse.[66]

If we think for a moment only about the effects that actions have in terms of helping and harming people, then other factors should be irrelevant. But in human moral psychology, other factors are always relevant. Another particularly striking one is that we tend to think of things as being worse if they are the result of *doing* something as opposed to *not doing* something. For example, if you do something that results in a human death (like strangling somebody), it's generally considered a worse moral action than if you fail to do something that results in the same human death (like failing to grab someone who's falling off of a cliff). We all feel this, so you're probably not surprised to hear about this "omission bias," but you might be surprised by how strong it is.

These biases show us that some factors that should be irrelevant to morality (like proximity) are used for moral decisions, and that some other factors that are relevant (like how many people died) are not. These moral inconsistencies are a natural, normal part of us. It's concerning that our moral judgments tend to be influenced by what we intellectually hold to be morally irrelevant factors.

A good moral theory is supposed to function so as to help us ignore these factors that lead our intuitive moral judgments to be contradictory and wrong. Moral theory is an attempt to create a coherent relationship between beliefs, feelings, and moral judgments. This theory, ideally, avoids inconsistency with itself and our most cherished moral intuitions, and tries to include only morally relevant factors, while excluding the irrelevant ones.

There are many other biases, but I wanted to point these out because they are particularly relevant to being a moral person today—helping or hurting beings at a distance, thinking in terms of numbers, and omission bias. They demonstrate why it is so important to have moral theories to help guide us, to keep us from letting our biases keep us from being the best people we can be.

Your Identity as a Charitable Person

Are you a generous person? Are you a fair person? Do you think of yourself as someone who gives a lot? Are you selfish, or do you give too much? Your answers to these questions reveal things about your moral identity.

You might expect that people who actually are good would think of themselves as good. But it's complicated. Dan Batson ran a study in which he had people decide who got a free reward—themselves or an unseen activity partner who they would never meet (and in reality didn't exist). Weeks before, Batson surveyed these people on a variety of moral measures, and some people reported being more concerned about things like caring and fairness than others. In the experiment, the participants were given a coin and told that most people thought that flipping a coin would be the fairest way to determine who got the reward. So people were given a reward, told that they could take it for themselves, or give it to someone else, and then told that flipping a coin might be a fair way to decide.

So what happened? Those who said they cared about fairness and caring on the survey were more likely to flip the coin. No surprise there. What's weird is that about 90 percent of coin flippers *and* non–coin flippers took the reward for themselves. That is, many people who flipped the coin basically ignored the result and took the reward for themselves anyway. So thinking you care about fairness sometimes makes one take a superficial step toward fairness, but when an actual reward was at stake, act selfishly in the end.[67]

These feelings of identity have other impacts on behavior, and they are not always good. For example, making a public declaration of how you are going to behave can make you (unconsciously) license yourself to do bad things, or fewer good things. Going public with your intentions, paradoxically, makes people less likely to follow through.[68]

When you think of yourself in a certain way, particularly a good way, you feel self-satisfied. This feeling of satisfaction can prevent you from doing behaviors that would increase your self-satisfaction. For example, suppose you are proud of yourself for not being sexist. You talk about not being sexist, and you see your views on gender equality as core to your identity. The problem is being so sure of yourself, because there might be parts of your mind that still hold bias, but because you think of yourself as a non-sexist, you're not on guard about it. Now, when you are faced with a situation in which you have to judge the relative qualities of a man and a woman, for example in a hiring decision, where a normal person might take into account their own sexism and question their gut feelings that the man is more qualified, the person who thinks of themselves as a non-sexist won't. "I feel that the man is more qualified than the woman. But this can't be because of bias because I'm a feminist." As a result, they might end up making sexist decisions based on their gut feelings, which still might be a little sexist, even though their cortex abhors sexism. The lesson here is that even if you don't think you're sexist or racist, you still should be on the lookout for prejudice in your own behavior, and don't let your identity pull your guard down.

One's moral identity can provide a satisfaction with one's behavior that is not always deserved. Here is how it can work: When people do good things, they think more highly of themselves. They get a warm-glow feeling, sometimes called "elevation." But people eventually stop going out of their way to be extra-good. They stop because they're satisfied. Whether they consciously think it or not, their minds have decided "I'm good enough, for right now." I think of it like a goodness thermostat that was set by your nature and upbringing at a certain level. When you're falling short, the heat comes on: you seek opportunities to do good things, and you are more vigilant about not doing bad things. When it's too hot, you stop going out of your way to do good things. This doesn't mean you will do *bad* things, just that you'll stop doing extra-good things until the temperature goes back below the set point.

So when you do good things, you raise the "temperature." But by how much? It's not that we have some kind of direct access to how much good

CHAPTER 4: WHAT WE THINK IS RIGHT AND WRONG   213

our acts *actually* do. We have to estimate, using cues from the environment. These perceptions cause emotions, which make us feel good or bad about ourselves. As we will see later, the actual amount of good done is often not reflected in our feelings about the act. Sometimes we might feel really good about something that doesn't do much good at all, or feel almost nothing about acts that do a whole lot of good.

What this means is that if you want to optimize being a good person, just like you would optimize your daily habits for productivity and happiness, you need to use evidence and reason to get more accurate estimates of the magnitude of goodness, and to overcome the moral biases you naturally have.

Doing Bad Is More Powerful than Doing Good

In many people's minds, being good simply means not being bad, as though avoiding doing bad things is all one needs to do to be a perfectly good person. When you say it like this, it sounds a little silly, but much moral theorizing seems to tacitly assume this. Moral codes often focus on "don't be bad," rather than "do a lot of good."[69]

Doing bad seems to be an almost bottomless pit of evil. Somebody who kicks a stray dog for no reason is a real jerk, but people can get really, really bad. Who's the worst person who ever lived? One way to look at it is in terms of human deaths caused. Historians' estimates of the number of deaths Joseph Stalin caused range from 15 million to 60 million people.[70] Mao Zedong is credited with the deaths of 40 to 70 million people.

But being good seems to be an even bottomlesser pit in the other direction, and it's fun to think about who *saved* more lives than anybody else in history. Edward Jenner saved 530 million lives by inventing the smallpox vaccine. Norman Borlaug helped to invent grain varieties that led to the green revolution, and by doing that saved over a billion lives. It's estimated that by inventing synthetic fertilizer, Fritz Haber and Carl Bosch saved 2.72 billion lives.[71]

The fact that you've probably heard about Stalin and Mao many more times than Bosch and Borlaug is evidence that people pay more attention to bad than good. Borlaug saved *twenty million times* more people than Mao Zedong eliminated. An argument could be made that Borlaug is over twenty million times more good than Mao Zedong was bad. The heights of goodness make the depths of evil look shallow in comparison.

These people are easy to see as heroes and villains because they worked hard for a long time to effect the change they did. Others have affected the lives of many with significantly less action. Stanislav Petrov was a Soviet military man who noticed that warning systems detected five nuclear missiles headed for his country. He decided not to return fire, figuring it was a computer glitch. He might have saved more lives than Haber and Bosch just by making one decision that prevented nuclear war between world superpowers.

On the flip side, Gavrilo Princip killed the Archduke Ferdinand of Austria. One murder is bad, of course, but it set off a chain of events that led to the First World War, which ended up killing about 37 million people.

All of these numbers are estimates and many are debated by historians, but for purposes of this book the exact numbers are not important. The point of all of this is that the good and bad you can effect in this world can be much, much more than how you treat the people you meet on the street, your coworkers, and your family. There's a whole world out there you can help out, or screw over.

THE NEGATIVITY BIAS

The human tendency to give more importance to bad acts than good ones is one of many ways that we think of the bad being more important than the good.[72] There is a well-documented "negativity bias" in many aspects of human psychology.

This makes us more afraid of loss than we should be. That is, we'll require much more money to part with an item we own than to acquire it in the first place (this is called "loss aversion" or "the endowment effect").[73]

People see the bad stuff as being worse than the good stuff is good. For example, studies show that the average person thinks that you'd need to engage in twenty-five lifesaving actions to make up for a single murder.[74] To keep love relationships stable, studies suggest it takes five acts of goodness to cancel one act of badness.[75]

Jonathan Haidt speculates that this bias is in part due to our brain structure. It takes time to consider the good and bad of something. Brain signals only travel at about 30 meters per second (about 67 miles per hour—think about that—you often drive faster than you can think). Because this processing requires lots of neuron communication, these appraisals take a second or two for your brain to process.

This just isn't fast enough to avoid many dangers in the real world. So your mind has a shortcut: the amygdala looks at the incoming stream of sensory information and makes a quick judgment about how dangerous it is. If the amygdala senses danger, it puts out the alert and ramps you up for reaction—fighting, running away, etc.—before the more thoughtful areas of your brain even get a chance to review what's going on. There is no equivalent shortcut for the good stuff, because they are not as urgent.[76]

When it comes to medical testing, this bias might be resulting in a gigantic net loss of life. We have a well-intentioned notion that drugs need to be safe before they are available. But our notions of safety are not often tempered with an appreciation of the damage being done by not having the drug available.

It's an empirical matter: would we save more lives if we allowed more "unsafe" drugs on the market? If a drug kills 5 percent of the people who use it, but saves 20 percent from a death they're heading toward, should the drug be legal? It's hard for us to say yes because of omission bias. The drug killing someone feels worse than allowing twenty people to die by not giving them a drug.

Your Brain on Morals

In the broadest strokes, human moral psychology can be understood as broadly being determined by automatic and controlled processes. We have unconscious moral tendencies, some of which we have as instincts, and others we learned so well that we don't have to consciously think through them anymore. Sometimes we use principles that we think about consciously, and we make moral decisions based on those principles. But even reasoning with these principles can, over time, become automatized and unconscious.

Much of our moral thinking is based on forces that are not only unconscious, but inaccessible to conscious inquiry. For most people most of the time, ideas of right and wrong are intuitive.

One of the most important things you can do, regarding self-understanding, is to realize where these intuitive notions come from. Intuitive "truths" feel true, but they do so, in part, because as humans, we evolved so that believing them led to a better propagation of our ancestors' genes. Just as there are things you can believe that might not be true, but make you happier and more successful, there are things that many of us believe that are not true, but the very belief in them made our species more successful. We inherited these beliefs.

Richard Dawkins gave a fascinating TED talk about what he calls "Middle World," which is the world we experience—not too big, not too small, not too slow, and not too fast, but right in the middle. Some things are true in Middle World that are untrue for the very large, small, fast, or slow.[77]

Lots of things we consider to be common sense are things that happen to be true only in Middle World. For example, the Earth appears to be generally flat. Turns out it's roughly spherical, just so enormous that it looks pretty flat when you're standing on it. Even if you were to walk around the entire world, it would still look pretty flat at every moment. Another example is relativistic effects of motion on time. We never move fast enough for ageing to slow down noticeably, so the idea that time slows down for fast-moving objects violates common sense. We perceive trees as motionless because their growth is too slow for us to see.

The reasons for this are that our sensory organs and perceptual brain systems evolved not to give us an accurate representation of the world, but instead to maximize our reproduction. It would not have helped our ancestors to perceive stones as being made of mostly empty space, because the way we interact with stones means that perceiving them as solid helped our ancestors flourish.

Other animals have different needs. Gravity is less important for buoyant fish. Liquid water's surface tension is important for some bugs, but not for us, because we can't walk on it. Much of what feels true does so as a result of our having evolved in Middle World. Even calling it "Middle World" reflects our Middle World biases, in that it assumes that what we perceive is what's centered, in the middle.

Now that we understand that our natural morality has some problems with it, and that we need a good theory to guide us, we can finally tackle the problem of what's *actually* right and wrong.[78]

Science and philosophy have done wonders to help us get a more objective view of the world than our natural minds allow. We invented ways of knowing that don't rely on our natural impressions. We have natural and cultural impressions of right and wrong, but we can see that these are inconsistent, and subject to change by variables that, when we sit back and look at them, are morally irrelevant. These make us search for something more, some underlying sense to all of it, which is why people look for ethical theories: structured ways of thinking about morality to guide us to make better moral judgments and actions.

Moral Theories as Detectors

I'll start with talking about theories in general: what they are and what their function is. What often happens over the course of scholarly history is that we start with an intuition and use careful thinking and science to try to make it more concrete. We might have a word, like "good," or a concept, like "kitsch," and then try to create a way to precisely define or measure it. I'm going to call this "creating a detector," because one of the functions of this kind of inquiry is to create some kind of objective, explicit way of classifying something.

Let's take the detection of light as an example. We know what light is, intuitively, and that intuition has something to do with our eyes and visual perception. We can see light. But that's not good enough for science, or even for professional photography. For a scientific experiment, we need a light detector, so we can get precise measurements. You can't just use your eyes to get an accurate enough measure of how much light there is. Some people are colorblind, some people are vision impaired, and people just aren't that good at assigning numerical magnitudes to things like the brightness of light.

So we make a sensor, a mechanical device that can tell when light is playing on it and at what intensity. It's a detector, and they're even used by professional photographers, who are experts at detecting and evaluating light.

Sometimes we create precise definitions out of intellectual curiosity. There are lots of philosophy papers, for example, that engage in what is called "conceptual analysis." This is no more than trying to figure out what a word or concept means in a thorough way. Defining something carefully is creating another kind of detector.

Let's take the concept of a marriage. Marriage appears in almost every culture that has been observed. But what, exactly, is it? How do we know if two people are married or not? In many parts of the world there are legal ramifications of marriage (inheritance, rights, and things like that), which means the culture needs a formal definition of marriage to know for sure who's married and who isn't.

In contemporary Canada, for example, traditional marriages need to be registered with the government. If you do this, you are considered married, by law, until that marriage dissolves, perhaps through divorce or death. This is a detector, and this detector gets used when, for example, you want to marry someone but suspect they're already married. You can use the legal definition of marriage to "detect" if they are married.

Sometimes you can define something prematurely if you don't understand what something is, but struggle to define it anyway. Then later, as you get new information, your definition can get in your way. For example, let's take anger. If one wanted to investigate anger, one might have an urge to define it first. The trouble is, how can one accurately define it before you investigate it? Shouldn't the definition be the result of an investigation and not created prior to it? Maybe one defines anger as a negatively valanced emotion caused by harm or a perceived threat of harm to oneself or something someone cares about. This doesn't sound so bad, and it's not a brain-dead definition to start with. But then suppose that you find something that sure seems like anger, but is not caused by harm or the perceived threat of harm—perhaps by direct brain stimulation, or as a side effect of a drug—where the anger isn't directed at anything in particular. By the definition one started with, it doesn't even count as an instance of anger. So your investigation of anger and your definition/detector of anger should change each other as your understanding matures.

How do we know if detectors are any good? Well, they have to match our original intuitions, of course! If somebody came up with a "light detector" that seemed to give random numbers given what appears to our eyes to be a constant amount of light, we'd think the detector wasn't actually measuring light. If it gave no change in readout at all, no matter how much we change the light we can see, we'd think it might be broken. Psychologists call this "face validity."

For marriage, we also have an intuitive idea of what it's like to be married. Usually, it's a couple, in love, living together, sharing finances, saying they're married, having gone through a ceremony, wearing rings, and so on. If the marriage registry didn't match up with these things much of the time, we wouldn't think it was a "marriage" registry at all, but something else.

If the detector matches up with our collective intuitions about a concept again and again, with really good reliability, then eventually we might come to trust the detector more than our intuitions. This is good, because that's the point of having a more precise detector in the first place. If the photographer trusted their judgment of the amount of light more than the detector, there would be no reason to use the detector.

Once we created light detectors, we found out that there was light we cannot see: radio waves, X-rays, etc. There's a whole spectrum of light, and it turns out we can only see a small portion of it. When a light detector says that there are gamma rays playing on it, we believe that there is light there even though we can't see it. We are going against our own intuition of light, but that's okay because we have come to trust the detector so much.

Likewise, suppose you meet a bachelor, living the bachelor life. Months later, you find out that he's "actually" married. That is, there is a person somewhere that he got married to, and they never divorced. We believe that he's still married, even though his life resembles not at all the life of a married person we have in our intuition, because we trust the detector society has made over the observable cues in the environment.

If you're wondering where all this is going, I'm going to bring it back to morality now. As we have seen, much of our moral thinking is intuitive. We have feelings of right and wrong. I'm going to describe my model of how morals interact with moral theories in terms of stages we go through.

Stage One: Going with Your Gut

People start trusting their intuitions without question. They have no ethical theory.

So if we have feelings already, why do we need some ethical theory? Because our natural morality is mysterious and inconsistent. This inconsistency leads to confusion and the need for guidance where intuitions contradict each other, for example, when a friend steals a stapler from work. One intuition says you should turn them in (fairness), the other says you don't rat on your friends (loyalty). We have moral issues arise in our lives that we don't have answers to, and we want something (a theory) to help us make sense of it all. This makes ethical theories attractive. They have two benefits: they promise to explain the moral intuitions we have into some coherent, rational framework, and provide guidance for moral issues we're unsure about.

Stage Two: Face Validity

What does it even mean for a theory to be an ethical theory, as opposed to a theory about something else? For any ethical theory to be even remotely plausible, to get any kind of a foothold in our minds, it has to correspond with at least some of our moral intuitions, just like a potential light detector has to match our experience with light. It needs face validity.

To take an extreme example, suppose I came up with an ethical theory that said that anything you do that involves spreading peanut butter on something is right, and everything else is wrong. Would you reject this theory? Of course you would. You would not reject it because it was internally inconsistent. Its consistency and simplicity are probably its only benefit. You reject it because some things you *know* intuitively are wrong are not considered wrong by the theory—for example, causing someone unnecessary, excruciating pain. It has no face validity. It's such a nonstarter ethical theory that some would say it's not even an ethical theory at all.

This is our starting point. When we look at different ethical theories, we try see how well they use some kind of underlying structure to make sense of our felt morality. If the assumptions of the structure are acceptable, for example, that harm is bad, and the theory can predict some high proportion of our most important moral intuitions, then we might start to accept that theory as being true or having some truth to it.

We can think of an ethical theory as a morality detector. These detectors take in situations or actions as input, and output the rightness and wrongness. They are like light detectors in that way. So we come up with ethics theories. Deontology. Utilitarianism. Virtue ethics. The Ten Commandments. And how do we evaluate them? Just like we do the others. We evaluate them based on how well they match up with our intuitions.[79] This is why many ethical theories end up with similar moral conclusions. They all say murder is wrong, they just have different justifications for it. They *have* to have similar moral conclusions, because any ethical theory that deviates much from our intuitive notions of right and wrong are dead in the water, in terms of people accepting them.[80]

Although our acceptance of an ethical theory is based on its rationality and how well it matches our intuition, the search for a perfectly *satisfying* moral system is fruitless. You can't create a consistent, rational moral code that satisfies all of our intuitions if the intuitions you're trying to match are inconsistent to begin with. Any rational moral code will necessarily entail moral judgments or rules that you don't feel right about. The conclusion isn't that all moral systems are flawed, but that our moral psychology is flawed, and should be jettisoned in favor of a good theory.[81]

Because no rational theory can perfectly fit our inconsistent moral psychology, *every* theory leads to ethical dilemmas or what feel like morally repulsive results. We can see that the search for an ethical theory involves balancing two things: what we intuitively, strongly believe is right and wrong, on the one hand, and reason, a coherent structure, and a way to generate moral decisions on the other. There's some sweet spot in there where the successful theories have to lie. If there is some true ethical theory out there, it will have to contradict some moral beliefs we cherish, because our natural morality has self-contradictions.

Endorsing an ethical theory is, in part, about choosing which uncomfortable conclusions you're willing to live with. When someone rejects a moral theory because it has some problem, what that means is that it has a particular problem that individual is unwilling to let slide.[82]

Some ethical dilemmas are due to problems internal to the ethical system—that is, the system itself says that the same thing is right and wrong when different parts of the system are applied. An example is the theory that if God tells you to do something, it's good. Here's a dilemma: God tells Abraham to sacrifice his son. Obeying God is good. Killing a child is bad.

But another kind of ethical dilemma is when the system's output violates our intuitions about what's right and wrong. Let's take a popular criticism of utilitarianism as an example (for now just think of utilitarianism as claiming that the right thing to do is that which does the most good for the most people). Suppose a doctor is in a remote clinic with five people who will die without organ transplants. A person walks into the clinic with a sprained wrist, but is otherwise healthy. This person has no friends or family. According to a (perhaps unfair) version of utilitarianism, the doctor should sacrifice the sprained-wrist man to harvest his organs to save the five people who need them. That's the most good for the most people. Yikes.[83]

Let's assume that there is some consistent, objective right and wrong in the universe, hard as it might be to figure out what it is. If our own moral intuitions are inconsistent, then some of our moral intuitions have got to be wrong. We might think carefully about which ones we are more willing to jettison.

Warts and all, once we accept a theory, it takes residence in our minds. It is installed like software. With a theory in mind, we can look to it to help us make moral sense of the issues we're unsure about. That is, we might not have a strong feeling that something is right or wrong, but we will endorse what the theory says, because we trust it, just like a photographer might believe a light meter's reading when she feels unsure about how much light is in the room.

Stage Three: We Believe the Theory When It Contradicts Our Intuitions

The believed theory becomes the lens through which we see every ethical issue. Eventually we come to trust the output of the theory more than some of the moral intuitions. This is a big step, because we only believed the theory in the first place because it *matched up* so well with our intuitions!

But just like for light and marriage, if someone becomes convinced that some ethical theory is the right one, it can then be used to replace intuitions. Someone might say, "I know it doesn't feel right, but I think this theory is correct, so it must be right." This is stage three, where one endorses the theory so strongly that one believes the theory's conclusions over their own feelings, like believing the light detector when it says X-rays are present, even though you can't see anything. And this is important. Although ethical theories serve to satisfy our curiosity for the justification of right and wrong, we also want them to lead us to sometimes surprising conclusions. If an ethical theory makes no surprising conclusions, then it has no practical use—we could just use our moral intuitions, and wouldn't need the theory, just like a photographer with superhuman eyes would not need a light meter.

CHAPTER 4: WHAT WE THINK IS RIGHT AND WRONG

For example, the bystander effect is when having lots of other people around makes you less likely to do the right thing. If you're the only person around when there's an accident, you're more likely to try to help. But when there are lots of other people around, there appears to be distributed responsibility, and people are less likely to do the right thing. Again, nobody endorses the idea that what is right or wrong depends on how many other people are standing around. And no matter how you ask them, everyone denies that the number of people around influenced their decision not to act. This is an example of a moral dissociation.[84] No serious ethical theory would propose that you should behave differently in these situations, and people don't endorse the bystander effect, but it's real nonetheless. This is where an ethical theory can help you think about what's right and wrong, allowing you to go beyond your instincts. After learning about the bystander effect, whenever I see a car accident I call the police, even though thoughts like "somebody probably already called" run through my head. I know that the bystander effect is a buggy piece of software in my head, and I've learned to ignore it.

Many times we don't even get to this stage. Often we *think* we are making a judgment based on rational principles, when we are actually only bending rationality and the "theory" to accommodate what we feel intuitively. Our deliberate, conscious mind often acts like a lawyer, arguing whatever case is given to it by our intuition, rather than acting like a judge.[85] Sometimes we try to force-fit the detector to justify our intuitions. This happens with those Christians who believe that the answers to all moral questions come from the Bible. They might get a moral feeling of something from the surrounding culture: that same-sex marriage is just fine, for example. Suppose a heterosexual Christian couple is friends with a lesbian couple. These lesbians are great, and the idea that their good God would send Janice and Julie to hell causes these Christians some cognitive dissonance. They scry the Bible for some passage that they can creatively interpret to allow for same-sex couples, or search for some favorable interpretation on the Web. Problem solved! The actual reason they thought same-sex couples were okay was because they got it from the surrounding culture, but they convince themselves they got this moral idea from the Bible.[86] In this way, they can avoid the cognitive dissonance of endorsing a theory that has an ethical ramification that they find uncomfortable. I don't mean to pick on religious people; this happens with secular ethical theories, too.

Stage Four: Changing Moral Intuitions

Stage four is when the ethical theory is so well integrated into their cognitive system that it changes the person's original ethical intuitions. I'll explain how this happened to me. When pirating music started to become popular with Napster in the late 1990s, like most people I just downloaded stuff to listen to and didn't have any ethical qualms about it. It didn't *feel* like stealing, for a couple of reasons: I wasn't facing the person or organization I was stealing from, I wasn't removing anything from anybody's possession. I was just making a copy.

Because people's notions of "stealing" are very ancient, they are kind of brittle. Taking something from someone, grabbing it off their person and running away as they yell, "Hey!" feels very much like stealing. Breaking into someone's house and taking something they love feels like stealing. But as the situations get less and less like these prototypical examples, the feeling gets less and less strong.

Taking a stapler from the huge company you work at feels a bit like stealing, but this feeling is attenuated by the fact that it's a large, faceless company and not some person in front of you. Some acts of stealing just don't activate our intuitive ideas of stealing, so we feel that they're okay.

But I started thinking about music piracy, and the more I thought about it, the more it looked like stealing. I wasn't taking any physical object, but I thought sneaking into a cinema to see a movie was a kind of stealing. I also thought that stealing someone's idea was a kind of stealing. I also knew that artists and labels are able to produce music (or as much as they did, anyway) because of the revenue from music sales. So I concluded that even though using Napster didn't feel like stealing, it actually was. So I stopped using it. (I still don't pirate, and if you pirated this book, shame on you. Go buy it.)

Over time, I told people it was stealing. I argued it. I started to think of pirates as thieves. I started to think less of people who did it. Eventually, I had an intuitive notion, rather than just an intellectual one, that piracy is stealing. My intuitions had changed. Now it *felt* wrong. The feeling was there when I thought about other media piracy, from downloading PDFs of books to watching music videos on YouTube that were uploaded by fans and not the labels that owned them.

These reasons, these theories we hear or come up with, get rehearsed: we repeat them in discussions with people, we listen to them at sermons, we read media that tends to agree with our moral positions. This rehearsal reinforces the original ideas and the framework the rationalization uses, leading to a change in our gut reactions. This is stage four.

Someone might experience this with respect to eating pork and conversion to Judaism. Let's say Marie is Gentile, and enjoys bacon. She decides to convert to Judaism, where she learns that avoiding the consumption of pork is a good thing, by God's decision. At first, the idea of eating pork feels fine, but she comes to believe it's wrong at an intellectual level. But years after her conversion, she actually might genuinely, deep down, feel horrified if she found out that she had accidentally eaten pork.

Again, we can compare the ethical theory to a light meter. A photographer, after using her light meter for years, eventually gets trained by the readings of the meter, improving her perception. She's looked at scenes and then read the light meter's output so many times that she gets better at predicting what the light meter is going to say. Her perceptions and intuitions about how much light there is any given situation become more in sync with the apparatus.

Let's go over these stages again for morality. People start with their intuitive feelings of morality, not explicitly questioned. I think many people never leave this stage for any of their morals, trusting their gut for their whole lives, inconsistencies tolerated or unnoticed.

Stage two is when you come to accept some explicit ethical *theory*, probably described in language. When you're unsure, you side with the theory. In stage three when your intuitions clash with the theory, you might side with the theory in spite of your feelings. Stage four is when the theory becomes so well ingrained that your intuitions change, and you don't need to go through deliberate thinking through the theory to determine what it says is right and wrong—you feel it.

Should we go through these stages? Yes, but only if it's for the right theory. The moral intuitions we evolved to have, and those we accepted through enculturation, are problematic, particularly in the modern world. So they need to be massaged a bit. But the ethical theory we go through these stages with needs to be the right one.

We need to find the ethics detector that accurately distinguishes right from wrong, accept it, and hope that we eventually feel it, because if people don't really feel the moral valance of something, and only have an intellectual opinion about it, the resulting change of behavior is unlikely to last, and, as we learned in the section on habit, will only result in moral behavior when the cognitive system is attending to it.[87] Just as in the productivity section, when your cognitive system is occupied, you rely on the habit and reward systems. As such, the fourth stage is important for moral development.

RELIGION AS AN ETHICS DETECTOR

Many people use religion to guide their ethics. From the detector point of view, this functions similarly to using a secular ethical theory. There's an old argument, originally from Plato, that tried to cast doubt on the use of gods for moral guidance. It goes something like this: is murder wrong because the gods say it is, or do the gods say it's wrong because it is wrong? Many people say the latter, because they don't like the idea of right and wrong being a matter of opinion, even if the opinion we're talking about is God's. But if the gods say something is wrong because it actually is wrong, then that means that there is some nonreligious rightness and wrongness that God or the gods are merely aware of. And if we think the gods are correct about morality, then we have some god-independent way to know what's good. So why do we need the gods at all?[88]

What many people don't realize, however, is that this same argument applies to any ethical theory. Is murder wrong because some deontological theory says it's wrong? Or does deontology say murder is wrong because it is?

I have a more favorable view of using religion as a moral detector, because I think it functions more like our secular ethical theories than it might appear at first blush. One accepts a particular religious ethical position because it aligns with many of our deeply held moral intuitions, same as secular theories (stage two). It also puts it into a larger structure that has coherence, same as secular theories. When that structure produces moral judgments that might not be intuitive, we are more likely to accept them because we accept the structure as a whole, same as secular theories (stage three). It's unfair to beat up on religion on this score, because all theories are similarly implicated.

Suppose you have a question about biology, and you meet a biologist and ask her the question. She tells you something. Is what she says correct because she says it is, or did she say it's correct because it is? Of course, it's the latter. We believe her because she is in an *authority* on biology. Using gods or a god is similarly treating them (or him) as a moral authority. Someone with whom you generally agree, but when you're not sure, you side with their opinion, because you trust they know better. An authority on a subject can function like a detector or a theory.

It's the same way with me and utilitarianism. If I have a moral question, I look at it from the point of view of utilitarianism, and accept the answer (or I'm more likely to, anyway), even if it violates my moral intuitions, because I trust the moral theory more than my own, inconsistent, incoherent moral psychology.

5
WHAT'S ACTUALLY RIGHT AND WRONG

I was driving around in the desert of the southwestern United States in the early '90s with a few people who would end up being notable scientists. We were all pre-PhD research assistants working at Los Alamos National Laboratory. On a particularly long stretch of road, I got to thinking about morality. I thought about how sometimes I did things I knew were wrong. My clichéd revelation in the desert was that this was unjustified. I resolved then to never do anything immoral again.

I announced it to the car. My buddies laughed at me.

But I actually was different from that day on. It wasn't that I was a menace before that day or anything, it's just that I became much more intolerant of minor ethical violations that I had let slide before. I felt like I was really a better person.

But about fifteen years after that I learned more about what it really means to be good. It's more than just not being bad.

So let's get down to brass tacks. What is the best ethical theory?

As we have seen, we want a theory that matches *pretty* well with what we intuitively believe is right and wrong, maybe the moral intuitions we hold most dear, or are most sure of, but one that irons out some of the inconsistencies. One of the glaring inconsistencies we talked about was our moral intuition's insensitivity to number. But when you're going to try to optimize your morality, and be the best person you can be, this is something that cannot be ignored. In more ethical decisions than you might think, *magnitudes matter.*

When many of us think of morality, we tend to think of it in terms of simple rules: Don't steal. Keep your promises. Be nice to the people around you. Don't betray your friends. If only one rule applies, no problem, you don't have to deal with magnitudes. The problem with any moral rule set is that you will run into situations where the rules will be in conflict.

Suppose your friend Andy has been spending too much money on going to the spa. It's causing some tension with his wife, Lorraine, who is worried about the household finances. Andy can't resist, though, and tells you over coffee that he's going to spend the day at the spa and is going to leave his phone home so Lorraine can't reach him and find out where he was. "Promise me you won't tell her where I went," Andy asks you, and you grant this promise.

Later, Lorraine calls you, frantic, because Andy needs a liver transplant, and a liver is available, but nobody can reach him. "Do you know where he is?"

Here we have conflicting morals: keep your promises, don't betray your friend's (Andy's) trust, don't betray your other friend's (Lorraine's) trust, help your friends when you can, don't lie. How do you navigate situations like this? In this case, the answer is pretty obvious. Andy needing a new liver is a life-and-death situation, and it pales in comparison to the promise you made to him about not telling his wife where he was. The promise just isn't as important as his opportunity to get a liver. Why is this decision so easy?

You come up with an answer by comparing magnitudes. How important is helping your friend in this instance? He needs a new liver; that's pretty serious. How serious was this promise, and how much is at stake? You might not explicitly think about *numbers* when you make this judgment, but you can bet that your mind is comparing magnitudes of some kind. That is, your mind is using the informational equivalent of numbers, even if you're unaware of what they are and experience it only as a strength of feeling or conviction.

In this case, I suspect most of us would break the promise to help Andy with his liver, because the liver transplant is so important, and hiding yet another spa trip from Lorraine is relatively unimportant. When values conflict, people weigh the importance of different factors.[1]

As I mentioned at the beginning, this book is for optimizers. I've talked about how to optimize your productivity and your happiness. Now I'm going to tell you what science has to say about not merely how to be a good person, but how to be the most good person you can be.

IS THERE A ROLE FOR SIMPLE RULES?

Rules and laws have to be uncomplicated to maximize compliance. So even though some rule or law might not be as morally justifiable in a theoretical sense as a complex, more nuanced one, due to problems of compliance by limited human beings, in the long run it's better for society to have the simple rule than the complex one. Individual violation of the simple rule for the greater moral good might be better in isolation, but the societal effects of ignoring the simple rules might lead enough others to break it such that in the future there is overall less good. So we treat rules as if they had no exceptions as an interface with the population of people trying to follow them. But we should not forget that laws are this way for practical reasons. We should not confuse the dumbed-down rules of the law with what is really right and wrong.

Being as good as you can be means thinking big, and magnitudes matter even more when we think big. Suppose you are considering where to put a homeless shelter. There are two candidate locations. You have lots of information to inform this decision, including costs to your organization, the proximity of the location to other services street people might need, and the impacts of the shelter on property values. You might have moral rules, like helping people when you can, and not letting people get hurt, but how do you use these rules to come up with a decision?

Well, putting a shelter in either location will cause harm to the property owners in the area by lowering their property values. Also, putting the homeless shelter here rather than there will help some street people and not others. So it appears that placing the shelter in either location is technically

CHAPTER 5: WHAT'S ACTUALLY RIGHT AND WRONG

immoral, if you're thinking purely in term of rules, as it's harming people no matter what you do. And if you try to avoid doing anything bad by doing nothing, then you don't help anyone at all, which (depending on your rule set) also breaks a moral rule.

When you're thinking about how you act in your day-to-day life, it's easier to think about rules that are or are not broken, rather than magnitudes of good and bad. This isn't all bad. The "user interface" of morality has practical importance. If mores are complex, and hard to understand and remember, compliance will suffer. We use rules of thumb to guide us, because figuring out what's right and wrong can be complicated.

This book tries to go one level deeper. In particular, when you think about what's the right thing to do when large numbers of people are affected, you more or less have to think in terms of magnitudes to find the moral way forward.[2]

Perhaps a more striking example is the moral justification of having a police force and justice system that is allowed to use violence or restriction of freedom. When police detain someone, or physically harm them, or a justice system imprisons someone, society is harming some individuals (the suspects and criminals) with the intention of helping society. Presumably, incarcerating dangerous people helps the other people in society more than it hurts the criminals.[3] In other words, you can't even justify a police force with moral rules that don't involve some kind of nuance regarding magnitude of helping and harming. People who claim that morality has nothing to do with magnitude are usually just trying to ignore the unexamined magnitudes in their heads that influence their moral judgments. We want these magnitudes out in the open so we can critique or change them.

Okay, when we want to maximize what is good or right, and minimize bad and wrong, how can we get more specific about what we need to maximize and minimize?

We'll start with what we're *most certain* is right and wrong, and what the most people agree is morally relevant: helping people is good and hurting people is bad. I say this is the most certain because every culture has a moral like this (even if they define "people" as only being members of their in-group). Every well-thought-out moral theory includes something like this. Just about every person (with the possible exception of persons with psychopathy) has this moral intuition.[4] When we try to step back from our cultural mores, get skeptical of any value of the difference from culture to culture, what is left? What moral value would someone have to be stupid or crazy to deny? For a lot of people there's only one answer: that things like happiness, pleasure, life satisfaction, utility, conscious good feelings, subjective well-being, preference fulfillment, etc., are good, and that harm, misery, pain, suffering, preference frustration, etc., are bad. Can we justify this moral stance? Most people cannot, but also feel they don't need to. That helping others is good and hurting them is bad feels self-evident.[5]

I've been sloppy about what care and harm means, so I'll take a moment to be clearer. On this view, there is an inherent good to things like happiness, pleasure, life-satisfaction, good feelings, not being hurt, and so on. Similarly, there is an inherent bad in feelings of misery, suffering, harm, preference frustration, pain, and so on. Technically, I mean positively and negatively valanced conscious mental states, and I'll refer to them simply as "good feelings" and "bad feelings." But know they are shorthands for complex phenomena.

Can *all* morality boil down to good and bad feelings? This is probably the most important question in ethics.[6] Some say no, there are other things that we care about that don't have to be justified in terms good and bad feelings. For them, these other moral things (perhaps knowledge, or beauty, or reparation) are also self-evident, and do not need to be justified in terms of care and harm.

Are they right? Perhaps. But acknowledging the existence of these other things is much easier than figuring out what those other things actually are. As we have seen, people disagree about many of their moral intuitions. People throughout history have held different opinions about the rightness and wrongness about abortion, infanticide, euthanasia, slavery, chastity, caste systems, cannibalism, eating meat, how many wives someone can have, the importance of religion, the importance of etiquette, and whether or not wearing a hat is respectful or disrespectful to religion.[7] If Elizabeth thinks that scolding other people's children is inherently wrong, and not just instrumentally wrong, and you disagree, there's not much to talk about. Because the moral is supposed to be self-evident, it can't have, or doesn't need, justification with anything more fundamental, such as helping and hurting.[8] With a zillion different cultures, with so many fundamental disagreements about rights and wrongs, and a zillion people in those cultures, each with their own, idiosyncratic moral preferences, how can we

CHAPTER 5: WHAT'S ACTUALLY RIGHT AND WRONG

be confident about what additional self-evident rights and wrongs should be endorsed?

I'm going to focus on good and bad feelings in this book, treating them as *the most important* moral factor, which most can agree on, even if they think there are other important factors besides. There are several reasons for this.

First, we have a high certainty that good and bad feelings are important moral ideas, and we have a high uncertainty about any of the others, which seem to be more subject to individual and cultural differences. It is telling that when diverse cultures come together to hammer out some kind of ethical agreement, their common ground ends up being weighing harms against benefits.[9] It is, perhaps, the one ethical thing that just about everybody can agree on.

Second, many moral ideas we have can be justified by appeal to good and bad feelings. For example, killing people and stealing things can be justified as bad because of the bad feelings that are likely to result. Why is it wrong to lie? Because it tends to hurt people. Why is it wrong to kill, or steal, or slap somebody? Because these things hurt people. Someone might argue that slavery is bad (or good) based on issues about how much people are helped or hurt. Similarly with helping, it's good to give to the needy, to help out a friend, to give a gift, to comfort someone when they're crying, because those actions help people. This trait is not shared by many other candidate values: not everything can be justified in terms of, say, beauty or knowledge.

Third, as we've learned in the section on moral psychology, all of us have the ability to think in this way. In this sense, we all agree with the concept of weighing the bad against the good, and all use it sometimes. Although we also all have the ability to think in terms of rules, our disagreement on what those rules should be (see reason one) makes this moot. There is little profit in agreeing that we should reason in terms of moral rules if we cannot agree on what those rules actually are.

Fourth, rule-based ethical theories don't tend to have much to say about *maximizing* goodness. I could write a book about how to minimize your chances of lying, stealing, and murdering, but that won't do anything for all the people who die of preventable disease, or animals living in the factory farms, or preventing the catastrophic effects of climate change. Popular versions of rule-based ethics do not have explicit guidelines for decision-making on a grand scale. They are simply not up to the task, which is why large organizations tend to use some form of good and bad cost-benefit analysis, rather than relying on rule-following without consideration of magnitudes. Can you imagine trying to make decisions for an entire country using only a few simple moral rules? However, if you are a die-hard deontologist, I recommend reading Peter Unger's *Living High and Letting Die* after you finish this book to see why deontology will bring you to many of the same conclusions as this book.

Fifth, I'm personally convinced by the philosophical arguments in its favor. When you understand how our moral psychology works (which, if you're reading this book front to back, you do), it looks like many of the objections to weighing benefits and harms stem from intuitions from the old brain, designed by evolution to propagate our genes rather than doing the right thing, or making the world a better place.[10] To my eyes, some of the "advantages" of rule-based ethics sound downright selfish.[11] Do you really want to trust a blind process that maximizes nothing but gene propagation as an arbiter on questions of morality? In regards to this reason, there are smarter people than I on both sides of this issue, so feel free to take this reason with a grain of salt.

But what I really want to communicate in this book is a way of *thinking* about ethics. When you're trying to optimize your goodness, you don't really have a choice but to think in terms of magnitudes. If you want to use more than care and harm, or you interpret care and harm in a way different from good or bad feelings, be my guest. But I encourage you to think in terms of magnitudes, not just good and bad as an all-or-nothing affair. That is, put numbers and weights to all you think is ethically relevant, and reason from there.

Can Science Tell Us Anything About Morality?

Thinking in magnitudes is not the only thing I want to encourage. Science and rationality are relevant, too, when it comes to right and wrong. I've talked about productivity, what makes people happy, and what people think is right and wrong, all through a scientific lens whenever possible. Now I'm going to talk about what's *actually* right and wrong. Can science even speak to that?

It's been said that science is, ultimately, without value. That science can tell us what is, but not what should be. That it is things like philosophy

CHAPTER 5: WHAT'S ACTUALLY RIGHT AND WRONG

and religion that describe value. Suppose Pat is sick, and Chris has medicine. Science can tell us what is likely to happen if Chris gives Pat the medicine, but what scientific experiment could you possibly run to show that Chris *should* give it?

The reason for this skepticism is because in ethics, if you keep asking why, why, why, you end up with some *assumptions* about what is right and wrong. Assumptions that can be supported with neither science nor argument. You have to start somewhere. These assumptions, sometimes called faiths, are thought to be antithetical to the scientific enterprise, which is based on evidence and reason.

However, anybody who thinks that science does not also depend on assumptions is seriously mistaken. I'll talk about a scientific experiment in a very simple form to make this point. Suppose you have some new medicine. We'll call it "curitall." You want to know if it will cure a sickness, which we'll call the heebie-jeebies. You get a bunch of people with the heebie-jeebies and give some of them the medicine and others a placebo. Suppose that the overwhelming result is that the patients who took curitall got better much faster than the placebo group. We conclude that the curitall effectively treats the heebie-jeebies.

But how do we reach this conclusion? Why are we justified in thinking that, just because curitall seemed to cure the heebie-jeebies in this experiment, that anyone else with it will react similarly? What do the trials in the experiment have to do with unobserved events, perhaps far away, perhaps in the past or future, on a completely distinct group of people?

We are justified because we believe that unobserved events will be similar to observed events. All of experimental science is based on this belief. Without it, we cannot generalize from what happened in the experiment to anything else, ever. We all believe that the way things work over here will work the same over there, and that the way things work now, or in the past, will work that way in the future, and worked that way even further in the past. Without this belief, experimental conclusions make no sense.

Is this belief justified? Or is it just an assumption? You might think, well, science has been going on for hundreds of years, using this belief, and it really works. When Galileo did his experiments with gravity, showing that two objects of different weights fell at the same rate, when people tried it later, it worked the same way. There are hundreds, thousands of examples of how what worked in the experiment worked later. Isn't that evidence that this belief, that unobserved events will work the same way as observed events, is justified?

There's a problem with this, though: the argument is circular, which means that to reach the conclusion, you have to assume the conclusion is true already. How? Because to say that unobserved experiments will resemble observed experiments because they have in the past requires us to assume that the future will resemble the past, which is what we're trying to find evidence for in the first place. If you don't accept this to begin with, the argument falls apart. This was articulated by my man David Hume over a century ago, and it's called "the problem of induction."

What's the way out of this? We just *assume* it's true, and don't try to justify it. And what is meant by assumption here is that we believe it without any rational argument or evidence. Sound familiar?

When we talk about ethics, we end up at core beliefs, or faiths, that are not justified by science or reason. They might be "don't cause unnecessary suffering" or "strive to make the world more like what God wants it to be." When we talk about science, we have assumptions and faiths, too: induction, the fundamentals of logic, and so on.

I'm not going to argue here that science is just another religion—it's not, because one of the great things it has over religion is that it tries to keep these faiths to a minimum, with the principle of parsimony, or Occam's razor. I want to argue that to the extent that we can use evidence and reason in science, we can also use it in ethics. Just because the *foundation* of ethics and science are ultimately assumptions doesn't mean we can't use logic, reason, and evidence to determine the answers to many—*most*—questions.

If we agree on some set of assumptions, like suffering is bad and that induction is okay, we can then throw the whole arsenal of reason, evidence, science, economics, and analytic philosophy at interesting problems that are relevant to how to live our lives and run the world.

When we ask about how right or wrong an action is, it's almost always a scientific question that reason and evidence can speak to.

Let's take a look at where we are so far. We're going to consider how to be as good as you can be, where being good means maximizing good feelings and minimizing bad feelings in the world. Further, we're going to use science and reason to try to figure out what are the best things to do.

The Limits of Empathy

As discussed in the moral psychology section, judgments based on emotions tend to be more deontological in nature (based on rights and rules), and judgments based on reasoning tend to be more utilitarian in nature. Psychopaths with low anxiety, who have social and emotional deficits, tend to be more utilitarian when judging others' actions, and people with disorders that reduce awareness of their own emotions do, too. People who say they often rely on their gut tend to make decisions that are less utilitarian. When you ask someone to make moral judgments while doing another task at the same time (putting their frontal areas under cognitive load), utilitarian judgments are affected, but not other ones.[12]

If we want to look at a utilitarian approach to bettering the world, then this suggests that moral judgments based on emotion are going to generally be worse than those made through reasoning. This makes the popular idea that empathy is the moral cure-all a very shaky position. If we want to be as good as we can be, we need to use thinking, not emotion.

6

A WORK IN PROGRESS

Here's where I'm going to try to introduce a healthy dose of skepticism about what you're about to read in this book. I've written it with the best, latest thoughts and scientific findings I could, but it's important to keep in mind *when* I wrote it. In this case, it was mostly written during 2019 and 2020. This means that any new insights that came to light after that are not here. This might seem obvious, but, depending on when you're reading this, it could be really important.

"Facts," as we often think about them, tend to expire. Science is a work in progress, as is philosophy and all of the other scholarly disciplines. What we think of as scientific facts are beliefs generally shared in the scientific community based on evidence and how well they fit into our structure of how the world works—that is, all the other facts. But the fact of one age becomes the falsehood of the next.

Samuel Arbesman's excellent book *The Half-Life of Facts: Why Everything We Know Has an Expiration Date* describes studies in which experts took old textbooks and counted what proportion of the facts described in them were no longer considered true. In general, you can be pretty sure that half of the facts in any textbook more than ten years old have been overturned since the textbook's publication.[1]

So you need to think about that when you are reading this book (or any other, for that matter). If it's after year 2031, about half of the facts herein probably need updating. Of course, knowing *which ones* expired is not obvious.

CHAPTER 6: A WORK IN PROGRESS

I am going to use a lot of estimates in this book, and some of these estimates are likely to change, as is their very importance. If the world turns to vat-grown meat or non-meat substitutes that are so good we don't need to raise cows anymore, then cow suffering might end up unmattering.

Given the great uncertainties with some of these numbers, I am sometimes hesitant to even include them. But I'm a firm believer that some number is better than no number. Once the reasoning is in place, the outdated numbers can be easily replaced, and the calculations redone. Why is this better than using no numbers at all? When you're trying to decide how to live your life, you're going to weigh things in your mind, and if you don't use an explicit number, you're going to use some implicit magnitude feeling in your head instead. Still, in some sense, a number, just as uncertain but not even recognized for what it is. Some of the numbers in this book are probably close to garbage. But without them, when you try to figure out whether to help climate change problems or current malaria problems, you're going to make a decision based on a feeling about how serious these problems are. We'd be able to calculate what the unknown number in your head is about the harm caused by carbon emission based on your choices. It's better to have these assumptions out in the open.

What I want you to take away, more than anything, is the thought process used in this book. The use of evidence, numbers, and calculation, even if rough, to help guide our life. That's more important to take away than the exact conclusions given by the result of the calculations.

Of course, even the best *way of thinking* can change, too. The sciences of productivity, happiness, and morality are relatively new—certainly on the scale of science itself, because scientific psychology was a little sleepy getting out of bed, but even on the time scale of psychology the interest in these particular areas have only recently blossomed.

Even ethics is a field that has changed rapidly in recent years. The effective altruism movement, which this book is a part of, only emerged in the late 2000s, and new philosophical problems in ethics are being discovered all the time. At some point, the facts, ways of thinking, and even values will change as we learn more about our world.

In contrast, our basic human nature changes much, much more slowly. Our instinctual morality takes hundreds of years to change, and the basics of what makes us happy are likely the same. Of course, our *understanding* of what human nature is like will change, but the nature itself is less of a moving target than our cultural values and ways of thinking, because genetic evolution takes so long.

So with all of these caveats out of the way we will start the analyses. The kinds of things I'm going to tell you, if you accept them, might make you a little uncomfortable. So settle in.

Having and Not Having Children: Moral Issues

From an ethical point of view, how many children should you have? Most people don't think with a moral lens at all about having children—they just think about what they want. But the potentially vast moral effects of having kids are often unappreciated.

First, we can consider the effect of having children on one's own happiness. As discussed earlier, the data on this is kind of mixed.

We also have environmental concerns. Having children means creating a self-replicating polluter and carbon consumer. That is, the child will consume carbon, but so will his or her children. If you estimate that carbon consumption over time, weighting the descendants according to genetic relatedness, one study found that having one child adds about 9,441 metric tons of carbon dioxide to the atmosphere, which is about 5.7 times an individual's contribution.[2] A single metric ton of carbon put into the atmosphere shrinks summer sea ice by about 32 square feet.[3]

These numbers are hard to grasp, so I'll compare them to the kind of things that people tend to talk about when they advise you on how to be good for the environment. Getting a more fuel-efficient car (say, from twenty to thirty miles per gallon) reduces 148 metric tons getting put in the atmosphere.

Let's say you recycle for your entire life. This saves a whopping 17 metric tons.[4] So, to offset the environmental damage of having children, you'd need to convince an additional 555 reluctant people to also recycle just to break even. *Per kid.*

When it comes to carbon consumption, most things you do or don't do seem negligible in comparison to reproducing. The lion's share of your environmental impact is based on how many children you have. You don't hear about this much, though, because telling people that they're hurting the world by having kids doesn't sell newspapers, so they focus on not using plastic straws.

IS *ANYBODY'S* LIFE WORTH LIVING?

There are some, like philosopher David Benatar, who argue that *nobody's* life is worth living.[5] That we vastly overestimate how much good feelings we tend to experience, and that the world would be a better place without any feelings at all: just a world of rocks and stars, with no beings that can experience anything.

Whether this idea is right or wrong, believing this is an evolutionary dead end and destined to die out in the intellectual marketplace. Suppose this idea was heritable, either genetically or through enculturation. The simple fact is that, over time, people who think reproduction is good will take over, and the people who don't endorse reproduction will die out. This is not a justification of their views, it's just how the world works. Our morals are, in part, a result of evolution, which means that, to some extent, we have the morals we needed to have to out-reproduce our historical competition.

Studies suggest that average happiness for human beings is pretty high, which suggests that the average human life is worth living. As for whether or not life in general is worth having around, we'll get into that later.

Even if having that child causes environmental damage, we might not be ready to say that, all said, the world is a worse place with the child than it would have been without. There are, of course, non-environmental issues that are relevant. Even a life that has a lot of hardship might be worth living. Moving on from environmental issues, we can also think about moral relevancy of the expected happiness of the child yet to be born. This has a degree of uncertainty, but there are some cases in which most of us feel pretty strongly that if the expected living conditions of the child-to-be are bad enough, it is not morally justified to bring that child into the world. There are a lot of things you might think are relevant: the probability of disease, poverty, a household with a history of abuse, living in a war zone, and so on. Also relevant, but even harder to estimate, would be the quality of life of all of that child's descendants.

We can imagine applauding a couple for choosing to avoid having a child that they want because they think the child's life might be miserable, or because it would make the world a worse place. In essence, we're concluding that the potential child's life is so bad that the world is better off without the child than with. Put another way, the potential child's life is not worth living. Although this feels reasonable, it has the strange property of caring for beings that don't and might never exist. That is, choosing not to have a child because they would have a bad life is making a moral choice to help someone that, by the very decision, does not and will not ever exist.

But if we are going to consider that quality of the child's life when deciding *not* to create a child, doesn't it make sense to consider the quality of the child's life when deciding whether one *should* create that child? That is, if we expect the child to have a great life, then yes, the couple should have the child.

But what if the couple doesn't really want the child? Do they then have a moral obligation to have the child, because not having the child would deny that child the good life they would have had?[6] And if so, how many children does this obligate the parents to produce? The logical conclusion of this is that a couple that could produce happy children would be obligated to keep having children until their resources were strained so thin that having one more child would render that child's life (and/or the sibling's lives) so bad as to be not worth living. Even with a suffering environment, isn't a life better than no life?

This feels alarming. I think of all of the ex-girlfriends I've had who were wonderful people. We chose to not have kids together, but I'm pretty confident that the kids we didn't have would have had lives *worth living*. At the same time, I'm very grateful I don't have a lot of children with all of my ex-girlfriends. My wife assures me that she is, too.

POPULATION ETHICS

Thinking about the moral implications of future people runs us into a branch of philosophy called "population ethics," and it's a doozy. It's about the ethics of people who do not yet (and might not ever) exist. What are our ethical obligations to them? Do people who *could* exist have the right to come into existence?

A central population ethics puzzle is this: is the world a better place with fewer people who are really happy, or with many people who are just kind of happy?[7] More concretely, suppose there were two islands, and the inhabitants of these islands were all infected with a deadly disease. You only have the resources to save one island—let's say you only have time to do a single delivery of medicine before the people die. On one island, there are a million people living decent, happy lives. They have enough food. On the other island, there are ten million people, but they live in poverty. They survive, but they're often hungry and are much less happy than the people on the other island, and the situations on both islands are unlikely to change, if they survive. Which island should one save?

The question of which island you'd send the medicine to is relevant to the ethics of having children. If you'd send the medicine to the more populous island, then that suggests you have *total* utilitarian leanings: the raw amount of happiness is what's morally important. If you'd choose the happy, smaller island, you have *average* utilitarian leanings: goodness means raising the average happiness. To a total utilitarian, it doesn't matter that by having a child you're making life a little bit worse for everybody, because you are creating a being (who might create other beings) that will probably have a life worth living.

Another argument for why we should have more kids is that the more there are, the more geniuses there will be to save the rest of us. Let's maximize the Einsteins. Of course, there will be more bad people, too, but it could be that the good that a single technological innovation the genius could make might be more good than the worst that a bad person could do, which would mean that if you're producing new good and bad people in equal numbers, the world will get better. Recall the discussion of how bottomless good and evil can be. The candidates for the people who saved the most lives saved more lives than the bad people killed.

Michelle Kline looked at people living on islands and how many different kinds of fishing tools they had. There was a linear relationship between the number of tools and the size of the population, suggesting that having more minds on the task, all learning from and building off of one another, generates more creative solutions to the problems a society faces.[8]

Anthropologist Jared Diamond's *Guns, Germs, and Steel* suggested that a part of the reason European and East Asian cultures dominate our modern world is because the layout of the continents—east-west, rather than the north-south orientation of the Americas and Africa—allowed for greater cultural transmission. This is because as one travels across longitudes, environments are similar, but as one travels north or south across latitudes, the environment changes enormously. This effectively creates larger populations that can effectively share and innovate.[9]

The ascent of these cultures happened before mass communication and relatively cheap air travel. Today we live in a globalized world, where these regional differences matter somewhat less. It could be that human prosperity will accelerate with an increased world population simply because we have more minds on the job. But there is the issue of finite resources that butts up against this—akin to the "one" vial of medicine vs. two islands of people.

So how many children should you have, all things considered? The environmental cost is high for each child, but every child you don't have denies a life that might be pretty good—and then we have to consider the lives of their descendants. The answer relies on answers to difficult population ethics questions, which are far from resolved. With this degree of uncertainty, I cannot confidently make an estimate about whether having a child makes the world a better or worse place.

Which is probably just as well, because I don't think arguments like these affect people's reproductive choices very much in either direction. So although science doesn't give a clear answer here, you now know enough to make people very uncomfortable at cocktail parties.

What seems more certain is that if you keep having children, there will come a point where you are stretched so thin (in terms of your finances and attention) that the happiness (or health) of you and your children will suffer. The emotional yen to have children should stand up to cognitive scrutiny.

Ethical Purchases

Many people try to improve the good they do for the world, or minimize the harm, through purchasing decisions. Like everything else, there are shopping choices that matter, unmatter, and antimatter.

Let's take sweatshops as an example. Sweatshops pay employees a very low amount of money—in Brazil, about $2,000 a year, in Bangladesh, $2 a day. Working conditions are atrocious. Because of this, many people try to avoid buying products that use sweatshops, and many companies abandon the use of them, and then advertise that they did so to impress potential customers.

When you look at the conditions of poor countries, however, it looks more and more like sweatshops look as bad as they do because we are comparing them to the jobs in the rich countries we live in, and, crucially, not the jobs that sweatshop workers would take otherwise.

Although some sweatshop employees might be working there because of coercion, most sweatshop employees choose to work there because, in their estimation, it is the best job they can get. In fact, many people leave their own countries, risking deportation or imprisonment, so that they can work in sweatshops in other countries. Why would they do this?

Because often the work they are leaving is far worse.[10]

Although $2 a day sounds grossly unfair to an American, many people in poor countries work for $1.25 a day. Now, you earning seventy-five cents more a day more might not feel like a big deal, but if you're only making $1.25, a seventy-five-cent raise means a lot. People who are against sweatshops might imagine that if a child is fired from a sweatshop, they will go to a well-funded school instead. They don't. Some people think that adults fired from sweatshops will turn to safer jobs with higher pay. They can't.

To work at a sweatshop, people are leaving work in agriculture, mining, and sorting through garbage. For these people, sweatshop work is the *cushy* job. This is the opinion of even esteemed left-wing economists. One well-intentioned act of the American Congress resulted in sweatshops closing. The result of this forced children to work in illegal sweatshops, where they made less money. Some even turned to prostitution.[11]

Sweatshops seem to be a phase that regions go through on their way to being a modern, industrialized economy. The West went through it during the Industrial Revolution, and Taiwan went through it in the 20th century. When the phase was over, the sweatshops went away, because there were even better jobs, so nobody wanted sweatshop jobs.

When you buy from a company that could use sweatshops but doesn't, you're paying money (usually more) to give jobs to comparatively wealthy people in richer countries. Working for minimum wage in Canada is far, far better than working for $2,000 a year in Bolivia.

So when you vote with your dollars by trying to put sweatshops out of business, you are contributing to putting people back into even lower-paying, more backbreaking work. A study by UNICEF found that after banning sweatshops, the children who worked at them mostly turned to stone-crushing, prostitution, and street hustling.[12] Choosing a company that proudly avoids using sweatshops antimatters. Don't try to stop sweatshops, try to end the poverty that makes those jobs attractive.[13] Ironically, patronizing sweat shops is an effective way to get rid of them, by making people rich enough to not need to work at them anymore.

The same reasoning suggests that buying local is also problematic, because you're contributing to the wages of people who are already wealthy, by world standards. Just to make it very clear, if you live in America or Canada, even if you have zero income, your access to shelters, emergency rooms, and soup kitchens means you effectively have more wealth than the poorest people in the world.

Is the welfare of someone in your own country more valuable than that of someone in a foreign country? I don't think so—living as an American in Canada makes me very aware of how arbitrary nationalism is. But even if you value your fellow citizens more than people in foreign lands, let's think in numbers. Ask yourself: *how much* more valuable are they? Because even if you think they are twice as valuable, it's probably still better for you to help people in poorer countries, because they are *more than twice as bad off* than even the street people living in your community.

Fair trade antimatters, too. This is because the countries that are rich enough to afford the required certification (like Mexico and Costa Rica) are ten times richer than other coffee producers like Ethiopia. Because your money goes so much further in a very poor country, you do more good for the poorest people by paying less for free trade coffee than you do by paying more for fair trade coffee. Furthermore, most of the extra money spent on fair trade coffee goes to middle men, and the farmers are not any better off. This has been found in academic studies as well as investigations by the Fairtrade Foundation itself, which admitted the limited evidence of a positive impact on workers, the very people fair trade was invented to help.[14]

Avoiding sweatshop products, and buying local and fair trade products antimatters.

Think Globally, Act Globally

In this book I'm going to explore ways to increase good feelings and decrease bad feelings, for all creatures that have feelings, until the end of time. When you put it that way, it sounds like a tall order. But the good thing is that, in our connected global world, we actually have more opportunities to help those who need it most than at any other point in history.

Think of our ancient ancestors. Doing good and bad for the world was simpler and smaller-scale, because there weren't that many people affected by the actions you took. If everybody in your group was doing okay, being good was a matter of being nice to them. If there were starving people on the other side of the continent, you wouldn't even know about their plight, much less be able to help them.

Historically speaking, it wasn't that long ago that human societies were beset with a more or less constant hammering by plague, famine, and war. Severe droughts commonly would kill off one out of every twenty people, and there wasn't much anybody could do about it.[15] Because this was the reality for humans for so long, our morals evolved to deal with that very small world we could know about and affect.

As we've seen, our stone-age moral psychology is made for that world. But today's world is very, very different. Rather than being at the mercy of inscrutable forces of nature, our species is largely in control of its destiny. Even the famines that happen nowadays are mostly caused by human screwups, rather than by unpredictable natural disasters.[16] And the actions individuals take have global effects. Nowadays, when we're considering being good, there are many people who can help in many ways, all around the world. The ethical ramifications of this are enormous and unintuitive for our small-group-of-hunter-gatherer moral brains.

One ramification is that many of the normal reasons we favor one person over another are becoming more and more ethically suspect. Let's take someone in Mexico who wants to optimize her positive impact on the world. Let's further suppose it is just as easy for her to help someone in Mexico as it is to help someone elsewhere, which, for many things, is true. Does it make any sense for her to prefer to help someone who is in Mexico, rather than somebody in a faraway country?

If the people are otherwise equally worthy of help, yet she favors Mexicans, she's acting as though Mexican people are more valuable than other people. As we have seen from moral psychology, this is a common belief—not that Mexicans are more valuable, but that people from one's own in-group (country) are treated as more valuable. People don't choose which country they are born in, so we end up treating some people better than others because those people got lucky enough to have been born in our own country. When we reflect on this, it seems to be a pretty flimsy criterion to use to determine a person's value.

So let's go on with the assumption that where someone lives, where they were born, what color their skin is, and so on, make them no more or less deserving of our help. This might suggest that it doesn't matter who you help—everybody's equally deserving!

Not so fast. We need to think about how *expensive* it is to help someone. Although at first blush this sounds crass, we all can feel the pull of cost-effectiveness. If it costs $50,000 to save someone's life in one country, and the same amount of money to save ten lives in another country, then it would be better to save ten people for the same amount of money. This is the nature of our world today. When taking a global view, it's much cheaper to help some people, and much more expensive to help others. Food, medicine, education, and just about everything cost a lot more in some places than in others.

Let's review. Doing something good has some kind of cost—we can typically think of costs in terms of time or money spent. We also might think of our willingness to do good as a depletable resource, too, because, as we've learned, as people do good they feel better about themselves. They eventually become satisfied with how good they are acting, and subsequently start licensing themselves to stop doing good things. If your goodwill can run out, then doing some good can keep you from doing other good, making goodwill a limited, precious resource. You only have so much time, so much money, and so much goodwill, so you want to use these resources so that they pack the most moral punch.

Furthermore, we can now affect people all over the world, and different people in different parts of the world vary a lot in how expensive it is to help them. A person's location alone is not a legitimate moral criterion, so if we want to maximize good, we should be helping those it's cheapest to help.

The conclusion, then, is that it's much more efficient to help the world by helping the people living in the cheapest places on Earth. The cheapest places tend to be where the poorest people are. And these people live far, far away from the rich people who can afford to make a difference. If you remember nothing else from this book, please remember this: the best way to optimize being good is to help the poorest people on Earth.

Let's look a little more deeply at poverty, which is a big cause of death and misery. Many of us see poverty in our day-to-day lives. There are street people at risk, homeless people, people who have trouble making ends meet. But the notion of poverty is generally thought of as relative to the wealth of the surrounding people. For example, in the United States, a person making less than $12,490 per year is classified as being "in poverty." That's about $34.69 a day. Alarming as that sounds, it's far more than what the World Bank considers extreme poverty. The World Bank needs to be able to consider what poverty means in some objective sense, so it can compare needs across diverse populations. For them, extreme poverty means earning $2.12 per day (2018 dollars)![17] Now a dollar goes a lot further in Chad than it does in Los Angeles. But buying power is taken into account. The World Bank estimates that it would cost $2.12 *in America* to meet your basic needs. Obviously, having a cell phone plan is not a basic need.[18]

What is extreme poverty really like? The World Bank interviewed 60,000 people in extreme poverty, across 73 countries, and asked them what characterized it. The most popular answers: you don't have enough food for most of the year, you can't save money for emergencies (putting you in debt when they occur), you can't afford to send your children to school, your house needs to be rebuilt often because it's made of mud or thatch, and you have no close source of safe drinking water.[19] In contrast, the poorest people in my city of Ottawa can get free health care at hospitals and clinics, free food at food banks (maybe not as much as they want, but they won't starve), free clean drinking water, and sometimes a bed in a shelter. Their lives aren't cushy, but they are nowhere as bad as those in extreme poverty when we look at it from the perspective of all of humanity. At least the homeless shelter doesn't have a straw roof.

By these measures, just about nobody in America, Canada, or the UK is in what the World Bank would consider extreme poverty. As we've discussed earlier, if you are living at the American poverty line, you're still richer than three out of every five people on Earth.

How can rich people (that is, people in industrialized countries) help poor people across the world from them? Their choices are really limited. Aside from purchasing choices, what they do locally doesn't make much difference for them.

For me, this is a startling thing to realize. What about being kind to your neighbors? What about donating to the local soup kitchen? What about not lying? As good as doing these things makes me feel, I have to admit that they are not saving lives, and when it comes to doing good, it's hard to beat saving lives. If we consider helping the people most in need as an important moral endeavor, which just about everybody agrees with, and the people who most need help are far away, then we get to the conclusion that being good to the (relatively rich) people around you doesn't matter as much as the poor people far away from you. The normal signals of being a good person, like being kind to the people you meet—helping old people get on the bus, visiting a sick person in the hospital—can't be directly applied to distant people. The way they can generally maximize their goodness is by donating to charities that have expertise and infrastructure to help the world's poorest people. For most of us, the only way to help distant people is through a middleman, like donating to a charity. Although we all kind of acknowledge that charitable giving is a good thing to do, it's not obvious to most of us that it matters far more than almost anything else we do.

I am not encouraging people to be jerks in their personal life. I'm just saying don't let being a nice person lull you into thinking that that's all you need to do to be a good person—let alone someone who maximizes how good they can be. With some exceptions, the lion's share of how good a person you are can be measured in how many lives you save, and for the vast majority of people that means donating to charities.

This is part of why it's hard to maximize your goodness: the most effective good you can do violates many of our natural moral psychology tendencies. The reason I spent so much time talking about moral psychology is so that when you learn about what it takes to be *really* good, you will understand why it doesn't *feel* right. It only *thinks* right.

But recall all we've reviewed about our moral psychology and where it came from. Rejecting evidence and reason in favor of our moral intuitions is like trying to make a theory of what we should eat based only on what tastes good. Someone might point to evidence that lots of sugar is bad for you, and it would be silly to counter that this has to be wrong because eating lots of sugar tastes good. We use evidence to avoid intellectual complacency we get from our moral intuitions.

So if we should help the poorest people, and one of the only ways to do that is to donate to charity, then we have a choice about which charities to donate to. How do people pick charities? Many people are opportunistic about charitable giving. They will often respond to someone at the door, on the street, or to a flyer in the mail. This method of charity selection results in giving to charities that happened to put resources into making themselves known, and not to the charities that are the most efficient ones. Most people give to charities to get a warm glow, and they don't look too closely at how much good their donation is actually doing. This makes a kind of sense, because the size of the warm glow you get is all most people have. There are no studies of this, but our insensitivity to numbers suggests it. Now, if your own personal happiness is all you care about, then this might be fine. But if you're optimizing your moral goodness, you'd want to do better.

It's in striking contrast with other investments, where they expect a personal financial return. Can you imagine being approached on the street by a twenty-year-old wearing a blue vest, being told over the course of thirty seconds about a great investment opportunity, and giving him $50 on the spot? Of course not. You'd want to see past rates of return, you'd want to compare the investment to other investments, and so on. You'd do your homework.[20]

But people seem to be perfectly willing to donate to charities without anything close to the same level of rigor.

Just as there are vast differences in the cost of helping one person over another, there are vast differences in the amount of good different charities can do per dollar. Which means that charity *selection* is just as important as how much money you give to them.

Even if you wanted to know how effective the charity was, in terms of numbers, how would you get this information? Until recently, the answer was: you couldn't. Sometimes charities would tell you, say, the percentage of money going to the CEO, advertising, operational expenses, and actually doing the good they're mandated to do, but this is a far cry from knowing how much the world is actually helped. The Charity Navigator website reviews charities to make sure they are not fraudulent, and they look at administrative bloat. But that's not good enough, because they don't measure the amount of good being done. A charity that gives carnations to people in hospitals might have a tiny administrative overhead and a poorly paid CEO and have no fraud, but still be a poor use of your donation dollars. No matter how efficiently you give carnations to people, it's not going to make a big difference. How many carnations would need to be delivered to sick people to do as much good as saving a twenty-year-old from blindness?

Using financial ratios as a proxy for effectiveness is the "overhead myth." To help understand why, we'll use business investment as an analogy. When most people make investment decisions, like when they buy stock in a company, they are interested in how much money they think the company is going to make. Now, if the CEO is getting paid more, this cuts into profits. If the company engages in advertising, that takes a cut of profits, too. But wait, you might say, don't those costs result in more profit in the long term? Maybe they do. The point is that what matters for an investment is the return. How much the CEO is paid, what the R&D and advertising budgets are don't *directly* matter. If a company makes more money by hiring a more expensive CEO, or by advertising, so be it.

There is an analogous thing going on with charities. Yes, some charities advertise, but they do so because (they believe) that the advertising brings in more money than it costs. Maybe one charity pays its CEO more than another, but that CEO might be really good. The point is that what makes a charity good or bad is its "profit" in terms of how much better it's making the world per dollar invested, and all the other details about how the organization is run are beside the point. You can have a charity with a poorly paid workforce, with zero advertising budget, but yet be *really bad* at effecting positive change in the world. You can also have a charity that spends a lot of money on advertising, and maybe more on overhead than you feel comfortable with, but dollar for dollar creates much more positive change anyway.

Looking at overhead and advertising budgets is easy, though, and we *believe* it correlates with effectiveness, but if we can measure effectiveness more directly, we can safely ignore those other factors.

Are there even ways to measure how much good is being done?

Given that there are so many moral considerations it is very difficult to compare things to each other. We have a serious apples and oranges problem. Economists are very good at reducing everything to dollars. But if what we really care about in this world is good feelings and bad feelings, money isn't really the best measure. We've seen that income, for example, has a nonlinear relationship with happiness. So we can't really equate human happiness with a certain number of dollars. What we need is a measure that is relatable, measurable, and quantifiable, and then try to put everything in those terms. The measure I'm going to use in this book is a year of quality human life.

7
THE (NUMERICAL) VALUE OF HUMAN LIFE

There are well-meaning people who say that you can't put a numerical or monetary value on human life. How far can we get with this idea?

Number of Lives Saved

Even if we don't put a particular numerical value on a human life, we can at least count the number of people who live or die. So a first pass might be that we can value human life in terms of the number of lives saved. Though not detailed enough for many analyses, it's not a bad way to think about goodness for the average person, simply because many good things we do clearly don't save any lives at all. For example, donating money to your local art museum is going to have an impact on the well-being of people, but nobody's life is in the balance. In this way, lives saved is a decent first pass to give you a rough idea of what matters and what unmatters.

In practice, saving lives often involves spending money for this or that intervention, which means that saving lives has some knowable financial cost. This brings us to the admittedly distasteful notion of putting a dollar value on human life.

If you are resistant about this, I get it. However, when anyone makes decisions about policy, charity, and other moral choices, one is *acting* as though there is some specific dollar value of a human life, even if one is *oblivious* to what that value is. That is, you can often look at someone's actions and infer what their value of human life must be for those actions to make any sense.

For example, the United States makes decisions about the highest allowed speed limits for motor vehicles. Speed limits have gone up over the years, which makes people get around faster, making the economy more efficient. But there are costs here—not just in terms of greenhouse gases (driving faster requires more fuel per unit traveled) but in terms of lives lost on the road. In 2017, for example, 37,133 people died in American motor accidents, but that is 1,934 higher than it would have been if we'd stayed with the speed limits we had in 1993.[1] The U.S. sacrificed almost two thousand people a year so that everybody could drive faster. Is it worth it? If we could calculate the increased speed limit's value to the economy, we could see exactly how much the U.S. valued the lives of those extra people who died on the road. (One analysis suggests that this isn't worth it, and that the social benefits of faster driving are two to seven times smaller than the costs due to death.)[2]

If we (perhaps grudgingly) accept that we have to make a decision about the monetary value of human life, we can investigate what is the best way to do it. For people who don't like to think in terms of numbers, this can be annoying, but it seems to me that there's no way around it. If you want to enter the best-person-you-can-be club, thinking in terms of numbers is the cover charge.

One way to look at it would be to try to estimate how much a person's life is worth *to them*. We can get numbers like this by looking at their choices regarding how much risk they are willing to take for a given amount of money. People take risky jobs, for example, and expect to get paid more for them. By looking at how much more, we can estimate the dollar value they place on their own lives. Various economists have made these calculations, and the value of life ends up being estimated at being around $9 million. In New Zealand and Australia, it's around $4 million.[3]

Government interventions also reveal how much they are willing to invest to save a life. Let's look at the government of the United States, which has many agencies in the business of life saving. It turns out that regulations vary enormously in the cost incurred to the government to expect to save one human life. A review showed that the costs of saving one life ranges from $100,000 (for the Childproof lighters regulation) to one billion dollars (for the Solid Waste Disposal Facility Criteria regulation). The average is around eight million dollars.[4] It's interesting that the revealed preference of the government for a human life is in the same ballpark (millions of dollars) as Americans themselves value their own lives.

Big organizations basically have to make estimates like this to function in anything close to a rational way. When governments decide what programs to invest in, they often compare the number of lives saved. As distasteful as putting a dollar value on a human life might feel, isn't it more distasteful to be spending a *billion* dollars over there to save one life, when over here you can save ten thousand people for the same cost?

As we will see later, even the low-end government cost of $100,000 for life-saving is about twenty times more expensive than saving the life of someone in the developing world—that is, if the United States saves an American life with the cheapest method its government has, a good aid organization can save twenty sub-Saharan African lives with the same money.

We've established that many people have an in-group bias. They feel that the people in their community deserve to be helped before people outside the community. I've tried to convince you that this is a morally irrelevant factor, but maybe you're unconvinced. Maybe you still feel that you'd rather help other people in your country than poorer people elsewhere.

If this is the case, let's put it in terms of numbers. Here's a tough question: *how many* foreign lives are worth one life of someone in your own country? Let's say you're American, and you value American lives over Nigerian lives. Let's say that, for the same cost, you have the opportunity to save one American life, or some number of Nigerian lives. How many Nigerian lives would you need to save to make this a difficult choice? This is a one-sided conversation. You don't need to actually tell me your answer, so be honest with yourself. You can be unsure of your answer, but make your guess precise—meaning, actually pick a number before you read on.

Let's say you come up with five lives. This would mean that an American life is five times more valuable to you than the life of a Nigerian.

When we look at charities, we can see how much it costs to save lives in America versus Nigeria, using the most cost-effective methods available. Evidence suggests that saving a life in Nigeria with the best charities costs about $5,000, and we just saw that saving a life in America costs about $100,000, if we go by the cost of the government's most efficient means. American lives, then, are twenty times more expensive to save. So even if you think American lives are *five times* more valuable as Nigerian lives, it still makes more sense for you to invest your charitable money to save Nigerians! Even at a moral exchange rate of 1:5, by saving Nigerians you'd still be doing four times as much good according to a moral stance that heavily biases in favor of Americans.

Now, maybe you picked a different number. But if that number was less than twenty, then the conclusion is the same: you'd be doing more good by saving Nigerians than your fellow Americans, even if your moral values say that one American life is worth, say, as many as nineteen Nigerian ones.

I would hope that you believe that an American life is worth exactly one Nigerian life, in which case you'd be doing twenty times more good saving Nigerian lives than American ones.

Some of you might be thinking that your currently preferred charity doesn't save any lives at all. Maybe you donate to a local kindergarten, food bank, or arts organization. The same logic applies to these efforts: the cost of education, the arts, and *everything* is far cheaper in poor countries than in yours! You could probably put on many plays in Nigeria for the same cost as a single production in your own country. So ask yourself the analogous question: how many *Nigerian theater productions* does it take to have the same positive impact on the world as a single American one? Even if you wanted to help the world's state of *theater*, it's still better to invest in the theater of a poor country.[5]

Another thing to consider is that improving health is one of the best ways to promote education and the arts, when a country is so poor that health problems get in the way of basically everything. Preventing kids from getting sick, and making sure they have enough nutrition, is more important than direct educational or arts investments, such as buying textbooks, computers, or lighting rigs for theaters—I'll discuss the evidence for this in a bit. So even if you ultimately really care about the arts (or education), you'd probably help the world of theater (or education) more by improving health than by funding the arts directly.

HELPING DISTANT PEOPLE AND DEONTOLOGY

I'm a utilitarian, but you can't avoid these problems simply by rejecting utilitarianism. All the major theories can lead to the same conclusions, with different justifications.

We might think of deontology as classifying acts into those that are forbidden, permissible, and required. Murder is forbidden, having a glass of iced tea is permissible, and feeding your children is required. Let's talk about helping others in the context of deontology. (Note that there are *many* different deontological theories; I'm just using a generic one.) We can assume that no reasonable deontology says that helping others is forbidden, so we are left with the two other categories: permissible and required. If helping others is required, then the conclusions of this book hold to be just as applicable as they would be under utilitarianism.

But many people believe that helping others, particularly those far away from us, is not required. It is permissible, but a good thing to do. They even have a term for this: "supererogatory acts." In this case, if you want to be better, and improve your goodness, then you should be doing not only the acts required of you, but more supererogatory acts, as well. In this case, too, the recommendations in this book also apply.

In short, if you want to be as good as you can be, deontologists and utilitarians would need to behave in much the same way.

Number of Years of Life Saved

Back to the practicalities of lifesaving. Utilitarianism typically says that everybody's life is equally valuable.[6] I think this makes sense when we think of race or nationality, but there are other considerations. A life saved is a simple concept that raises a lot of important questions. For one, how long is the life saved going to be? If you save the life of a forty-five-year-old female, you might be adding 38.72 years of life to that person's life. If you

CHAPTER 7: THE (NUMERICAL) VALUE OF HUMAN LIFE

save a male infant, you are expected to add eighty years.[7] If, for the same cost, we can save the life of an eighty-year-old or a twenty-year-old, it seems to most people that we should save the younger one. This suggests that a better way to look at lifesaving isn't in terms of number of lives saved, but the number of years of life saved. This would suggest, for example, that adding twenty years to one person's life might be better than adding five years to two different people's lives. It would follow that saving a life in America, where the life expectancy is 79.3 years, is better than saving life of a person of the same age in Angola, where the life expectancy is only 52.4—assuming that the cost of saving a life is the same in both countries. But this assumption is false, because the cost of saving lives turns out to vary widely. If you can pay the same price to save more years of life, then it would make sense to save an American over an Angolan. But because it is much, much cheaper to save an Angolan's life, utilitarianism suggests that, on the basis of the cost of saving one year of human life, we should help the Angolan.

This is progress. We've moved from simply saving lives to being able to think about the number of life-years saved. When we save someone's life, we can subtract their current age from the life expectancy to estimate how many years of life we are actually getting them.

But we need to go even deeper because we have the problem of how good that person's life is going to be. Extending someone's life so they can continue to be miserable is not as valuable as extending someone's life who will live happily. And if someone's life is so bad that it's not worth living, then adding years of life only increases suffering, making the world a worse place. Not only do we have to account for the number of years saved, but the quality of those years.[8] Can science help with this?

It's not straightforward to do, because, as we've discussed, human life has several ways it can be good. There's health, long life, happiness, etc. A good candidate measure is known as the Disability-Adjusted Life-Year, or DALY. It's the metric used by the World Health Organization to think about the global burden of disease. It tries to combine loss of years of life and the cost of living with a disease into a single number. One DALY is one year of healthy life lost. If a healthy person has a life expectancy of eighty, and they die from an accident at seventy, then they have ten years of life lost, or ten DALYs. According to this metric, averting DALYs is a way to decrease human suffering.[9]

EXAMPLE OF CALCULATING WITH DALYS

Alzheimer's disease and schizophrenia are not preventable, but let's suppose they were. Let's say that for the same cost, in England, we could prevent a sixty-three-year-old man from getting Alzheimer's or a twenty-year-old woman from getting schizophrenia. Which one should we prioritize?

For an individual, we calculate DALYs by adding two things. First is how many years of life are lost due to an early death caused by disease. The second, "years lost due to disability," is how many years before death they lived with the disease, multiplied by the disability weight.

Let's say a man gets Alzheimer's disease when he is sixty-three, and will die seven years later at seventy. Let's assume that life expectancy for most men is eighty. This means that with Alzheimer's he will be dead ten years earlier than he would have without Alzheimer's. But he is also suffering with Alzheimer's for seven years. Having Alzheimer's for a year is estimated as being 0.67 as bad as being dead for a year,[10] so we multiply 0.67 by 7 and get 4.66. That is, having Alzheimer's for seven years is about as bad as losing 4.66 years healthy life. Ten life-years lost plus 4.67 "years" lost due to disability is 14.66 DALYs worth of suffering.

A woman gets schizophrenia when she is twenty. This disease shortens a woman's life by an estimated 13.6 years.[11] If we assume a healthy woman will live to 83, this means our patient is expected to die at 69.4 years of age, which means she'll be living with schizophrenia for 49.4 years. The disability weight for schizophrenia is 0.58, which gives us 28.45 additional "years" lost due to disability. We add years lost (13.6) to this and get 42.05 DALYs.

Preventing disease in the man would avert 14.66 DALYs and preventing disease in the woman would avert 42.05 DALYs. The choice is clear: prevent schizophrenia in the woman.

The WHO uses this kind of calculation when looking at health interventions for thousands of people at a time.

Of course, being "diseased" shouldn't be looked at as an on-off switch, because some diseases are worse than others in terms of how it affects one's quality of life. Scholars use a variety of methods, including attitude surveys, to come up with numbers to account for this, too. Living for a year with blindness, for example, isn't as bad as not living that year at all, but it's not as good as living that same year with good vision. It's estimated a year lived with blindness is 60 percent as bad as not living that year at all. So if losing a year of life is 1 DALY, losing a year of your life with blindness is 0.6 DALYs. Another way to look at it is that this measure says that if someone had to choose between losing three healthy years of life and living five of their life's years with blindness, we would expect it to be a tough choice, because they represent the same loss of happiness.

How did scientists come up with these numbers? They asked people who were living with the disease to rate how bad it is in five ways: mobility, self-care, usual activities, physical pain, and mental pain. They then gave these numbers to people who weren't sick, and asked them, "Suppose you had this illness for ten years. How many years of healthy life would you consider as of equivalent value to you?" From these answers they calculate the weighting.[12]

People who have never been on dialysis estimate their life would be 39 percent as good as if it were healthy, but people actually on dialysis say it's 56 percent as good. In general, the public thinks that having this or that medical condition is going to be worse than it is.[13] This is likely due to the impact bias, which makes us overestimate the emotional impact of future events.[14]

On the other hand, asking sick people how happy they are sometimes gives us unintuitive answers because of adaptation. That is, after a few years, with some diseases, people aren't any less happy (there are exceptions). One interview study of 150 people showed that about half of them reported having a good or excellent quality of life—this is in comparison to people without disabilities, 80 percent of whom reported a good or excellent life.[15] Most people would expect the differences to be greater. These two forces are working against one another, so in some sense they cancel each other out. But research on how much they cancel out has not been done.

Although people overestimate the happiness effects of physical disease, healthy people vastly underestimate how much *mental* pain affects happiness. They think it's worse to have diabetes than to be depressed, and they are very wrong about that.[16] So healthy people overestimate the happiness reduction due to physical illness, and underestimate the happiness reduction due to mental illness.

The DALY measure takes into account disease, which means that the normal variance in happiness isn't accounted for at all, unless it's manifested in a disease, like schizophrenia or depression. But we can get ballpark idea about happiness by looking at their measures for depression. A year of mild depression is 14 percent as bad as being dead, moderate depression is 35 percent, and severe depression is even worse than being blind, estimated to be 76 percent as bad as being dead (for one year that you normally would have been expected to live). From this we might extrapolate. Is being mildly clinically depressed, say, *twice* as bad as being generically miserable for a year? If so, I estimate that a year of being generically miserable is 7 percent as bad as being dead for one year.

DISABILITY AND FAIRNESS

One critique of the use of DALYs is that it feels like it is unfair to people who already have a disease.[17] Here's how it works: suppose we have two groups of people in danger of getting diabetes, and you have an intervention that can prevent this. The first group is otherwise healthy. The second group is exactly like the first group in every way you can measure, except that they are also blind. Which group should you help?

The paradox is that we have two different, sensible conclusions. First, we should help the blind group, because they are already suffering and helping those *most in need* sounds morally justified. On the other hand, if we calculate the value of, say, ten years of life for a blind person, it's only got a value of averting six DALYs, not ten, because of the 0.6 disability adjustment for blindness. So a naive DALY calculation points us toward helping the otherwise healthy group because they would enjoy the rest of their lives more than people in the blind group would.

I'm not the only one who finds this disturbing. However, I should note that this is really only a theoretical problem at this point—actual analyses using adjusted life-years do not take into account preexisting medical conditions, so in practice this problem does not happen, for now. However, it's an interesting theoretical problem that might need to be solved in the future. Similarly, nationality, race, education, etc., are also not considered, only age and gender.[18]

Is there a way out of this? I have an idea, but it might be wrong because it hasn't been studied. Remember how money is worth more to poor people than to rich people? That is, giving a poor person $1,000 gives them ten times more happiness than giving it to someone ten times richer? Could the same thing be going on with health? That is, could it be that the loss of happiness or utility is greater if you're already suffering?

To my knowledge, this has not been studied. I did an informal poll of my friends on Facebook about this: "Lets say you were going to be blind for a few months, and have a broken hand that would take a few months to heal. I'm taking a poll—click 'like' (thumbs up) if you would rather have them at the same time, and click the heart if you'd rather have them one after the other."

Thirty-seven people responded, and it was split right down the middle. This seriously needs scientific study.

Nor does the DALY take into account extreme happiness. The value of one healthy year is the same as the value of an ecstatically happy year. (Even the flip side of the DALY, the Quality-Adjusted Life-Year, or QALY, considers a healthy life as good as it gets. But is living an extraordinarily happy life for one year worth *more* than 1 QALY?) So the DALY isn't perfect. However, it has the advantages of being in common use, which allows us to compare interventions.

What is more tragic, the death of an infant or a twenty-year-old? According to the standard DALY measure, it would be the infant, simply because they'd have more years taken away from their lives. But there are reasons to believe that a twenty-year-old's death is more tragic than an infant's: society has invested more into their capabilities, they have a network of people who care about them, they have memories, and a story that ends. Some have argued that all intuitive moralities agree that the value of a life increases through pregnancy and childhood, and peaks in adulthood—that is, a death at twenty is *more* tragic than the death of a child.[19]

For different reasons, some think that living from age twenty to twenty-one is more valuable than living from eighty to eighty-one. Maybe all people are *created* equal, but then their value changes as they age, first up, then down. To reflect these ideas, some DALY measures use age-weighting.

Another relevant question concerns the value of lives now versus lives in the future. We tend to value things more now than in the future with something called "temporal discounting." Economists disagree on what the discounting rate should be (e.g., the rate of it and whether the shape should be hyperbolic versus exponential), and there does not, as of yet, seem to be any way to empirically know what it *should* be. For what it's worth, when we test regular people (non-economists) in the lab, they seem to behave as though they are using hyperbolic discounting.[20] But we also know the dangers of blindly trusting our intuitions.

We touched on this in the section on whether or not to have children and the value of people yet to be born. Is averting twenty DALYs next year more valuable than averting twenty DALYs ten years from now? On the one hand, it feels a little arbitrary, and many scholars believe we shouldn't put a discount on the value of future lives. But we might defend temporal discounting on the grounds that the future is so uncertain that treating future people like people now would be foolhardy—we don't know what will change. When we look at what problems people of the past thought were going to be problems today, they are sometimes laughable. Another reason for discounting (money) is that we assume that future generations will be richer, and therefore get a lower value from whatever good we are allocating.[21] Also, in a richer future people will probably live longer, and will have access to better medical technology. Without discounting future lives we should always choose future people over present people, because a year of their lives saved would have more happiness than the life-years of today.[22]

Lots of scholars use temporal discounting for a very practical reason: if we count far-future lives as valuable as near-future lives, it does screwy things with the math. There are just so many more potential people than currently living people that treating them as equally valuable would lead to the conclusion that *all* of our resources should go to them. For example, according to this line of reasoning, we should spend money on trying to cure malaria, and spend no money at all on treating people who currently have it, and no money to prevent current people from getting it. As providing bed nets is such a reliably effective way to save lives, it seems a little discounting is in order.

THE WEIRD WORLD OF TEMPORAL DISCOUNTING

The notion of a discount rate for the future makes intuitive sense. We would rather have a dollar today than a dollar in a year, and it does seem sensible that we would regard the lives of potential, future people with a little less value than people alive right now. Unfortunately, this kind of temporal discounting looks a little bit strange when we apply it to the past, and when we think about the far future.

For example, if we go far enough back into the past, there was a point where a single individual person in the past is worth as much as every single person alive today, because relative to back when they were alive we, as future people, would be discounted. For example, the human population in 2019 was about 7.55 billion. At a 3 percent discount rate, we find that a single person living around 770 or 780 years ago has the same moral worth of everyone living today. Similarly, if we look into the far future, people seem to have very little value at all. Put another way, a single person living today would be worth 7.55 billion people born 780 years from now.

We can actually put this back into money if we want to, but we have to be careful. Money is less valuable to people who already have a lot of it. But if we are looking at how money can be used to avert DALYs for very poor people, we don't have to worry about that so much, because all the money would go to preventing DALYs in poor people. Some organizations are really good at it, and others not so much. We'll talk about what are the most effective organizations, and assume that the cost of averting a DALY is that of the most effective charities. So how much does it cost to avoid a loss of a year of human life for the most effective charities? $78 USD.[23]

IMMORTALITY

This book explores how to maximize good feelings and minimize bad feelings. One obvious fact is that, for biological beings, a requirement for feeling good is simply being alive. Although we might argue that some people have lives so bad that they are not worth living, we tend to think that these people are rare, and in any case it's not for us to decide that they should die. In general we think that lifesaving can be expected to increase good feelings.

Assuming a year of life is worth living at all, having more years of life is better than fewer. Almost everybody believes this, because almost everybody uses medical interventions to help prolong their lives and the lives of those they care about. Living a long, happy, healthy life is something that really matters.

But are there limits to this? Although most people are happy with the idea of living longer, the idea of living forever makes them a bit uncomfortable. Why?

The first thing many people say is that living forever would be horrific because you'd get more and more weak in body and mind. But it's not as though medical science is trying hard to extend life without also extending general physical and mental health. Perhaps instead of "immortality" we should use the phrase "eternal youth" to clear up this objection. Of course, it's not really eternal; you could get killed by buffalo trampling or poisoned brussels sprouts. What we mean is that you would not die of old age, disease, or for any reason having to do with your own biology. About half of people say they'd be fine living as long as their health was still good. People generally feel better about living longer if they expect old age to be a good experience.[24] This suggests that as medical science improves, and people have more rosy outlooks on what life will be like when they're older, they will want a still longer life.

The resistance to eternal youth, then, probably boils down to the naturalistic fallacy. The idea of living "forever" feels unnatural, and many things we label as unnatural we also feel is morally wrong. This is worth questioning. If people should not have eternal youth, then how long is enough? The answers people give tend to be based on their own life expectancy. In rich societies today it's not unusual for people to live well into their eighties. This was much more uncommon in the past. People are less comfortable with the idea of living to be 400 than they are with the idea of living to 120.

But people's opinions on how long life should be aren't going to affect whether or not we get continued life extension, because our species has made an implicit, collective decision to live as long as possible. Each medical scientist is working in their own little niche, but collectively this means that science has and will continue to extend our life expectancy. There will never be a point where all scientists will collectively say, "Okay, we're good. No need to try to prevent death anymore!" Scientists build on each other's progress, but they are not a coordinated group. Even if most of them wanted to put on the brakes, there is no organization to arrange the solidarity.[25]

The effect of this is that science will continue to increase life expectancy as much as possible, *even without any overarching plan to do so.* As they do, their ideas of what is natural and unnatural will change, making living longer and longer gradually more acceptable in people's minds.

Humanity might get eternal youth without even having an explicit goal to get it.

We started by talking about how to be a good person, and we ended up talking about saving lives and fighting disease. Is this okay?

One assumption I'm making in this book is that doing good means maximizing good feelings and minimizing bad ones. In our day-to-day lives, in industrialized countries, this brings to mind being nice to people and not acting like a jerk. But on the world's stage, hurting people's feelings with curt remarks doesn't matter that much when you can do things to save people years of life from disease.

Not everybody agrees that we should be focused on improving good feelings and reducing bad feelings. Many scholars, for example, like to focus on preference satisfaction. But if we're looking at how to make this world a better place, this nuance makes little difference. Because just about everyone agrees that a person living a longer, healthier life both increases good feelings and satisfies preferences more effectively. I have my own philosophical take on these matters, but these conclusions hold even under different views.

All of this talk about numbers and money might feel like you're swimming in foreign waters. It certainly was to me when I first heard about it. Isn't being good more about how nice you are to the people around you, and gifts you give, and not being mean at work? That's what our monkey brains would suggest. The monkey brain that evolved to only deal with a hundred and fifty or so people around us. The monkey brain that evolved to keep us alive and reproducing, and not necessarily maximizing the good of the people around us. This is why, in the modern world, when we have better information to understand what it really means to be good, and in a globally connected world where we can make a difference to people far away from us, thinking in terms of numbers and money is an adjustment, a difficult thing to swallow.

Being nice to the people around you is easy for some people, and more difficult for others. Some people are naturally irascible, negative, and lash out—not the kind of people their dogs believe they are. How much effort should one take to be nice if it doesn't come naturally to them? If it takes effort, it very likely takes effort away from more effective efforts at being good. However, being nice to people around you has instrumental value. If you're nice, people might look up to you more, and be more willing to emulate your goodness that matters. They might help you with your goals, and donate more when you hold a funding drive for buying malaria nets.

8

CHOOSING A CAREER

Let's look at some advice you often hear about career choice. One is "Follow your passion." The benefits of doing work in a field you're passionate about is that you have a higher chance of loving what you do. You would also be less likely to burn out, and probably more effective at the work you do. But there are problems with this advice.

One is that lots of people are passionate about the same things early on in life, and other fields get neglected. Lots of people just leaving college are passionate about playing video games, comedy improv, and watching sports. Among students, 84 percent had passions, but they tended to involve sports, music, and other arts.[1] But there are far more people who want to do these things than there are jobs. Only 3 percent of jobs are in sports or the arts. This means that most of the people who follow their passion will not be successful in working in that area. It also means that the jobs will be highly competitive, and thus lower paying.

A good modification of the "follow your passion" rule doesn't sound as good, but has much more wisdom. Follow one of your passions that balances how passionate you are about it with how many other people are passionate about it, and how much people are willing to pay for that skill. If, for example, you are kind of passionate about working with numbers, risk, and probability, but what you *really* love is composing country music songs, you just might have a better life becoming an actuary and doing country music in your spare time than the other way around.

A further modification has to do with the idea of passion at all. In the Western world we tend to think of passion as something you're born with, or have to discover. Some people seem to be like this. Famed athlete Wayne Gretzky seemed like he was destined to be a hockey player since he was three years old.

But this isn't always true, and other cultures recognize that. Sometimes you become passionate about something that you get good at. Your parents might force you to practice piano when you're young, and when you're young you might hate it. But when you get really good at it, you just might love it in adolescence. Even Gretzky had a father who encouraged him and pushed him to succeed.[2]

What you're passionate about and what you're good at are usually, but not always, the same thing. You can be passionate about singing, but not have a particularly good voice, or a voice that tires easily.

The lesson is that the ideas of what you're passionate about when someone is young and not good at anything yet should not be treated as some indication of a person's authentic self. The self is something that grows over time. You might think of your passions as something that are core to your being, perhaps even genetic. But much of it is probably a result of your culture—as historian Yuval Harari puts it, "a combination of nineteenth-century Romantic myths and twentieth-century consumerist myths."[3] Your passion for travel is a modern value that you got from your culture. This is not to say that culture isn't a legitimate part of you, it's just to emphasize that culture is learned, and you should remember that what you'll learn as you age will change who you are—what your authentic self really is.

Keep in mind what you want to do, what you're good at, and how hard that job is to get.

Studies also show that job satisfaction has less to do with matching your passion to a job and more with how good the job is in a more or less objective way. The best jobs allow you some independence, have variety, give a sense of completion and positive contribution, and make it easy to know if you're doing well or not.[4]

Most of us have to work to earn money to live, and most of us spend around forty hours a week doing it. But just because you need to earn money to live doesn't mean that work can't pay off in other ways.

No matter what kind of employment people have, studies have found that workers have one of three basic mindsets about them: it's a job, it's a career, or it's a calling.[5]

If you look at your work as just a *job*, then you see it only as a means to money. You might look forward to the weekends, and have other ambitions in your life, like mastering calligraphy, volunteering, or completing your Beastie Boys record collection. If you see your work as a *career*, you're more ambitious about it. You see your job as a part of a larger plan in your life in which you advance. You seek prestige and promotions. Those who see their jobs as *callings* find it intrinsically fulfilling, because they are doing something they genuinely want to do, or feel is important.

I'm lucky in that my job is my calling. I think I'd do cognitive science even if I wasn't getting paid to do it. But you don't need to be a university professor to have a calling. Amy Wrzesniewski found that there are people with these attitudes toward their work in almost every profession. Even among custodial staff, there were people who were just punching the clock, and those who saw their work as an important part of a larger organization. A hospital worker, while cleaning up vomit, might see her work as a part of healing people, and find it very fulfilling.[6]

To the extent that you are able, you might be able to change your mindset so that the work you do is more of a calling. Assuming that your job involves doing *something* good for the world, and you're not working as an advertiser for a cigarette company or something, you might be able to nurture the idea of your work as a calling in your mind. Think about what your field is doing for the world, and how you are a part of a larger force for good.

Admittedly, this is harder for some professions. Journalists, for example, often feel that in the workplace they are unable to live by the ideals that brought them into the field in the first place. It's the sad truth that many people go into environmental law to help the environment, and refugee law to help refugees, but most of the jobs in these areas are defending companies who are trashing the environment, or working to keep refugees out of the country. In those professions, the jobs that work toward the values of the employees are rare, coveted, and as a result pay much less.

It's an unfortunate fact of our modern world that there is a correlation between doing good for the world and getting paid less. Jobs where you get a visceral feeling of helping people or the environment, like teaching, social work, and so on, tend to be low-paying jobs. This is one of the reasons women, on average, are paid less than men: women tend to choose meaningful jobs where they are helping people, and men are more likely to do unmeaningful work that they might hate so they can get a bigger paycheck. We end up with the situation where women get paid less, but are happier with their jobs.

If you can't *make* your work a calling, it might be that another job would be better for you. It's possible that some people would not find their work to be a calling no matter what they did. But if you have a calling, you'd have a better life if you had work that channeled it.

There are some jobs that seem great because of the effect they have on the world, or because they are associated with some ideal, but the day-to-day activities of the job aren't very pleasant. This mismatch between what people think a job involves and what a job actually involves cause a lot of people to get into areas of work that they shouldn't. They think that wedding photographers spend most of their time snapping pictures of happy couples, and don't realize that the vast majority of their time is spent alone, editing photos on a computer. Luckily, the cure for this is simple: chat with a few people in the profession you're targeting and ask them what their day-to-day work-life actually is. This will wake you out of your television-induced delusions of what certain careers are like.

Truly fulfilling work involves flow states for the things you do day-to-day, as well as a feeling of meaningfulness about what the work is doing to the world. These two things correspond to the two main aspects of happiness: pleasure and life satisfaction.[7]

Having your work be a calling is a win-win-win: you make money, you are happy, and you're doing good for the world.

Choosing a Career on Moral Grounds

A good starting point for choosing a career on moral grounds is to decide *how* you want to make the world a better place. There are two basic ways to think about this. The first is that you want to create something that is good. Maybe you will try to cure malaria, or work to encourage good governance.

The other is that you choose a problem and try to reduce its magnitude. There is no shortage of problems in this world. But we can break the most important problems up into a few classes: one is human suffering, another is animal suffering, another is prevention of future catastrophes, and the last one is environmental protection.[8]

Of course, these problems have fuzzy boundaries and overlap with each other. Preventing global climate change, for example, has ramifications for human and animal suffering. But one of the reasons to break it up this way is that it's difficult to compare a charity that helps one to a charity that helps another. For example, is it better to save rainforests, which helps prevent climate change and increases species diversity, or is it better to save people's lives with malaria nets, or is it better to prevent factory farming so that animals don't suffer? These apples and oranges problems have yet to be solved when it comes to occupation choice.

The bright side of this is that you can choose whichever one you like the best! Let's say, for example, you're particularly interested in preventing global pandemics.

At some point, a pandemic might wipe out large amounts of people. No good. What kind of things can you do to try to prevent global pandemics? Well, one consideration is your skill set. You can do research to try to find cures for infectious diseases. This might be a career path you can choose if your skills are in science and biology. You can also work for a nonprofit that creates programs that prepare communities for these events, or distribute vaccines. This is the kind of thing that someone with social and organizational skills can do. If you're good at writing, then perhaps you can be a journalist, or be in some kind of a communications job where you can raise awareness for the problem so that other people can help. If you're into politics, say, a lobbyist, you can try to influence government policy so as to reduce the probability of devastating pandemics. If you're not good at any of these kinds of things, then you can try to get a job where you earn a lot of money, and then just donate money to the charities that are trying to prevent pandemics in all the ways described above.[9]

The lesson here is that even though the problem of global pandemics is in some sense a biological and sociological one, it doesn't mean you need to be an expert in biology or sociology to be able to help with the problem. Lots of problems are multifaceted, and involve research, government, communication, and so on. You can be creative, and try to figure out how you can best use your skills and interests to work toward solving the problem of your choice.

Furthermore, it's really important that you enjoy your work. When you're twenty-two years old, you might be full of vigor and want to force yourself to do something for the good of the world that, fifteen years later, you might not like anymore. Recall that burnout is a real problem. If you're unable to do your job because you're so miserable, then you're no good to anybody. So if you choose a job that does good for the world, make sure that you're reasonably happy doing it, or else it will be unsustainable.[10]

Let's look at a job everybody thinks is good for the world: being a doctor (not the kind of doctor I am, I mean the kind that actually helps people). If you add one doctor to America, for example, there will be an estimated four lives saved over the course of the doctor's career. Is that a lot? Well, to me it sounds low, but part of the reason it's not higher is because there are already a lot of doctors. Does this mean that if you work as a doctor, you'll save an estimated four more lives? Actually, no, because your becoming a doctor does not mean that there will be one more doctor, because if you had not become a doctor, someone else would have. Medical schools are competitive places with only a few spots. If you hadn't taken one of those spots, someone else would have. Let's assume that the selection criteria these schools use is pretty good. This means that if you got the spot, you're going to be a better doctor than the person who would have taken your place. Fair enough. But when you're looking at the lives saved by becoming a doctor, you need to look at the difference between the number of lives you would save versus the number of lives the person behind you line would have saved. When you do this calculation, we find that becoming a doctor only saves one or two additional lives than would have been saved by someone else if they had become the doctor instead.[11]

Being a doctor has benefits for you, such as prestige and (often) a high salary. But if we're thinking only about the good contribution to the world, is all the stress, sleep deprivation, and up-front financial cost worth one or two additional lives saved? As we will see, the best way for even doctors to maximize the number of lives they save is by donating good chunks of their salary to effective charities.

Earning to Give

When people want to do good for the world, they often think of careers that help people directly: teachers, doctors, social workers, and working for nonprofits. There is also an idea among do-gooders that earning a lot of money is distasteful.

But another way to look at it is that you might be doing more good by earning obscene amounts of money and giving it to effective charities. Is it better to work for very little as a social worker, probably helping people in rich countries one at a time, or working in the financial sector, and donating tens of thousands of dollars to charities, and saving several lives a year? The concept of "earning to give" turns many ideas of doing good for the world on its head.[12]

Philosopher William MacAskill introduces a hypothetical newly educated doctor, Sophie, interested in helping the world, which I have adapted for this book. She has two tempting job offers. One is to work as a doctor in the developing world for a charity Doctors Against Sickness Greatly Undermining Death, or DASGUD. This job would pay her $50k annually, and let's estimate that if she took this job she'd treat lots of people and save the equivalent of ten lives per year, because more lives are at risk in the developing world, and doctors there can do more good than doctors in rich countries. Her other offer is working at a hospital in Atlanta, giving some of the richest people in the world nose jobs, for $150k per year. Let's assume that the medical work she would do in Atlanta saves no lives. Because she wants to maximize her goodness, if she were to take the high-paying job, she'd live on $50k and donate the other $100k to charity every year, which would save twenty lives yearly. Let's assume that she'd be equally happy in either job, and we're only comparing these jobs in terms of the magnitude of her positive effect on the world.[13]

In both cases she'd be living on $50k per year. Which job should she take?

There are some good reasons for her to take the high-paying job. One is that the $100k she'd be able to donate would pay for two other doctors if she donated that $100k to DASGUD, which, you will recall, pays their doctors a $50k salary. So right here we see that she'd have *twice* the positive effect with the Atlanta job, even if she spent her days doing nose jobs for rich people.

Another reason is that the donated money can go wherever she thinks it would be most valuable, and changing her mind about where to put her money is very inexpensive. Let's suppose that when she enters the workforce Sophie thinks that being a doctor is a great way to help the world. If, five years later, she decides that she now thinks investing in education is more effective, or helping the environment, or helping animals, or doing medical research, she can easily start pouring her $100k somewhere else. If she took the charity job, she'd have to switch jobs, which is quite expensive in terms of time. She's also a doctor, and because of her skill set might not be able to contribute very much to, say, an environmental cause.

A third reason is the "replaceability" argument. Charity jobs are highly competitive. This means that if she doesn't work at DASGUD, someone almost as qualified as her will. Let's say Sophie working at the charity job would save ten lives per year. How many lives would be saved if the next best candidate, Candace, had the job instead? We can assume that, on average, the best person was hired for the job, so the next best person for the job would be slightly less effective. So let's say that Candace would save nine lives per year instead of Sophie's ten. Still pretty good! If Sophie doesn't take the Atlanta job, someone else will, too, but this person will probably not donate $100k of their salary to charity (most people donate little, and what they do donate usually goes to relatively ineffective charities).

This isn't something that people think about a lot, but it's important: not only should you think about the effect of your actions, but you need to think about what will happen as a result of your *not* choosing the other options (opportunity cost). The added good to the world Sophie would do in the charity job is not ten saved lives, but one: the difference between her taking the job and Candace taking the job. In this case, not choosing the charity job results in a minor difference in lives saved (Candace would cure one less person than Sophie would), but not choosing the Atlanta job results in a big difference.

What you might think at first is that the charity job saves ten lives and the Atlanta job saves twenty, but when you consider replaceability, you look at the difference between what Sophie would do in one job versus what her replacement would probably do. When you do this, her taking the charity job only would save *one more* life per year, and taking the Atlanta job would save *twenty more*. On this reasoning, she'd be doing twenty times as much good in the Atlanta job.[14]

The same goes for my own job. I'm a university professor—a highly competitive job. I could pat myself on the back for doing science and educating people, but if I didn't have this job, somebody almost as good as me would. Or someone better than me, as I sometimes think when I'm suffering from imposter syndrome.

At the beginning of your career, it's more important to get job skills and credentials than to try to make an impact right away. This is because you will still be learning about what you want to do, and working will help with that. People change jobs every few years, so don't think of the first job you take as set in stone. Also keep in mind that it's easier to move from for-profit to nonprofit than the other way around.[15]

Being a doctor is something people generally think of as a do-gooder job. How does this reasoning apply to morally problematic jobs? Let's suppose Kurt has a job offer at a factory that raises pigs for consumption, and the pigs there are treated terribly. Kurt's job there would be to make the company more money, which would, in many cases, be bad for the pigs. Kurt cares about animal welfare, and could also work at a nonprofit that tries to encourage companies to have more ethical practices.

The meat company pays much more. If he donated a substantial portion of his earnings to the animal welfare nonprofit, he could potentially do much more good for animals by working at a factory farm, as weird as that sounds. I'm not saying he necessarily would, just that it's possible.

I'm sure some of you are getting your backs up over this. Recall from the section on moral psychology how we're very harm-focused, and we are prone to be morally harsh on people who cause harm, and it takes a lot to make up for causing harm. That is, saving a life doesn't make up for murder, in most people's minds. Although we can try to resist this bias, it's tough. And the person actually doing the working has to live with herself.

MacAskill suggests a real-world example to help us feel these situations to be more acceptable: Oskar Schindler. Here's somebody who's not working for a factory farm, but *manufacturing bullets for the Nazis so they can kill innocent people and try to take over the world*. But he used the money he made from it to save the lives of an estimated 1,200 Jews. He's on everybody's list of heroes.[16]

Still, it can be hard to work for a terrible organization, even if you can rationalize that you're doing more good than harm by doing so, due to your donations. One way to feel better about it is to bring back the principle of double-effect. I brought this up earlier in the context of moral psychology. I don't know whether it's something we *should* hold as a moral principle, but you and I both probably do. It's the belief that doing harm isn't so morally bad if it's a side effect of what you're trying to do, rather than the intention of what you're trying to do.

An example from philosophy is the "tactical bombing" case. Murdering 200 innocent people is pretty bad. Luis Garavito, the worst serial killer in history, was responsible for 138 proven victims, and probably killed more.[17] So any action that results in the death of 200 innocent people is comparable to the effects of most of the world's worst serial killer. Suppose a nation could end a bloody war by bombing a weapons facility, but we can pretty confidently foresee that 200 innocent people would die. But ending the war would save thousands. In the tactical bombing case, the nation is not killing the innocent people *to end the war*. The intention is only to destroy the weapons, and the innocents who die are a side effect of an otherwise moral action. Many people would say that the tactical bombing case is morally justified for this reason. But if a doctor killed one innocent to harvest organs to save five others, we feel differently about it, because the innocent is being killed to save others.

If this difference is meaningful to you, then it can help you feel better about harming the world in your job so you can earn to give. The point of taking the job with the mining company isn't to destroy the environment, and the point of taking the factory farm job isn't to hurt animals—that's a side effect. The reason you're doing it is to earn money so you can do philanthropy.

Some large companies donate money to charities, and some of those offer matching donation programs, such that you can donate a part of your salary and the company will match it.[18] This is an effective way to amplify the impact of your own donations, but it must be tempered with consideration of the effectiveness of the charity the company chooses. If it's a really ineffective charity, even given the amplification it might not be as good for the world as just donating your money to an effective charity without any matching program. If you work at a company, you might try to start a matching program if there isn't one, or try to steer the company toward more effective causes.

The judgment of what job you should take depends on, as always, the numbers. If Kurt is good at his factory job, let's assume that it means some number more pigs a year have terrible lives. By working at the nonprofit, he would contribute to saving some number of pigs. By donating, he's saving some different number.

Coming up with these numbers in an individual's life is often not easy. Years ago, it was impossible.

Luckily, there's a website dedicated to helping you choose a career based on ethics: https://80000hours.org, based on the estimate that you will work about 80 thousand hours over the course of your life.

What about research? Many of the advances that makes us capable of doing good today are the result of people who did not spend all of their time trying to feed the world's hungry or keep them from getting malaria. If nobody did anything but try to help the world's poor directly, we'd have to postpone every other endeavor of value. If we had done this from the beginning, we'd have never invented writing, or high-productivity crops, or smallpox vaccines.[19]

This suggests that if you have the capacity to be a researcher, then you might well consider becoming one. What should you research? Although it's tempting to focus on practical problems, most of the great advances in science were the result of what we call "basic" research, which is research that is not aimed at solving any particular practical problem. For example, in the 1800s Boolean algebra was invented because George Boole thought it was kind of interesting. A hundred years later it became the basis of all computing technology. It turned out that a key to understanding the Southwestern fever depended on a deer-mouse population study that was conducted because the scientists were interested. They had no idea of the practical applications later down the road. If you wanted to find a cure for disease, would you fund a researcher looking at mouse populations in Arizona?[20]

Working to find solutions to long-term world problems is something that very few people are capable of doing. If this an option for you, it's probably the best choice of career you can make. For everyone else, thinking about the good done in the work you're doing, tempered with the amount of good you can do through donation is the best way to choose what field to go into.

9
MEASURING GOOD DONE

Now that we have an inkling of how to measure goodness in terms of healthy human life years saved, we can return to the question of how one knows the effectiveness of our actions. If we assume that things like race, religion, and nationality of a person don't make people's lives any more or less valuable, and all we care about are numbers of healthy human life-years saved (more specifically, DALYs averted), then we can look at the charities we might or might not contribute to, and see where we can get the most bang for our buck. Just as a wise person would do their homework before making a financial investment, we can choose charities that prevent the most DALYs.

On the one hand, the charity with the most money has the potential to avert the most DALYs. But we can't assume that all charities are working at the same efficiency. What we want to know is our return on investment, so to speak. That is, we want to know how many DALYs are averted for every dollar we put in. (I'm going to speak in terms of American dollars for this discussion, but the reasoning is the same for any currency. I'm choosing American dollars because I assume most readers can more easily mentally convert their own currency into USD than any other currency.)

Charity's Dark Ages

In the past, this information was completely unavailable. I'll call this charity's Dark Ages because we were really in the dark about what good they

CHAPTER 9: MEASURING GOOD DONE

were really doing. This is not to say they did no good, we just didn't know how much. Not only did the giving public not know, but the charities themselves didn't know, and they didn't take any steps to find out.

Measuring the good the charities actually do is an embarrassingly new concept. Before the early 1990s, nobody had done it at all.[1] This might sound unbelievable, but their reasoning wasn't completely brain-dead: finding out how effective you are takes resources that could be spent on actually helping people! So they preferred to have faith that what they were doing helped, rather than investing time and money into finding out how much (or indeed, if) it did.

Once, I was talking to a scholar who was (and still is) one of the world's experts on computer science education. He had a teaching method he advocated and used in his university classroom. I asked him once if he wanted to run a study, where he taught one class in a traditional way (the control group), and another class in the new way. Then he could look at learning outcomes and see how much better the new way was. He said he didn't want to do that, *for ethical reasons*. He was so sure that his way was better that he thought it wouldn't be fair to the students in the control group to teach them in the traditional way.

He had a cold at the time, so I sold him some leeches to help him get over it.

Such is the reasoning behind why charities don't want to measure what good they're doing. The big problem with this is that when you do make the effort to measure the amount of good done, it turns out that some charities matter (they are doing some good), other charities unmatter (they're doing nothing at all), and some charities even antimatter (they're doing harm).

The measurement of the effectiveness of charities was started by Michael Kremer and Rachel Glennerster around 1993, who worked with a charity to try to improve education in Kenya. If you wanted to improve education for Kenyan kids, what would you do? Think about what your ideas would be before reading on.

Got a list? Whatever's on that list, these two probably tried it. They tried providing textbooks. But that only helped the students who were already doing well. They tried hiring more teachers to reduce class size. That didn't make any difference. Over and over, the things they tried turned out to unmatter when they measured the outcomes. I want to emphasize here that these were all interventions that common sense suggests would be really good ways to improve education.[2]

Okay, but what about the charities that actually antimatter? What are these antimatter charities and what kind of monsters are dedicating their lives to running them? They're not just charities working toward something you fundamentally disagree with (such as pro- or anti-abortion charities), but ones that are trying to do things that are kind of universally agreed-upon as good.

A pioneering team decided to try to measure whether their own charity was doing any good. The famous Cambridge-Somerville Youth Study looked at a treatment that seemed very promising: they would find children at risk and mentor them. These mentors helped with psychiatric problems, financial problems, health, etc. What was new was that they compared these treated kids with a control group—kids in similar situations who got no treatment. This had never been done before. They looked at forty-three different ways the kids' lives might have improved over the course of the next thirty years, including health measures, criminal activity, and family issues.

The program did not work. The treatment group was no better than the control group on *any* measure. The shocking finding was that not only did the program not work, it did harm: for seven of the measures, the treatment group was *worse off* than the control. That is, the poor, disadvantaged students who got mentored and helped by people ended up worse off. People are still debating exactly why, but there is little doubt that the well-intentioned, hard work done by these people did more harm than good. This is sobering. The peer-mentoring, as it was done, antimattered.[3]

This is not an isolated incident.

One development group built a fish factory to try to improve the lives of the hunter-gatherers who lived on Lake Turkana. The lake got overfished and the fish stock collapsed.[4]

"Scared Straight" was another well-intentioned intervention that tried to get kids to have better lives. It involved having incarcerated people come talk to kids at school, telling them about why they should not follow the same path they did. It backfired. The criminal life was romanticized and the kids who were "scared straight" had increased delinquency.

Scared Straight was not based on sound criminological principles. The Adolescent Transitions Program, however, was. It promoted self-monitoring, communication skills, and pro-social goal-setting. Yet even in this program, delinquency was increased three years later, presumably because grouping problematic people together undermines the treatment.[5]

It's hard for a charity to know what to do, and it takes a lot of extra work and money to see if whatever they're doing works. Sometimes, for educational, criminal, or health interventions, these measurements must be done years after the intervention.

I hope these examples have seriously undermined your trust in any charity that doesn't do evaluation. Doing charity well isn't easy.

Let's get back to the Kenyan schools. Kremer and Glennerster finally found something that worked: based on a tip from a friend at the World Bank, they tried something that didn't seem to have anything to do with education: deworming. It worked really, really well. Not only did it improve education, but it generally helped with the health and economy of the area.[6]

What I love about Kremer and Glennerster's story is that this solution, which was very cheap, and had such amazing benefits, isn't something that most of us would think of when we try to imagine improving education. (Was de-worming, or any health intervention, on your list of ideas of how to help the education of the Kenyan children?) People who study education often don't learn much about disease, and people who study disease often don't learn much about education. But in this case all the common-sense interventions unmattered, and preventing disease mattered.[7] Do you care more about health or education? In this case, it didn't matter: helping health was better for both causes.

These new evaluations show that the differences in effectiveness between the best charities and the merely good ones can be huge—the best being sometimes *hundreds* of times more effective.[8]

We are living in an exciting new era where we now have the tools and the motivation to do better than using our instincts about how to make this world a better place. When it comes to donating money to charity, the way is pretty clear: choose a charity that, according to evidence, averts the most DALYs per dollar. What charities do that? You can always look at GiveWell.org for their top recommended charities. At the time of this writing, some of the best charities are the Against Malaria Foundation, GiveDirectly, and Evidence Action's Deworm the World Initiative.

But humans aren't the only creatures that can have good or bad feelings. Let's turn to animals, and see what we can do there.

10
ANIMALS

Animals are incredibly diverse. Agreeing that animals have *some* moral standing is one thing, but do they all get the *same* moral standing? A 2018 survey of Americans showed that 47 percent of them believe that animals should have the same rights as people in terms of being free from harm or exploitation.[1] I'm not doubting that people respond that way when asked directly, but I'm also pretty sure that even those people tend to be more upset by a moose dying than a goose.

When we think about what kinds of things have moral standing, one characteristic that's pretty much universally agreed upon as important is consciousness. That is, most people and ethicists believe that having some ability to consciously experience pleasant and unpleasant mental states (feelings, as I'm referring to them in this book) is sufficient for some kind of moral consideration.[2] This is the reason we think that a rock doesn't have moral standing, but a penguin does. When we look at humans, deer, mice, and flies, many of us get a strong intuition that as we go down the list there's less suffering capability. How does science weigh in?

Consciousness

Let's assume that suffering requires conscious experience. That is, if something is harmed, or its goals are frustrated, it doesn't count as suffering unless it actually feels like something—or, as we say in cognitive science, has some subjective experience.[3] Even if we might say that a Roomba has

CHAPTER 10: ANIMALS

a "goal" to clean the floor, we don't think it's an ethical violation to prevent it from doing so, because the Roomba can't actually *feel* frustration. At first blush, consciousness seems to be an on-or-off kind of thing. "Creature consciousness" is the question of whether beings of a particular kind are capable of being conscious of anything at all. The idea is that some entities, such as peanut butter, do not have creature consciousness, and others, like humans, can be conscious of things. This raises the question of which entities have creature consciousness and which do not. Do mice? Do flies? Paramecium? Most relevant to this book is the capacity of a creature to consciously feel positive and negative mental states, which is often referred to as "sentience."[4]

But when we look closer, it's clear that even for a "fully" conscious human being, there are lots of brain processes going on that she's unaware of. It's not enough to say that she's conscious. It matters what she's conscious of and what she isn't. She might be conscious of her feeling that she wants to sit down, but unconscious of the edge-detection processing going on in her early visual system, or how her mind puts words together to make grammatical sentences. This is "state consciousness": you're conscious of some states of your mind, and not others.

Most cognitive scientists, though, don't think of consciousness of even a particular thought or feeling as being simply on or off. Our consciousness of any particular thing is graded. There are things that we are barely conscious of, and other things we feel very conscious of, suggesting that there are levels of consciousness. When we're anesthetized, we can experience reduced, but not always eliminated, conscious states. Perhaps state consciousness is a continuous variable—more like a dimmer than an on-off light switch.[5] So even if we think mice have creature consciousness, we have the further question of *what* they can be conscious of, and *how* conscious of things they can be.

Although we cannot get into the heads of other creatures, we do have several reasons to think many of them are sentient to some extent. First, the feeling of pain and pleasure solves an evolutionary "design problem" that almost all living creatures share: getting the organism to do things that are good for it. How do you make a being eat food that's nutritious for it? Make it pleasurable to eat it. How do you make a being reproduce? Make sex pleasurable. How do you make it avoid harm? Make bodily harm painful. Some animals, like mussels, are not able to move later in life, so we assume that they don't feel pain. The sea squirt begins life like a fish and swims around until it finds a good place to secure itself, then it sticks to a rock and eats its own brain, presumably because it doesn't need it anymore. For a sea squirt, pain has an adaptive value only for the swimming part of its life. So although it *might* feel pain later, without a brain, it's unlikely, because maintaining the ability to feel pain would incur the kind of needless nutritive cost that evolution tends to be good at eliminating.

The fact that so many different kinds of creatures need to avoid some things and approach others suggests that some mechanism for motivating behavior would be, evolutionarily speaking, very old. This means that it's likely that conscious feelings of pain and pleasure probably evolved a long time ago, and most creatures alive today who descended from this ancient population of critters probably inherited sentience, too.

And indeed, the brain structures, neurotransmitters, and brain behavior related to feelings appear to be somewhat consistent across many species. For example, the brain activity and structures that are correlated with pain in humans match up pretty well with other creatures. Finally, creatures of many species *behave* the same way when put in situations that we would expect would cause pain and pleasure. They cower and rest when hurt, and get energetic when well-fed, healthy, and rested. Mammals even have similar facial expressions to human beings in different emotional states. Anesthetic drugs (painkillers) make creatures of many different species, even insects, behave as though they are in less pain, putting in less effort to avoid harm.

Some people take extreme views, saying that the suffering of any human or animal is equivalent, or, at the other extreme, that human capacity for suffering is so great that animal suffering is negligible or nonexistent. But most of us are in the middle somewhere. But let's take a look at the possible views. I break them into four categories. The Unconscious Animal Theory holds that nonhuman animals are conscious of nothing at all. The Muted Animal Theory holds that animals have consciousness, but experience the same event with less intensity. The Same Pain Theory holds that the same event, like breaking a femur, causes the same exact amount of suffering in a human as any other animal. And finally, there is the Tinker Bell Theory, that nonhuman animals actually feel *more* suffering than humans given the same event.[6] Let's talk about them.

Unconscious Animal Theory:
Nonhuman Animals Feel Nothing

We might prefer a human life to a pig life, but would we prefer a human life to 100 pigs? Or 1,000 pigs? Infinite pigs? If animal lives are worth nothing, it could mean two things. First, it could mean that one believes that animals actually can't suffer. That is, they are morally equivalent to rocks and water: you actually *can't* hurt them in any moral way, because they cannot suffer.

The other thing it could mean is that animals are capable of suffering, but we shouldn't care about them morally.

Thinking about a human *death* is a little extreme, so let's think about a less severe form of suffering, say, stubbing your toe so that it mildly hurts a little for a minute or so. If one's view is that animals cannot suffer, then one can test how sure one is about this view by asking: would you rather have the experience of a human stubbing their toe or the suffering of 10,000 pigs slowly starving to death? Believing that animals cannot suffer means that anything done to animals to cause the slightest bit of happiness for a human is worth doing. For example, if a psychopath gets a thrill out of torturing dogs, this position would have to conclude that, aside from upsetting other people who hear about it, there is nothing morally wrong with it, as it is giving happiness to a human, and the dogs feel nothing anyway. If these ideas makes you hesitate, then maybe you're not so sure that animals don't feel anything after all.

Although believing animals cannot feel pain has a long intellectual history, it is certainly out of touch with modern sensibilities. But there are still a few scholars who believe that nonhuman animals feel nothing at all.[7]

Muted Animal Theory:
Simpler Animals Feel Things Less Intensely

This means that the simpler the animal is, the less intensely it feels anything. Simplicity can be defined in several ways. I'll talk about this at length later.

Same Suffering Theory:
Nonhuman Animals Feel Just as Intensely as Humans

This idea, which seems very plausible to some, holds that the capacity and intensity of animal consciousness is the same as that of human beings.

We might think of an equivalent action, like having a leg torn off, and speculate that a human, cow, or spider would feel the same amount of suffering.

**Tinker Bell Theory:
Nonhuman Animals Feel *More* Pain than Humans**

One thing we don't often think about is how simpler animals might have the capacity to feel *more* pain than complex ones. How could this be possible?[8]

For one thing, people can contextualize their emotions. For example, being in prison might suck, but knowing you'll be in prison for a lot longer is an additional source of suck. On the other hand, if you've been in prison for a long time, and you're getting out tomorrow, you might be happier than many people who are free![9]

Suffering (and happiness) is more complex for human beings. Being in pain because you are assaulted is more distressing than being in the same amount of physical pain in a medical procedure.

Chickens can't contextualize anything. For better or worse, they are always in the moment. A wild animal captured so we can help it is just as terrified as an animal trapped to be killed and eaten.[10] Suffering might be worse for animals, overall, because they can't attenuate it with any sense-making. They are always in the moment, suffering, with no hope of rationalization.

JAINISM AND ANIMALS

The Jainism religion cares about animals a whole lot. Some Jainists famously sweep the sidewalk in front of them as to not kill any bugs while they walk. This comes from their philosophy of nonviolence, ahimsa, which puts restrictions on causing injury to anything alive, from humans to animals, plants, and microorganisms. Many Jainists do not eat potatoes, for example, because digging them up kills microorganisms.

But even the Jainists don't think every being suffers equally. They break living things into categories, roughly according to how many senses they have. Plants, they believe, only have one sense, touch, and as such have less capacity for suffering as creatures like ocelots, who can see, hear, and smell.

In fact, plants actually are sensitive to light, as demonstrated by how your plants will grow in the direction of light. They also are sensitive to chemicals in the air (a primitive sense of smell), know up from down, and can detect temperature changes.[11] Recent studies show that some plants use sound to find water, and others put defensive toxins in their leaves in response to the sound of munching caterpillars.[12]

So even if science doesn't hold up Jainist perceptual psychology, we generally agree with their ranking of beings' capacity to suffer: killing animals is morally worse than killing plants. For Jainists, one should kill *only* plants, and ascetics should avoid even injuring those to the best of their ability.[13] Fruit, for example, is food you can harvest without killing any organism, as are some vegetables that can be taken from a plant without killing it.

Another way to look at it is that when simpler animals experience something, their very simplicity makes that feeling all-consuming. This idea is expressed in the stage directions of the play *Peter Pan*, in reference to Tinker Bell's mental states: "She is not wholly heartless, but is so small that she has only room for one feeling at a time." Think about having a pain, like a sprained ankle. You can distract yourself with movies, talking to people, reading, and thinking complex thoughts. We all know that the subjective suffering when in pain is worse when you are focusing on it. But if you didn't have the cognitive capacity to understand language, and appreciate stories, and to focus on intellectual problems, how would you distract yourself from the pain? Wouldn't the pain be inescapable, all-consuming? If your mind was simple, the pain would account for a higher proportion of the overall thinking that was happening, which could, subjectively, be a much *worse* state of suffering. Could it be that a stegosaurus feels *more* pain than an elephant, or a snail feels more pain than a human?[14]

What we would need to be sure about this are good theories about what parts of brains are capable of conscious feelings. Scientists are still working on this, and there is substantial disagreement. The harder problem is knowing how to compare the intensity of the conscious experience across beings.

So should we assume that all animals, regardless of complexity, feel the same intensity of feelings, or that they are graded in some way, or maybe they feel nothing, or more?

Given the uncertainty about this, it's fair to acknowledge the different points of view and give them all some probability of being true. I'll do this when feasible. But in going deeply into reasoning about ethics, I'll continue on the theory I endorse, the Muted Animal Theory, which holds that simpler animals feel less change in welfare than complex ones, given equivalent events or situations.

HAVING PAIN, FEELING PAIN, BEING IN PAIN, BEING BOTHERED BY PAIN

Human pain is complex, and we can talk about distinct aspects of it. There is, for lack of a better term, biological pain, having to do with nociceptors, aversion, and learning. We share this with lots of other animals. It's possible that some beings have this kind of pain without conscious feeling of it.

Which brings us to feeling pain, which is the conscious perception of pain. This can happen even without activating the biological nociceptor cells (the cells in your body that detect harm), like when you feel pain in dreams or under hypnotic suggestion.

Then there is what you might think of as "being in pain," which is when you focus on the pain and it's all you think about. This might sound like splitting hairs, but sometimes Buddhists talk about trying to get to a state where you can *have* pain without *being* in pain. You are supposed to acknowledge that a part of you is in pain, but try not to let it become all-engrossing. You are able to get some distance from it.

Then there is being bothered by pain. In sexual contexts, masochists might take sexual pleasure in pain. Some people like the pain of getting a tattoo, perhaps because they are focusing on the accompanying endorphins. Whatever the reason, these people actively seek out pain because, in some sense, they like it. They're experiencing pain, but not suffering from it.

Certain nonhuman animals might have biological pain and feel pain, and be in pain, but probably don't have the ability to not be bothered by pain.

In mammals, the sensory feeling of pain (location, intensity, type) is done in a different part of the brain (primary and somatosensory cortex) than being bothered by pain (which happens in the cingulate cortex). People who have their cingulate cortex removed can feel the pain but say they don't mind it—a condition similar to what people on morphine say. Rats behave in a way that appears to be similar to this. Some have suggested that we could genetically engineer farm animals so that they still feel pain, but don't suffer from it.[15]

One study found that very experienced meditators could reduce the awfulness of pain by 57 percent, but the pain intensity by only 40 percent, further supporting the idea that pain and the awfulness that usually comes with it are two different things.[16] How this dissociation relates to animals is unknown.

Levels of Consciousness vs. Moral Value

When we think about these issues, it's helpful to distinguish some concepts. Let's take an example of getting stabbed. Painful, right? How might we compare the stabbing of a human with the stabbing of a mouse? We need to think about three things: the event, the feeling, and the moral value.

First, the *event*. The event is the thing happening in the world that is externally observable. Getting stabbed, eating a piece of cake, dying, being hungry, having your feelings improved by a compliment, and so on. When we're comparing different creatures, we need to take care in what we consider the "same" event. Getting stabbed with a switchblade has about one fourth to one third of a chance of killing a human being.[17] Although there is no data on this, I assume that getting stabbed with a switchblade is much more lethal for a mouse! So maybe the equivalent event for a mouse would be being stabbed by a mouse-sized switchblade, scaled down appropriately.

Second, the event causes some *feeling* (a conscious pleasant or unpleasant mental state), which might have some intensity and duration. Here is where we might look at whether or not the same events result in the same feelings. To take a stabbing example, is the amount of pain felt by a mouse and a human the same amount of pain, given the same (or equivalent) external event? Maybe. We just talked about several theories with different answers to this question.[18]

Third, this feeling might have some *moral value*. Even if we found two events that caused the exact same amount of pain in a mouse and a human, some people think that the suffering in the human is of more moral worth than the same suffering in a mouse. That is, we should try to prevent suffering in the human over the mouse even if the intensity of suffering felt is the same.[19]

Let's assume that two animals, say a mouse and a human, are suffering an equivalent amount. If animals have muted consciousness, then the mouse might need a greater *event* to cause the same amount of suffering. If animals feel the same intensity as humans, then it might be the same event. Now we can consider the moral value of this pain. How much more important is it to relieve the human of suffering than a mouse of the exact same amount of suffering? The answer to this question reveals differences in the moral importance of changes in welfare across different species.

Extreme speciesism holds that animals get no moral consideration at all. A way to favor humans over animals would be to acknowledge some nonzero discount for animal suffering, but to simply not care about it. That is, one might acknowledge that animals suffer, but simply not be concerned with it. To maintain this position, one needs to abandon the notion that happiness and suffering are the ultimate moral goods, and qualify it to something along the lines of happiness and suffering *of human beings*. This strikes me as what some consider speciesist: that is, like racism or sexism, but for species.[20] Caring about happiness and suffering is something that appears to be self-evident, but restricting it to a particular species feels somewhat arbitrary. It's hard to argue for it, and many arguments you could make might equally apply to sub-groups of human beings, which should make everybody uncomfortable: if one accepts this arbitrary addendum, one

CHAPTER 10: ANIMALS

cannot, in principle, oppose any other arbitrary addendum, such as caring only about people with a particular religion or skin color. It's likely that in societies with slavery, many slave owners believed that slaves could suffer, but just didn't give a shit. There are some smart people who hold this view, however, so it cannot simply be dismissed.[21]

The view of "Equal Consideration" holds that the same amount of suffering has the same moral value, no matter who is feeling it. The "Unequal Consideration" view is that some beings get a higher moral consideration than others.[22] Perhaps humans at the top, mice below them, and insects of lower value still. A possible view is that animals get *more* moral consideration than humans, but this view, if anyone at all holds it, is very rare.

Understand that this is a different set of theories from how much consciousness different animals feel. The uncertainty of both of these transitions needs to be acknowledged: how much feeling in a particular being is caused by a given event, and how much moral weight do we give that feeling? Sometimes people agree that a human is more important than a chicken, but differ in their reasons why: one says that chickens feel less, but that all suffering is of equal moral value, and others say that chickens feel the same amount of suffering, but humans have more moral worth.

Let's go into some more detail about the Muted Animal Theory, and why I think it is the most plausible. According to this idea, if a human breaking a femur has an experienced pain of ten, maybe a mouse's broken femur would have an experienced pain of five, or two. Acknowledging that there is a difference is easy. The hard part is knowing what this weighting is.

Can we use brain size to estimate the capacity for feeling? It makes intuitive sense that if a brain is too small, then it would be less complex, and a certain amount of complexity is required for conscious states. In human beings, bigger brains positively correlate with intelligence.[23] Scientists disagree on what part of the brain is involved with conscious experience, but several theories have something to do with high connectivity. And high connectivity is more likely with more neurons.

So maybe bigger brains are more conscious? This leads to some conclusions that are not intuitive. For example, it would mean that whales and elephants are more conscious than humans, and that men are more conscious than women. The average brain volume of an adult human male is 1,345 grams, and for an adult human female is 1,222 grams.[24] (In practice, nobody thinks that small differences in brain size or complexity between individuals of the same species should be used in law or ethics to favor some people over others. As a practical matter, there is general agreement that all healthy humans should be treated as being equally conscious.) Some scholars suggest that babies are *more* conscious than adults, because adults have seen so much that they don't notice the things they are used to, and babies notice *everything*.[25]

A better estimate than raw brain size is brain-to-body ratio. That is, brain size relative to body size. A blue whale needs a bigger brain just to run a huge body, but that doesn't mean it's smarter or more conscious. The brain-to-body ratio has the comforting advantage of putting human beings at the top of the list.[26] I'm not going to get into details, but there are many examples in human history of the group with power over others endorsing a moral system that perpetuates that power by choosing criteria that give the ones in power more moral consideration. Just something to keep in mind!

But what does seem true is that the amount of suffering a being is capable of is not *linearly* proportional to brain size or number of neurons. That is, just because a human has about four times as many neurons as a chimpanzee[27] doesn't mean that getting stabbed with a pencil feels four times better for the chimp than for a human.

Of course, it's not like our brains get smaller when we lose consciousness, either. A better way to estimate consciousness would be to look at the activity of those parts of our brains, or patterns of functioning in our brains, that have to do with conscious processing. This is problematic for a couple of reasons. First, scientists don't know (or think they know but don't agree) about what those parts of the brain are. The second is that we don't know that other animals would necessarily use the same brain areas as humans. An octopus, for example, might create consciousness in a different way.[28]

Let's compare a stegosaurus, which had a tiny brain, and an elephant, which has a much larger brain. The stegosaurus weighed about four and a half metric tons, but its brain weighed about eighty grams—about the same as a dog. An elephant weights about six metric tons, but its brain weighs about five kilograms—that's over sixty times larger! Both are large, herbivorous land animals, so what on earth is the elephant doing with all that extra brain that the stegosaurus didn't need to thrive for over five million years?

Although it's tempting to think of animal evolution as directly optimizing things like eating, reproducing, and staying alive, over the eons many species evolved more and more complex brain structures to *indirectly* get to these things. Elephants are particularly social creatures, so they need more brain processing power to be able to handle social interactions (primate brain sizes are proportional to how social the species is).[29] In this example, an elephant can experience loneliness in a way that is likely much less in a stegosaurus.

CAN PLANTS FEEL PAIN?

We tend to assume that plants can't feel anything, but this is, in part, because they don't move. Many children don't even understand that they are alive.[30] But plants actually *do* move. Some, like the Venus flytrap, we can see move, but all plants move by growing. It's just too slow for us to perceive, leading some to suggest that when you take the long view, plants are slow-motion animals. They grow toward light and away from things that touch them, for example. So if a plant moves toward things good for it, and away from things bad for it, isn't that pain and pleasure?

Not necessarily. Scientists understand that not all behavioral responses to pain require conscious feelings. The bodies of people with spinal cord injuries, for example, will sometimes react to harmful things with motion, but with no conscious feeling or awareness at all. Nociception is the general term for responses to harmful stimuli, and only some are accompanied by conscious experiences. We also tend to think that the plant responses are simple enough to not *need* the complexity we assume is necessary for consciousness.

We better hope this is the case, because if plants feel pain then it might be nearly impossible for any large animal, including humans, to live an ethical life.

But a great deal of mammalian brain volume is dedicated to running bodily systems more effectively, such as digestion and the immune system. These processes are unlikely to be relevant to feelings. That is, the brain functions you have to optimize your immune system probably can't feel happy or sad. Bigger brains also mean some redundancy, and more in-brain infrastructure so that distant parts of the brain can communicate with each other.[31] These communication channels also might be morally irrelevant. This gives us reason to think that some, but not all, differences in brain size correspond to increases in the potential to have good or bad feelings.

Pain, however, is likely equally important for a stegosaurus and an elephant. They both have a strong evolutionary incentive to act in a way that avoids bodily harm, starvation, and so on. So in terms of physical pain, these two animals have the same needs.

So the suffering of big-brained animals might be more complex. A human can feel loneliness, anxiety about the future, and existential angst as well as pain. Nonhumans are "merely conscious."[32] But to make decisions about how to treat different kinds of beings, we would need more than just a vague idea that more complex creatures can suffer more.

Just using our imagination is likely to lead us astray. Studies show that when a person experiences some pain and suffering, and some other person experiences the same pain and suffering, a person will believe that their own pain was worse than the other person. For example, a study by Cathy McFarland found that when people estimated pain that would be felt in film clips about getting treatment for serious wounds, they estimated (on a scale of 1 to 9) that they themselves would feel pain at a level of 7.36, on average, but that other people would feel it only at 6.32.[33]

Brought to the scale of a species, it might well be that we think humans can suffer more than other creatures simply because it's easier to imagine what the experience of another human would be than that of an ocelot or a cow. This is a reason to be skeptical of our own intuitions.

Can we rank animals? Or even better, can we come up with numerical magnitudes to adjust feeling intensities for each species of animal?[34]

To reiterate, we just don't know to what degree nonhuman animals can suffer. It would help if we knew how to detect consciousness in animal brains, but we really don't, and some say that consciousness isn't anywhere in the brain, because it's merely our central executive system attending to some other part of the brain's functioning.[35] But let's assume that it has to do with the complexity of the cortex (also a rough measure of intelligence). If we compare cortical neuron counts (where data is available), we can calculate an estimate for how much suffering a member of some nonhuman species can experience relative to a human.

CHAPTER 10: ANIMALS

Does this make any sense? I believe that it does, as a rough estimate, and I'll give you a reason to think so. Scientists have come up with a way to measure the level of consciousness of human beings in different states using brain measurements. It involves stimulating a part of the cortex and measuring with an EEG the complexity with which that stimulation reverberates around the rest of the cortex. This measurement technique comes up with a single number, between 0 and 1, that is intended to reflect the degree of consciousness that brain is in at the time of measurement. This number is the perturbational complexity index, and if it's higher than 0.31, then the brain is at least a little conscious. Why should we trust this number? Because it predicts the levels of consciousness we would expect for states like being awake, being in deep sleep, or being under deep anesthesia. What's interesting about this consciousness measure is that it shows that consciousness might be, or might be a result of, a lot of active connections across the cortex. If the cortex isn't widely communicating with itself, there doesn't seem to be any consciousness.[36]

How does this relate to animals? Well, let's assume that the level of connectivity in the cortex of a species scales with the number of neurons in the cortex—probably nonlinearly. Then there will be far fewer connections in a smaller cortex than in a larger one. As such, even if an animal is awake and at peak brain capacity, maybe its consciousness is diminished, relative to human beings, simply because their small brain cannot support the same intensity of inter-neuron communication as humans.[37]

In what is probably the most important table in this book, here are moral values for humans and the animals we eat (or could eat):[38]

Suffering Discount by Species, by Cortical Neuron Count

Species	Consciousness adjustment multiplier
Human	1
Cattle	0.035
Pig	0.027
Chicken	0.0038
Salmon	0.0012
Shrimp	0.0000012
Cricket	0.0000029
Mealworm	0.00000029

Crickets and mealworms are included not because people eat them very often (they don't), but because they are sometimes suggested as alternative possible food sources.

If a human year of suffering counts as 1, a cow year of suffering counts as 0.035. This estimate says that it would take 28.57 cows suffering for a year to be just as bad, suffering wise, as one human suffering for a year (1/0.035).[39]

This reflects the idea that humans have a greater *capacity* for conscious feelings than other animals. That is, our joys are happier and our pains are more miserable. It assumes that simpler brains have muted feelings.

This does not mean that an animal can't have a better life than a human. Imagine a Chihuahua who lives with a retired human and is doted on for their whole life. Now imagine a human who suffers from crippling depression and is shunned by society, and never recovers. Humans can suffer in all the ways (or in equivalent ways) animals can, and in many other ways besides. So it's not that human lives are always better than those of other animals, it's that they are more intense, and can be much better or much worse.

ARE WE BEING SPECIESIST?

There are several ways we could measure the relative suffering of different beings, and doing so is important because we don't want our natural prejudices to get in the way of it. Counting cortical neurons sounds sensible, but is it?[40] It puts humans on top, which we have an incentive to do. Are we being speciesist?

One danger is that we are fooling ourselves into thinking we are avoiding speciesism because we are measuring the amount of suffering that a being can experience based on some objective measure, and not explicitly basing it on the fact that it's not human, or doesn't have a face. But speciesism might be sneaking in the back door in the sense that we might be choosing the measure itself based on our speciesist intuitions about how much things suffer!

The way natural history worked out, the descendants of our ancestral primates became the dominant life form (us). But what if it had worked out differently? Octopus arms can think independently.

Even when cut off from the rest of the body, they continue to search for food. Human arms can't do that. What if, instead, the dominant species had come from the ancestors of octopuses? Then this book, written by some octopoid-like Jim Davies, might say something like, "Well, obviously the great apes don't suffer as much as the various species of octopuses, because apes, like all mammals, have all of their cognition in their brains. Their arms can't even think for themselves. The intensity of feeling, the pleasures and sufferings, of any creature with a mental system more distributed throughout the body is obviously going to be greater than any creature that uses only its brain to think."

One analysis lists fifty-three traits we might plausibly care about when trying to determine whether an animal can suffer—and neuron counts is just one of them.[41] By choosing to focus on cortical neuron counts, are we choosing the variable that makes humans come out on top? If we picked something else, like raw brain volume, then whales, elephants, and dolphins would rank higher than humans. If we used brain-size-to-body-size ratios, then mice and humans come out even, but small birds come out ahead. Birds, because they have to fly, require very light bodies, but that seems to be irrelevant to how much pleasure and pain they would be able to feel. The point is that we are only tempted to put in caveats like this when human superiority is threatened.

That's its own form of speciesism.

You might doubt the entire notion of muted feelings for some species. But note that even among normally functioning human beings we have variation in pain sensitivity and emotional response. Even if you see consciousness as an all-or-nothing affair, we still have the problem of identifying *which* beings in the world have it and which do not. In this case you might want to consider the *probability* of each creature being conscious, rather than their *degree* of consciousness. And on what basis would we estimate this probability? The same ones we'd use to estimate the degree of consciousness. These numbers would also be between 0 and 1, and the moral conclusions we'd draw would be similar.

We can use this table to find equivalent amounts of suffering. If we give some number to how much pain a human would feel if we cut off their arm with a chain saw, we multiply that number by 0.035 to find how much suffering a cow would experience for the equivalent event. This means that we'd have to saw off the legs of twenty-eight cows to get the same amount of raw suffering as sawing off one human arm (sorry for the grisly descriptions).

But now we have to decide on the *moral value* of this suffering. I believe in Equal Consideration, so for me, suffering is suffering, and well-being is well-being, and it doesn't matter who is experiencing it. That is, I give the same moral value to the same amount of pain and suffering, even if it takes different events to generate it in different animals. If we assume that cortical count is the best way to come up with these weights, then not only do we need twenty-eight harmed cows to create the same amount of suffering of one human, but that these harms would be morally equivalent. Not everyone agrees with this.[42]

Looking at this table (and accepting the assumptions behind the calculations that were used to make it) makes it look like shrimp are the most moral animals to eat, in terms of direct animal suffering (unless you can stomach crickets, but a friend of mine says they don't taste very good).[43]

This might be true if all animals have the same amount of meat on them. But you only have to picture a chicken and a cow in your mind to see that that isn't the case.

If you think that suffering is equally bad for every animal, then the suffering of one chicken is ethically equivalent to one cow. If you think that cows can suffer ten times as much, then maybe you think that the suffering of one cow is equivalent to the suffering of ten chickens. Before we go on, what number do you pick? How many chicken sufferings are equivalent to one cow suffering?

It takes about 200 chickens to get the same amount of meat as one cow. So if your number was less than 200, then it's more ethical to eat beef than chicken. The smaller your number, the more ethically problematic it is to eat chicken than beef. If your number was one, then eating beef is 200 times more ethical than eating chicken, when it comes to animal suffering due to death (the environment is another matter, because farmed cows contribute a lot more greenhouse gases). That would also mean that somebody who ate beef every other night for a year would be morally better (in terms of animal suffering due to death) than a vegetarian who slipped up and ate chicken once.

And what about whales?

If we're talking about eating these animals, though, the most relevant way to compare them is the number of calories one can get from one animal. Calories per kilogram of meat vary from animal to animal, but range between about 850 to 2,760, with beef, pork, and poultry all being above 2,000 kcal/kg.

But the animals we eat suffer more than just the pain at their moment of death. We also might want to consider the number of years taken away from its expected life span, and the conditions under which it lives those years. I'm not going to bore you with all the math, but you can read it in the papers I reference.[44] But when you mix up all that stuff, the winner is (drum roll, please) . . . salmon. If you want to reduce the suffering you cause to animals, avoid poultry and eggs, and focus on salmon, pork, beef, and milk. Even though crickets and shrimp are really small, and possibly feel so little pain, it takes so many of them to make a meal that it's ethically worse to eat them.

Here is where we encounter a big problem with estimating suffering in a being's life. Lives can be better and worse, and just about everybody agrees that at some point a life is so bad, so miserable, that the creature would have been better off not existing at all. If we think of the feelings of a creature, over the course of its life, in terms of utility, we might say that if a creature's life is bad enough that it has an overall negative utility. Although it's easy to agree with this common-sense notion, it turns some ideas on their head. Let's say, for example, that the life of a chicken raised in a batter cage is so awful that its life has an average negative utility. This is entirely possible, given the terrible life conditions of battery hens. I have a friend who worked at a chicken factory. I asked him, "Would you rather have a life as one of the chickens in that factory or not have any life at all?" He only had to think for about two seconds before he said, "No life at all."

Under this condition, ending a being's life earlier is better, not worse, because you're putting it out of its misery. We would not count years of life lost, because those years of life it might have had would be torturous.

I acknowledge that many readers will not agree with my views on animal welfare. This is to be expected, because even experts disagree. By how much? Can we look at what the entire *field* thinks about it, given all of the uncertainty?

WAIT A MINUTE, DID YOU SAY CRICKETS FEEL PAIN?

Note that there is no evolutionary pressure for humans to have evolved a moral reaction to killing and harming insects. We might feel for the animals we do because they look a bit like humans. Mammals even have similar facial expressions for the same emotions.

But a cricket has no facial expressions, and is so small and unrelatable that it's no surprise that we have little intuitive notion that they might feel pain. Here is where we can't trust our instincts, because evolution had no adaptive reason to give us an accurate picture in this case.

Although it's tempting to think that bugs don't feel anything at all, you should know two things. First, we don't have very convincing places to draw the line at consciousness and nonconsciousness. A famous philosopher of consciousness, David Chalmers, thinks it's possible that even a thermostat has some very small amount of it. If David Chalmers, who's thought very hard about this for a long time, thinks a thermostat might be conscious, maybe you shouldn't be quite so sure that a cricket isn't.

Because we cannot get into the mind of a bug and experience what getting hurt might feel like, if anything at all, the best we can do is look at their behavior and neural systems, and make an estimate.

Crickets, fruit flies, and larval moths will try to get away from harmful things, so they definitely have nociception. Crickets are slower to avoid dangerous heat if given morphine. There are studies of creatures even smaller: fruit flies, animals so small we can barely see different parts of them with the naked eye. Fruit flies are pretty far from humans in the evolutionary tree of life, but they use similar neurotransmitters in response to pain (reduction of GABA). Even more disturbing, injured fruit flies show behavioral and neural responses similar to those of vertebrates, such as human beings. A broken leg will cause the fruit fly to go into a state of increased vigilance. Their muscles get primed for escape, and they become more sensitive to stimuli in general. Flies have an "escape circuit," a fast-acting way to avoid looming dangers that bypasses their normal motor responses—this is why it's hard to swat a fly with your hand. The escape circuit works even faster when the fly has suffered an injury. These effects remain for the rest of the fly's life, even after the wound has healed, suggesting that flies can have chronic pain. Worms have endorphins. Many insects have serotonin—the thing in Prozac that makes you happier.

Because flies and vertebrates are so distant in the evolutionary tree, and we have such similar responses to pain, both in terms of behavior and neural response, we can infer that pain evolved a very long time ago—perhaps 500 million years ago, when the first multicellular creatures evolved.[45] If we view consciousness as a continuous thing, rather than on and off, there is a very good chance that fruit flies and crickets have *some* amount of conscious experience of pain.

If you still think that bugs have no conscious experience and are therefore exempt from moral consideration, it simply means that you think that the discount rate is 0.0, rather than the 0.0000029 estimate we get from cortical neuron counts. If you're really confident about this, then you should be less willing to experience a stubbed toe as a human than a thousand years of experiencing the life of one caterpillar after another being eaten alive from the inside out by wasp larvae (this really happens to caterpillars).[46] If your answer isn't an immediate, unqualified yes, then maybe you're not as confident as you thought that the caterpillar feels *exactly nothing at all*.

I collaborated with economist Richard Bruns on a paper that tries to grapple with the uncertainty in the field.[47] What we did was estimate the percentage of scholars who believed this or that theory. Ideally we would survey experts, but for now we estimated that 3 percent of scholars endorsed the Unconscious Animal Theory, 62 percent endorsed the Muted Animal Theory,[48] 30 percent endorsed the Same Suffering Theory, and 5 percent endorse the Tinker Bell Theory.

Further, we estimated the field's endorsement of the various ethical theories. We estimated that 10 percent endorse the Extreme Speciesism view, 60 percent endorse the Equal Consideration View, and 30 percent endorse the Unequal Consideration view.

The nice thing about all of these views is that they can be expressed in terms of numbers. For example, if you think that a dog has muted consciousness, and feels only 60 percent of the suffering that a human would for the same event, then you would express this as a multiplier of 0.6. So if a human would experience 100 points of suffering from an event, a dog would experience 100*0.6=60 points of suffering for the same event. With the Unconscious Animal view, the multiple is 0, and for the Same Suffering theory, the multiple is 1 (in other words, there is no change). There are similar multiples for the ethical theories. If animal suffering doesn't matter at all (Extreme Speciesism), then the multiple is 0, meaning any number you multiply it by turns to nothing.

We assume that a theory's probability of being true is equal to the percentage of experts in the field who endorse it (for example, if 62 percent of people endorse the Muted Animal Theory, we assume there is a 62 percent chance it is correct). We made a simulation that randomly chose an animal suffering theory according to the chance of it being correct. Then it chose an ethical theory in the same way. You multiply the two associated multiples together to get a measure of how much hurting an animal morally matters. Our simulation was run 10,000 times (this is called a Monte Carlo Simulation), giving us a spread of uncertainty in the field, *independent of what any individual thinks is true.*[49]

The results of these simulations, which we ran for crickets, salmon, chickens, pigs, cows, and elephants, was that the uncertainty was enormous. If we respect the views of people in the field, for all of these animals there is a 90 percent chance that the suffering multiplier is between 0 (it doesn't matter at all) and 1 (each individual animal's suffering is as intense and matters as much as a human's).

I will return to this point in the conclusion, but what I take from this is that this degree of uncertainty is so high that we are better off putting our efforts into helping the world in ways for which there is more expert agreement. Specifically, it's more effective to treat human health than putting the same effort into helping the conditions of animals.

Is Life Suffering?

Recently I read the novel *Watership Down*, an animal fantasy about some rabbits. When I would tell people I was reading it, they would often say something like, "That's really violent and brutal, isn't it?" This reaction stems from an unrealistically sanguine idea of what the lives of wild animals are like.

Suppose you're sitting on your back porch on a fine autumn evening, and you see a rabbit come out of the woods to eat some clover growing around the grass of your back lawn. You smile and watch the rabbit, peacefully munching away, until its ears perk up, and it hops out of sight, back into the forest. It's so beautiful! Then the mosquitoes start to annoy you, so you retire inside your house.

What you didn't see is that this rabbit had recently been forced out of its den by other rabbits around midday, and is smarting from a wound on her leg that she suffered in the scuffle. You didn't notice this because she behaves in a way that hides it, because predators target wounded animals. She's been hiding all day, stressed out about going outside during the day, and waiting for the evening, when she can still see but is less likely to be seen by predators. She desperately needs to create another den somewhere safe, as she is pregnant. But she's practically starving, which is why she dared to venture out onto your lawn, where she is more visible to predators. Including you. She knows you are there, and her stress levels are high, knowing you can see her. Finally, she sates her hunger enough so that she's not quite so desperate. But she hears something that might be a predator, and hops away to the relative safety of the woods, where she is plagued by mosquitoes for several hours as she tries to find a place for a new den. She doesn't. She gets mauled by the fox she heard earlier, who brings the rabbit carcass to her kits. The foxes, similarly famished, devour the rabbit. They will survive at least two more cold nights.

Animal welfare often focuses on the way human beings treat animals. But let's not be under any illusions that animals in the wild are living in some paradise. How bad is it? The first thing to talk about is bugs, because, compared to the number of bugs, large animals like rabbits practically don't exist. There are an estimated ten quintillion insects alive at any given time. That's a 1 with 19 zeroes after it. That's 300 pounds of insects for every human.[50]

"Bug" is not a perfect term. It's a nonscientific, functional term for small animals we want to keep out of our food and off of our bodies. A more accurate term is "small invertebrate animal," but that's clumsy, so I'll use the term "bug," and what I mean are insects, spiders, worms, and things like that.[51]

Can bugs feel pain? Probably (see the sidebar "Wait a Minute. Did You Say Crickets Feel Pain?"). Recall that the cricket, for example, has been estimated to feel 0.0000029 the amount of consciousness as a human being.[52]

So nothing to worry about, right? Unfortunately, there are so many bugs in the world that the amount of suffering that bugs are capable of, taken as a group, is staggering. I discussed above the implications for using things like crickets as food, but even that problem is dwarfed by the problem of wild animals suffering.

PROPORTION OF LIFE SPENT DYING

Another thing to consider in the death of an animal is how much of a proportion of its life is spent dying. It seems that a painful death is worse if it's a greater portion of your life. In the following table, we have the expected human lifespan, and the age of animals we eat when we put their lives to an end.[53]

How Many Seconds Humans Tend to Live and How Many Seconds Food Animals Live Before Slaughter

Species	Seconds of Life
Human	2,100,000,000 seconds
Cattle	47,300,000 seconds
Pig	14,500,000 seconds
Chicken	3,629,000 seconds
Salmon	63,070,000 seconds
Shrimp	13,140,000 seconds
Cricket	5,256,000 seconds
Mealworm	2,628,000 seconds

If we assume that dying takes, on average, ten painful seconds (I just made this up), we divide ten seconds by the expected lifespan (in seconds) and come up with the proportion of each animal's life spent dying. What's important here are the orders of magnitude. Humans are in one category, cattle, pigs, shrimp, and salmon in another, and chickens are in a category with mealworms and crickets.

Proportion of Life Spent Dying

Human	0.0000000048
Shrimp	0.00000076
Pig	0.00000069
Cattle	0.00000021
Salmon	0.00000016
Mealworm	0.0000038
Chicken	0.0000028
Cricket	0.0000019

Wild Animal Suffering

The scholarly field of "wild animal suffering" endeavors to understand how much suffering happens in more-or-less natural environments, without human interference. A few things are relevant to this discussion. One is the difference between K and r strategies in reproduction. The K strategy is the one that humans and elephants use. It involves having very few offspring and investing a lot of parental care into each one of them to ensure their survival. The r strategy, on the other hand, involves having a giant brood of offspring, sometimes tens of thousands, with little or no parental investment. It works because even though the vast majority of them will die due to starvation or predation, enough survive to keep the species going. Different species occupy different points on this K-r continuum.

Another relevant fact of ecosystems is that for every large creature there are many smaller creatures.[54] For every large animal like a moose, there are many more medium-sized animals, like raccoons. For every medium-sized animal, there are many more smaller animals, like mice. Every small animal, there are far more bugs. For every bug there are thousands upon thousands of microorganisms, which consist of only a single cell. This relationship, that the smaller an organism is, the more of them there are, is true of all ecosystems.[55]

This is important because the smaller the animal is, the more likely it is to use an r reproductive strategy. If most animals are small, and smaller animals tend to be more on the r side of the spectrum, then the vast majority of animals that live in the world have lives that are very short, followed by a painful death. If an organism's population is relatively constant over time, and broods are numbered in the thousands, then that means that thousands of creatures (all but two, for the two parents of the next generation) will die a possibly painful death (from starvation, thirst, or predation) every single generation. It appears that, in evolution, there arose a reproductive strategy that caused pain and very short life spans for the vast majority of creatures that come into existence. Would you rather be born as an average fish, where you are more than likely to be devoured by a predator in the first few days of living, or not live at all?

Here's another way to look at it. Suppose you had two options: die by being hit painfully by a car, or die peacefully in your sleep, but earlier in life. That is, you can have a longer life if it ends in a painful car accident. How much earlier would you be willing to die so that you could avoid the car-accident death? Is your answer more than a few weeks? If so, then you might think that the life of an average animal isn't worth living, because the majority of them live less than a few weeks before dying horribly of predation, starvation, or thirst.

The answers to these questions are important, because they help answer the next question: On average, do animals enjoy a positive net welfare, or negative? That is, if you look at the *average* animal, is their life worth living, or is it not worth living? If your answer to the car-accident question was more than a few weeks, then you'd probably rather not live the life of an average animal, because the average animal is small, is a member of an r-reproducing species, and dies a painful death that is a large portion of its short life. The pain of death would be too high a proportion of your life in total.

Let me reiterate here that because of the vast numbers involved, if bugs feel pain, then the sufferings and pleasures of the large animals, humans included, is negligible. Even with the vast discounting of conscious experience that we're assuming for smaller creatures, their sheer numbers mean that bug welfare is where all the moral action is.

THE POTENTIAL SUFFERING OF ALL INSECTS COMPARED TO ALL HUMANS

Let's compare insects to human beings. There are 7.53 billion people in the world at the time of this writing. Their suffering is not discounted, so let's assume their potential suffering is 7.53 billion units. There are ten quintillion insects estimated to be alive today, and if we apply the discount on the assumption that they experience muted feelings, the number is 29 trillion, much higher than 7.53 billion. This suggests that the welfare of insects, their potential suffering and pleasure, is over 3851 times "bigger" than human welfare, even with the steep discount.[56]

And that's not even counting the other bugs, like spiders and worms.

So the vast majority of creatures in Earth's ecosystems are living very short lives with deaths involving suffering, and it just might be the case that it is worse for them to be alive than to have never lived at all. I really hope this is not the case, because it means that a universe without life at all would be less miserable than a world with life. Imagine some utopia, with blissfully happy humans living in it. It's a disturbing thought that to ecologically support all of this human happiness requires cosmic amounts of suffering, paid for by the smaller creatures dealt a terrible hand by evolution with r-reproductive strategies, or for whatever other reason live lives of net negative welfare.

This line of reasoning has been used by some to justify habitat destruction. I know this sounds bonkers, but the reasoning is consistent, whether or not you agree with its premises. If more life means more suffering, on net, then fewer ecosystems means a better world. Such is the view of Brian Tomasik. Let's say there were two identical islands, both thriving with life. One gets plowed over and covered with blacktop—we'll call it Tomasik Park. The other stays a wilderness. Which island is preferable, from a moral point of view? If the net suffering is higher than the net good feelings for bugs, then Tomasik Park is better. Yikes.

Looking to the future, on this view terraforming other planets will be an ethical catastrophe, because it would cause more suffering than well-being. Just like life does on Earth.

Why We Should Preserve Natural Habitats

I'm going to tell you why I think it's a bad idea to destroy animal habitats in an effort to improve animal welfare, in case you need any convincing of that.

One reason is that we're really not sure that the average animal experiences net suffering. We assume that creatures that can't move don't have pain, because we understand that the creation of a conscious pain system is reproductively expensive. That is, a species will not evolve to have conscious experience at all unless it has a strong adaptive reason to do so. Some have argued that the expensiveness of conscious experience suggests that creatures who live with extreme r strategies are less likely to feel pain, or at least feel reduced pain. The (suggested) trade-off is that the more suffering the individuals of a species have to experience to reproduce, the less each individual organism will actually feel the suffering. If this trade-off is real, then we simply don't know if average animals enjoy a positive net welfare, or negative.[57] Our confidence in knowing whether bugs' lives are worth living or not is low.

Some have argued (rather persuasively) that ray-finned fish can't even feel pain.[58] It would be a strange world indeed if a fruit fly felt pain, but a fifteen-foot tuna did not.

We also have good reason to believe that for many other reasons, ecosystem preservation is a really good thing, to say the least! That is, even if bugs *probably* feel pain, the known cost of ecosystem destruction is so great that we still shouldn't destroy them. We should be *very* cautious when entertaining the idea that we're better off without ecosystems than with them.

Should we intervene with wild animals' suffering at all? People offer reasons why the answer to this question should be no. The one that comes to many people's minds is the idea of nature being somehow an intrinsic good—that we shouldn't be messing with nature, because there's something about the natural world that should remain pure; that the biological world and the processes it uses to keep going are as they should be, so we shouldn't change them.

This is a difficult position to justify, and in fact philosophers and psychologists have a word for it: the naturalistic fallacy. Although we tend to focus on the beautiful parts of nature, a lot of it's really nasty. Just about nobody thinks that we should have let smallpox just run its course, or even let sudden oak death destroy forests, simply because pandemics are a natural part of the world. Even if we ignore bugs, the lives of wild animals are brutal.

Perhaps we're tempted to think that the way evolution has created ecosystems is in some sense optimized, and thus good. There's a saying in biology: evolution is smarter than you are. How exactly is evolution smart? In a nutshell, evolution is really good at optimizing reproduction.

There are exceptions to this: the trait of green fur never arose in mammals. We have snakes and bugs that are green, suggesting that it might be a good camouflage color in green environments. *I* thought of green fur, but it didn't arise in evolution. Cats never evolved to figure out how to stay downwind of prey, which means that they often get smelled while stalking.[59] So in many ways, living things are not really even optimized for reproduction. But in general, evolution does a damn fine job of reproducing as efficiently as possible, and modifying organisms so that their reproduction is maximized.

So if a scientist came along and said that he had a plan to create or modify an organism so that it would reproduce better, we should be skeptical of that, because when it comes to optimizing reproduction, evolution is almost always smarter than we are. However, if someone were to say that the world that evolution has created is necessarily the most good world, in a moral sense, we have very good reason to be skeptical, because in the domain of generating moral goodness, evolution is very likely *not* smarter than we are. Evolutionary biologists agree that evolution is optimizing reproduction, not the happiness of the individual members of species involved in it.

Evolution Doesn't "Care" About Happiness

Remember, evolution doesn't actually care about anything, it's just something that happens. When we talk about evolution caring, it can make it easier to think about, but keep in mind that it's only metaphorical.

But allowing this metaphorical language, all evolution "cares" about is reproduction. Everything else it cares about is only instrumental. That is, if a species has evolutionary pressure to be able to run quickly, evolution only cares about running quickly because it leads to more reproduction. Even survival itself is merely instrumental to reproduction, which is why, in most species, evolution doesn't optimize survival of beings that can no longer reproduce.

WHO SUFFERS MORE, LIVESTOCK OR WILD ANIMALS?

Our ignorance of exactly how much pain and suffering another being can experience forces us to use what we hope are correlates of it. For animals, the general guidelines are access to food, water, shelter from the elements and predation, health, and the ability to express natural behaviors.

For farmed animals, they get plenty of food, water, shelter, temperature control, and protection from disease. They lack freedom of behavior and movement. They are protected from predators, though sometimes chickens will try to kill each other in cages (but farmers burn off their beaks so they can't do this, so . . . problem solved?).

In the wild, though, only freedom of movement and the ability to engage in natural behaviors is a guarantee. They struggle with hunger, thirst, competition with conspecifics, predators, and the elements. Living in the wild is no picnic. Or maybe it is a picnic, but the kind of picnic that lasts three days, during a snowstorm, and there's only one sandwich for twenty people.

I've seen animal rights activists ask, "How would you like to live your whole life in an airplane seat?" This is a comparison to the confinement that pigs have to endure in farms. But when we're comparing domestic, farmed pigs to wild pigs, perhaps the question should be: would you rather live your life in an airplane seat, with nothing to do, but not be too cold or too hot, not suffer disease, have enough food and water, and be killed painfully at the end, or be free to do what you like, but be frequently hungry, possibly suffer from some terrible disease, be often very cold, and die of either starvation or killed painfully by a predator?

That choice isn't so easy.

If asked whether I'd rather be in an airplane seat the rest of my life, bored, uncomfortable, and depressed, or put out in the North American wilderness to fend for myself, starving and freezing to death, I'd be like, "Can I get the whole can of Diet Coke, please? No? That's okay."

Now, you might be thinking that this isn't a fair comparison. That a professor living a cushy life is unprepared for living in the wilderness. You would be correct. I am domesticated. Similarly, livestock is also domesticated, and putting a broiler hen into the wild could very well mean a worse life for it, because such a creature no longer has any meaningful natural environment, and is utterly unprepared for living outside of captivity. So the idea of releasing farm animals into the wild is a preposterous idea, in terms of relieving animal suffering.

A more objective way to look at this is in terms of stress levels. Here we find that wild animals have much higher cortisol (a biological correlate of stress) than domestic animals. This is partly due to the effects of breeding on domestic animals, but their very captivity has an effect, too, as shown with experiments done on wild animals caught and kept in captivity. Their cortisol goes down.

And (modern) zoo life is even better. Given the choice between being an animal living in a modern zoo or the wild, I'd choose the zoo for any animal, anytime. And I think most of us would rather be a domestic pet dog than a wild dingo.[60]

It just might be that wild animal suffering is, all said and done, *worse* than living in captivity.[61] This does *not* mean that we shouldn't care about livestock suffering. We should, in part because it's something we can actually do something about.

The only reason we shouldn't *also* intervene in the wild is because we have no idea what we're doing.

What is conspicuously missing from evolution's optimization is happiness. Evolution doesn't even have any intrinsic goal of happiness, and the existence of so many of Earth's creatures that use an r strategy for reproduction is evidence of how little evolution cares. Remember, if a reproductive strategy arises that involves enormous pain and suffering, but turns out to be a better way to propagate the species, *evolution will take it.* There is no law of biology that says that good feelings have to outweigh the bad. Feelings are only there at all to aid in reproduction.[62]

Let's look at the agricultural revolution that happened to human beings with cultural evolution. Before farming wheat, hunter-gatherers had to live sparsely. The same area of land could support people living tightly together, if they farmed wheat. But those people would be more poorly nourished and more susceptible to disease. Why, then, did the agricultural way of life take over the world, pushing the hunter-gatherers out? The same area of land that might support a hundred hunter-gatherers can support about a thousand people. So even though they were worse off at the time, there were so many more of them that they took over.[63] More people that were (probably) less happy.

If evolution is not optimizing goodness, we cannot use its products as a moral standard. That is, just because some system evolved does not mean that it's good, or is something that should exist, from standpoints of happiness, goodness, morality, or anything like it.

Even people who believe that nature is sacred have their limits. They probably advocate, at least in principle, for wildlife rehabilitation: medical help for hurt animals, and euthanasia for animals that are too far gone and just need to be relieved of their suffering. They want to intervene in nature in these circumstances. It's telling, though, that the wildlife rehabilitation world's number two problem is burnout.[64] That's how much suffering is in the natural world. (I assume their number one problem is animal suffering.)

This leads us to the position that perhaps we *should* intervene on nature, where it is reasonable to do so. The problem with this is that most of the time we have no idea what we're doing, and our attempts to change natural patterns end up doing more damage than helping at least half the time. Rabbits were introduced to Australia, which was a disaster. There were too many rabbits, so they introduced foxes to hunt them. Now there are too many rabbits *and* foxes, to the detriment of just about every indigenous animal species on the continent.

In my opinion, and opinions of many other scholars, we shouldn't try to improve ecosystems simply because of our ethical, economic, and ecological understanding of these systems is too primitive to be confident that we're going to be doing more good than harm with any given intervention that we come up with.[65]

So given that we really don't know whether life is, on average, good or bad when it comes to pain and pleasure, along with our ignorance about how the world works that would make interventions unlikely to be successful, it makes sense for people concerned with expanding happiness in the world to support the conventional views with regard to environmental conservation: preserving habitats, reducing pollution, conserving resources.

That said, people intervene with nature *all the time*, and when we do we should consider the welfare of all of the sentient beings involved. The kind of thinking I'm advocating here can be used when deciding between, for example, two different interventions that will affect nature.

To summarize, whether bugs have lives worth living is very possibly the most important question there is. But our uncertainty about its answer, and the vastly different and hugely consequential results of doing anything about it are also so large, that it's best that we don't destroy Earth's ecosystems just yet, turning all the habitats into Tomasik Park.

Okay, so when it comes to animals, what matters and what unmatters? Even though wild animal suffering is kind of the only thing that matters, we are in such a terrible position to do anything about it that the best we can do is focus on what we can more safely control—how we treat livestock, deal with fish, and preserve natural habitats. It's not much, but it's all we can do.

For now.

Helping Animals

Figuring out how to most effectively help human beings, complicated as it is, is relatively simple, because we generally think that all human beings have equal value, with adjustments for age and a few other things. But when we talk about nonhuman animals (who I'll refer to simply as "animals"), there is enormous variety in size, biology, cognitive ability, and how much people care about individual species. Even experts don't agree on some fundamental issues.

When I talk to people about helping reduce the suffering of animals, people often suggest I help out at the local humane society. When people think about abuse to animals, they think of dogs chained in a backyard with no water, or animals used in laboratory experiments.

The majority of human-caused suffering for large animals is in catching fish and raising animals for eating. If we add up all the animals killed in shelters, sacrificed in laboratory experiments, and killed for fur, we get

25.5 million animals per year. Not a small number, but it's dwarfed by the number of chickens we eat every year: 139 million.

How many land animals are killed to feed humanity? About 60 billion per year, which is about nine animals per human.[66] The global capture of wild fish is estimated to be between 1 and 2.7 trillion fish every year, or about 150 per human.

It takes land to raise animals, and it takes land to grow crops. But don't forget that we need to grow crops to feed many of the animals we eat. So not only are we using land and other resources to raise the animals, but we have cropland needed to make food to feed them. This leads to inefficiencies: only about one-tenth to one-quarter of the calories are retained for human consumption by the time the meat is eaten. That is, it's much more efficient to get the calories directly from crops.[67]

As I mentioned above, animal protection laws are often different for pets, livestock, and sea animals. If you go to a restaurant and eat chicken or bacon, that animal was probably treated in a way that could get you arrested if you did it to a dog. But for the animals we eat, it's perfectly legal and socially accepted. Although I find animal welfare advertisements' focus on charismatic animals like cats and seals a little annoying, I recognize that for some they are gateway animals that get people interested in animal welfare in general. As people look more closely, they might then realize that suffering cats and seals are a negligible problem in the world today.

Changing How You Eat

Given that one of the biggest effects our behavior has on animal welfare is due to farmed and fished animals, it's worth talking about which animals we should eat. The subject is one of the most fascinating to investigate, because eating can be evaluated through so many lenses. There's taste, nutrition, cost, environmental effects, animal welfare issues, national business values, and cultural values associated with eating.

Staying alive requires eating and, for most animals, to eat means to kill. Humans, in particular, require consuming living cells to live, be they from plants or animals. With very few exceptions, every living thing gets its sustenance from the death of other living things. As a result of this, almost every living thing has evolved measures to try to prevent it from being eaten. Animals have claws, shells, camouflage, and the ability to run away. Plants have toxins, bark, and thorns.

Fruit, for example, is made of living cells, but eating fruit does not harm the plant it came from. For one thing, plants don't feel pain, so far as we know, and for another thing, the whole *point* of fruit is to be eaten.[68] Fruiting plants evolved to create food for other animals, who, by eating it, help the plant reproduce. But even this has nuance: hot peppers plants are better spread when birds eat them than mammals, so they evolved to be hot in our mouths to keep mammals from eating them. But birds eat these fruits freely because they can't feel capsaicin.

Fruitarians go to the extreme, with some people only eating fallen fruit, so that no plants are harmed due to eating. The fruitarian diet is not recommended by nutritionists, though, and several children have died as a result of having this diet imposed on them.[69]

The consumption of most plants, however, involves killing the plant. Even though, from an evolutionary perspective, most plants don't "want" to be eaten, at least they feel no pain when they are cut down and killed, as far as we know. Killing plants for food is generally viewed as morally justified for this reason. As discussed above, eating less meat causes fewer beings overall to suffer. This is generally true, but there are some interesting nuances.

What about avoiding meat that's going to be thrown out anyway? Let's suppose you go out to dinner with a couple of friends and one ordered a gigantic plate of food that she couldn't finish. She is going out of town and can't take the leftovers home. Should you eat the meat or throw it out?

Eating this meat will not increase demand because it has already been purchased. So eating it does not increase animal suffering in the near-term. However, if you're a vegetarian, eating meat might make you fall off the wagon: if eating the meat might tempt you to eat more meat, it's probably best to avoid it. Another possible downside is that people who know you might see you eating the meat. If you're just a reducetarian, like I am, this doesn't matter much, but if you're vegetarian or vegan, people might see you as being inconsistent, or see your apparent giving in to temptation as evidence that vegetarianism is unsustainable.

Meat served at restaurants is typically from factory farms, so let's assume that the meat was raised unethically. As bad as it is to harm animals in this way for human consumption, I think it's much worse to harm animals in that way for *no reason at all*, which is what happens if that meat gets thrown out: the animal suffers and dies for nothing.

But the larger issue is that wasting food is bad. Composting only recovers a fraction of the energy needed to create the food (and most restaurants don't compost). If the leftovers are eaten, it means less food that needs to be purchased in the future. So who should take it home? The ethical answer is: the person most likely to buy meat in the future. If the other friend at the table is a carnivore, give it to her because she'll have it for lunch the next day, and not have to buy a pastrami sandwich.[70]

But most people eat meat, and that means, at the very least, killing beings that have feelings. Just like pet cats, humans have, in practice, "harmful needs." We might accept that animals have to die to keep other animals alive. In this case, we should strive to kill as humanely as possible. For some animals, this simply isn't possible. Whales, for instance, are too big to kill humanely while keeping intact the reason to kill them in the first place—harvesting oil and meat.

But even for cases where the killing is humanely carried out, there is still the important question of needless suffering on the part of those animals as they live in captivity, and how it can be reduced.

Let's look at chicken meat raised in a typical farm in an industrialized nation like the United States. There's so much ammonia in the air that it hurts the eyes and lungs of the birds. Bred to be heavy, sometimes they collapse on their own weight and die of thirst.

Eating eggs seems better than eating chicken meat, because you don't have to kill the bird. But laying hens are kept in cages so small they can't even stretch their wings if they were in there alone. But they're not kept alone; they're in there with three or more others. The most aggressive bird in the cage sometimes pecks the others to death (this is where the term "pecking order" actually comes from). But this problem was solved by slicing off their beaks with a hot blade. Their beaks are full of nerves, and this is done without anesthetic.[71]

I'm not going to go into more detail, because I'm afraid that I'll lose my readers. It's just too horrible to read. I'll just say that it's also really bad for pigs.

So what can be done about it? Well, you've got choices. The obvious one is to reduce the amount of meat that you eat. If you live in America or Canada, you probably eat much more meat than people in other countries, so your idea of what a normal amount of meat to eat is rather skewed.

ANIMAL DEATH AS SIDE EFFECT

The process of getting animals to your plate involves death of even more than just that animal and the animals it's fed—lots of death happens on the way. Animals die from stress in transport. Animals might get crushed by other animals. Chickens are raised to be so plump that their legs can't lift their bodies up to get to water. The system isn't perfectly efficient, so for every animal successfully raised, slaughtered, and eaten, there are animals that died and never made it that far.[72]

Also keep in mind that for fish, lots of fish are caught and die before they are able to be fed to any human or animal. The average American, through eating, causes the death of 224 wild fish every year, either by eating them or by eating animals that were fed with wild fish. Also consumed are 186 shellfish, mostly shrimp, eaten every year.[73]

So when you eat an animal that eats animals, you're contributing to the death of, usually, many animals.[74] How many? The typical American causes the death of between 262 and 406 animals by eating their thirty animals.

The mildest version of eating less meat is being a reducitarian, where a person tries to reduce the amount of meat they consume. The most extreme is veganism, where no animal products are consumed, including gummy bears, which include gelatin, an animal product.

Suppose someone wants to eat less meat so that they are doing some good for animals. How many animals suffer due to your eating? By some estimates, the average American eats thirty land animals every year, twenty-eight of which are chickens.[75]

If you eat a chicken, then that's one chicken. Obvious, right? If you eat steak, you're only eating a fraction of the cow, but that's still easily measurable. What if you eat pork loin, or salmon? Here things start to get complicated, because pigs and fish are sometimes fed meat, and when you eat an animal that was fed meat, you are responsible not only for that animal, but the animals that animal ate. For example, only about 10 percent of the fish caught for food are directly eaten by human beings. The rest of it goes to feeding larger animals we'd rather eat. Pigs are fed sardines, fish are fed chicken and fish. Most fish are carnivorous. To get a kilogram of salmon, you need to feed it three or four kilograms of other fish to fatten it up.[76]

THE BEST PROTEIN TO EAT

In my opinion, the most ethical animal protein to eat is the meat of mussels. They have a lot of things going for them.

They are really nutritious. They are high in protein, almost as much per gram as steak or chicken, but with less fat.[77] They're also high in vitamin B12 and heme iron, nutrients vegans have to be particularly careful about making sure they get in their diet.

These animals don't feel any pain, as far as we can tell. Their nervous system is very simple, they do not react to tissue damage, and they cannot locomote (which makes them more like plants in that they have no evolutionary reason to feel pain).[78]

They are filter feeders, so they actually *clean* the water they are raised in, particularly by removing nitrogen. Most fish farming makes the water dirtier. But mussels make the water cleaner. This might worry you, because aren't you eating the crap they clean out of the water? I asked a fish farming scientist about this, and was told no, they are so small, and reach maturity so quickly, that the bioaccumulation is negligible, unlike large fish, like tuna, where bioaccumulation is more of a problem.

Mussel farming also contributes very little to climate change. To generate a kilogram of protein from mussels puts 6 kilograms of carbon dioxide into the atmosphere. That's not only lower than any other meat, but it's lower than any other *food* except for peanuts. Lamb puts out 160 kilograms of carbon dioxide and even lentils generate 22 kilograms of carbon dioxide for a single kilogram of protein.[79] This suggests that eating mussels might be an even more ethical source of protein than protein-rich vegetables! Who would have thought that eating mussels is three times better than lentils for the climate?

Unlike other meats, and even most vegetables, mussels require no extra land or water use, which means that these resources are not taken from other animals.

They don't need antibiotics, and don't need to be fed to other animals.

One downside is that they cost about three times as much as chicken (per gram of meat) at my grocery store.

I think they taste great, they're nutritious, they don't suffer, and they're good for the environment. Is there any other way a food can be good?

Creating Demand

Does reducing meat consumption do any good? There's an argument that it doesn't, and this argument feels particularly tempting if you love to eat meat, like I do. Let's consider the effects of buying a chicken breast. The grocery store (or restaurant) buys meat in bulk—let's say, they order a number of cases of chicken breast at once. One person buying or not buying a single breast is probably not going to affect how many cases are bought next week. This makes it seem like your eating a chicken breast won't make any difference to how many chicken breasts your grocery store buys—let alone change the number of chickens raised by the farm.

However, once in a while the store will be in a position to buy an additional (or one fewer) case due to a single extra chicken breast being (or not being) purchased by a customer. If you happen to be considering buying chicken at that time, then you *will* influence demand in a real way. This will only happen once in a while, but when it does, your buying the chicken breast will cause a *great deal more* demand for chicken, far greater than a single breast. Your purchase (or decision not to purchase) will cause the buying (or not buying) of an entire case of breasts! So the expected effect of buying chicken breasts, over a lifetime, will be that you cause demand roughly in proportion to what you purchase. Some weeks it's you causing more or less demand, sometimes it's somebody else. This is why an individual refraining from buying meat results in fewer animals being farmed.[80]

Later, when I talk about environmental ethics, I will return to the subject of meat-eating, because cows are worse for climate change than chickens. My estimates suggest that even taking into account the suffering due to climate change, it's still far worse to eat chicken than beef.

Hunting

People can get bent out of shape about hunting, but there are reasons we probably shouldn't be too upset about it when considering animal suffering. First of all, hunting accounts for a small amount of animal death relative to fishing and the farming industry. In the United States, only 5 percent of people hunt at all, and the decline is expected to accelerate. Further, hunters tend to target animals that are full-grown, and have lived a full life of the kind that most animal welfare advocates would like animals to have—a life in the wild. Although death is not always quick for hunted animals, it's probably less suffering than what's experienced in a factory farm. If someone hunts a deer and eats that meat, that's better for animals than that same person eating the same amount of meat from the grocery store. What I'm saying applies to the suffering of the hunted animal, but of course there are other factors to consider. People hunt endangered species, which is bad for different reasons. But putting your energies into protecting non-endangered animals from hunting is a less effective use of your resources than trying to protect animals in factory farms.

Hunting makes people upset because of the vivid, violent image it generates in people's imaginations. But you might be surprised to hear that much of the American success of bringing species back from the brink of extinction was paid for by hunters. Sixty percent of the revenue from wildlife conservation groups comes from the sales of hunting licenses and permits. And these agencies are facing dwindling resources as hunting goes out of fashion. Wildlife conservation is one area where people who are otherwise quite different—say, hunters and women's groups who love birds—can come together to create positive policy change. The Audubon Society was started by a hunter, George Bird Grinnell. So not only does hunting do a only a small amount of damage to the animals killed by hunters, the activity of hunting has had an overall positive effect on wildlife and environmental conservation.[81]

All of this discussion of hunting has not included industrialized fishing, which is an altogether different enterprise, and one that actually matters a whole lot. Wild fish get to live out their lives free in the water, but their death due to fishing is pretty brutal. Fish get caught in hooks and are dragged for many hours until the net is pulled up. Then they suffocate in the net.[82]

Fish are quite alien, so it's easy to ignore the possibility that they feel pain just like other animals. But they react to harm similarly to other animals, learn to avoid harm, and their behavior changes when they're drugged with painkillers. However, there is currently a debate in the literature over whether fish (specifically ray-finned fish, unlike sharks and rays) can feel pain at all. There are a few arguments against the idea that fish feel pain. One is that pain seems to involve the cortex in humans, but when the fish equivalent is removed they still try to get away from harm. Further, the behaviors they take when trying to avoid harm can be explained by reflex actions, similar to those that a human will still engage in even after a spinal cord injury that renders them unable to consciously feel pain below the neck. Finally, when their fins are harmed, they don't favor them like mammals do—if you harm a rat's paw, for example, it will lick it and try not to walk on it. Damaged fish don't seem to react this way. Other scientists dispute these findings and interpretations of them, so the science is inconclusive.[83]

Vegan Diets

Although a vegan diet means that animals are not killed so that you can eat meat, we can't fool ourselves into thinking that veganism does not result in any deaths at all. When you harvest plants as a part of industrial agriculture, there are deaths as a side effect. Mice, rats, and voles can get chopped up or trapped in compacted tunnels, crushed by tractors, or have their burrows destroyed by plows. Poisons kill rodents and insects. When you plough a field, predatory birds follow the plough to feed on all the small mammals, lizards, and snakes killed in the farming process. Beetles live in the flour processing plants and must be killed. Small amounts of their crushed bodies are in our bread—it actually makes the bread a little more nutritious.

Some think that these deaths are so numerous that they could potentially outweigh the animal suffering associated with eating pasture-raised cattle. The idea is that when you raise a cow on a field, it doesn't kill that many mice or insects. When you eat beef, you're only killing the equivalent of a tiny fraction of a cow. This means that to the extent that you eat plant-based food from industrialized agriculture instead of pasture-raised cattle, you are doing a disservice to animal welfare. This view is *the new omnivorism*.

A problem with thinking about this issue is that the data on which how many animals are killed in harvesting are rare and problematic in many ways. Getting an answer is tough for many reasons, including the following: different plants have different numbers of animals living in and around them, different climates have different kinds and numbers of animals, it's hard to know which animals merely fled when the harvester came versus those that died because of it, it's even harder to know how those animals fared after being displaced, some animals die of increased predation due to the clearing of a field, pesticides have runoff that has effects on other animals, including fish in the waterways, and some animals actually have increased populations due to agriculture.[84] So few scientists care about the issue that almost no work is being done on it.

A very, very rough estimate suggests that there is one rodent death for every hectare of cropland harvested.[85] Okay, so not so many rodents killed. The vast majority of animals killed in agriculture are bugs killed by pesticides.

In a previous draft of this book, I tried to calculate the moral impact of pesticides, but eventually decided not to include it, because the numbers we have to use to try to estimate it are very uncertain, and often missing entirely. Without a much more in-depth analysis, and more data collection, I am not confident knowing if plant agriculture is good or bad for animals. Here are some of the relevant factors we'd need to consider for, say, a hectare of land: the number of bugs in a cultivated field, how often the bugs die by pesticides, how much bugs enjoy their lives, and how much they suffer when they die, how much that matters, how much food is generated by the crops, and how much utility humans get from that, and so on. But even if we understood this, we'd have to compare that land use to what would be done with it otherwise: the counterfactual. Would it be wild? A parking lot for a clothing store? To understand this we need an estimate of the moral value of a field of crops versus what else would be done with the land, and there's just too much uncertainty here. I'm afraid I'll have to reserve judgment on this until more work is done.

The new omnivorism argument is controversial and considered debunked by many animal rights activists. But even if it were true, it only applies to pasture-raised beef. Other meat, like chicken and pork, requires growing food to feed the animals, so you get all of the drawbacks of farming food anyway.

As it looks right now, reducing meat consumption appears to be better for animal welfare if the choice is between eating a lot of meat or eating none at all. But some have argued that pasture-raised cattle have fairly decent lives—lives that would not have existed if we didn't eat beef. So by eating beef you're *giving* a pretty good life to a cow. Again, though, the moral value of this depends on what the counterfactual is. What would we be eating in place of beef? What would the pasture have been used for otherwise? What animals would have lived if those cows hadn't, and how happy would they have been? We don't know.

Cultured Meat and Plant-Based Meat Substitutes

Eventually, raising animals for their meat might become outdated and all but disappear because the alternatives are a better value for consumers. I say "all but disappear" because there will likely always be a market for meat from animals. Some might see it as more authentic. Perhaps the reduction in farm animals will be the kind of magnitude decrease that the world experienced when cars replaced horses—there are still horses around, but far, far fewer.

There are two ways to create meat alternatives: making plant-based products that taste like meat—your veggie burger and the like. I'm glad to see that, at the time of this writing, several plant-based burgers are not only on the market, but getting popular with consumers.

The other way is to create meat without having to raise an animal. This involves growing cells to create a piece of meat (or simulated ground meat) that is actually made of animal cells. But the cells are created in a vat, and didn't require raising and killing an animal to make it. This latter method has a host of difficulties—for example, a piece of meat growing in a vat doesn't have an immune system. But if it works, people could eat meat without contributing significantly to animal suffering.

Right now meat substitutes are not competitive with meat in terms of attracting the business of non-vegetarians, but this will change. When it will change, though, is highly uncertain. Estimates range from just a few years away to seventy years.[86]

Keep in mind, though, that because of insect death, these plant-based burgers are possibly not as ethical to eat as pasture-raised beef burgers, but are probably better than grain-fed beef burgers.

Activism

The individual actions you take in your life, like what you choose to eat, or whether or how you adopt a pet, are fairly small potatoes compared to what you can do in terms of donating to effective animal charities. The situation is similar to being good to people—actions that aren't donations don't add up to much good or bad.

The most effective animal welfare charities are listed on the Animal Charity Evaluators (ACE) website.

I'll focus on one, the Humane League. The ACE estimates that for every dollar donated to the Humane League, between -6 and 13 animals are saved.[87] This big range shows how much uncertainty is in this calculation, so we have to be humble about what we think we know for sure. The midpoint of -6 and 13 is 3.5, so you can expect to save 3.5 animals for every dollar you donate to the Humane League, mostly chickens. Think about this for a minute—if you donate a dollar to the Humane League, you're probably saving 3.5 chickens from factory-farming. That's a very good deal. If you gave $365 a year (a dollar a day) to the Humane League, you would save 1,277.5 chickens from factory farms. Even if you ate a chicken a day, if you donated a dollar a day to the Human League the world would have 912.5 fewer chickens in factory farms than if you didn't exist.[88]

The most efficient thing you can do for animal welfare is to donate to an effective animal-welfare charity. Given that the typical American diet results in the death of about 300 animals (the thirty we eat and the rest as side effects[89]), donating even $100 might completely cover you for eating meat, and donating $200 is far more effective at helping animals than even being vegetarian. Because it's so cheap to save animals in this way, let me be perfectly clear: it's far easier and more efficient to help animals through donation than by changing your diet.

But if you are still interested in changing your diet to help animals, eat mostly plants, and when you eat meat, try to eat mussels, salmon, and pasture-raised beef.

The study that concluded that salmon was good to eat did not put into its estimations the other morally relevant aspects of eating animals: environmental impacts, nutritional value (aside from caloric content), or the pleasure we get from the meal.

The nutritional effects of eating meat are contested and complex.[90] Some studies show that moderate amounts of eating meat has no effect on mortality, and other studies show that eating too much meat increases risk for cancer. Obviously, if you're starving, eating meat lengthens your life, not shortens it, so there is not one number that says how much your life will be shortened or lengthened by eating a kilogram of beef, for example. It's some unknown nonlinear effect based on how hungry you are and how much meat you eat in general. It also matters what diet you're turning vegan from. A meta-analysis of studies showed that going vegan from an American diet has more health benefits than turning vegan from a Taiwanese diet, for example, because the normal Taiwanese diet is more healthful than a normal American one.[91]

For people who have enough to eat, however, it looks like being vegetarian adds about 3.6 years to your life, with many effects occurring twenty years after you become a vegetarian.[92] It also saves you about $11 per week in terms of food cost.[93]

It also does not take into account how delicious meat is, either. The pleasure of eating the meat could be measured, converted into life years saved, and brought into the calculation.

We Raise Them to Be Eaten

When we often think about the ethics of eating animals, we tend to think about an animal being alive, and then, if we kill it, it loses its life. This is what actually happens for hunting and catching of wild fish for food. But a moment's reflection reveals that for farmed animals, the only reason they are alive in the first place is because farmers brought them into existence for the sole purpose of being eaten. Is this morally relevant?

If you refrain from eating ten chickens, it does not mean that there will be ten chickens who would have otherwise been killed will now live happy chicken lives. It does not mean a better life for ten chickens. Refraining from eating ten chickens means that (roughly) ten future chicken lives *will not be brought into existence in the first place*.

Looking at it this way, we might ask the question differently: when we eat a farmed animal, looking at its whole life, including its death, is that animal's life worth living, or would it have been better for it to never have existed?[94]

When framed this way, different intuitions about right and wrong come into play. Let's take, as an example, a human being who is suffering a lot. At age thirty, they have had chronic pain for their whole life, they have depression, and no social life. In fact, they are scorned by their society because of some superficial physical trait and live a lonely, miserable life. Although we are comfortable saying that this life is far from great, and that they experience a lot of suffering, it's another thing entirely to say that this person's life is not worth living; that it would have been better if that person had never existed at all. Curiously, even when a human's life involves more bad feelings than good, we have trouble assigning an overall negative moral weight to their lives. We want to *help* such a person, but stop short of saying that they're better off dead, or that it would have been better had they never been born.

But things get weird when we apply this to animals. Our intuitions about the horrendous conditions of factory farms are saying just that: that the animals in there suffer so much that it would have been better for them never to have existed. And this is why refraining from eating factory-farmed meat is a moral good: it prevents painful lives from existing in the first place.

It's interesting that even for human beings, our intuitions are different if we're talking about potential human lives of the future or present human lives that already exist. For example, suppose a couple is considering having children, but they live in a war-torn, famine-stricken area, and on top of that the couple is in extreme poverty and suffer from addiction problems. Suppose this couple decided that the life that they expect their child to live would be very difficult, and for this reason they refrain from having a child. Most of us would applaud this as a responsible, moral decision: don't have kids if you can't take care of them. And the reason we think this is a morally good decision is because of the suffering that the child would experience. But once that child is actually born, and becomes an actual person, and not just a potential one, our ideas about the moral value of their life changes. We still might think it was irresponsible to have had the child, but we wouldn't want to say that the child would be better off dead. Our intuitions resist the notion that, if that child were to die in her sleep, the world would be a better place. We'd still see that as tragic.

The ethics of eating farmed meat, then, is relevant to the lives of *potential* animals of the future. And when we think about potential lives, we are comfortable weighing the pleasures and pains of that future animal to decide whether or not it would be good for it to exist. On this framing, the question of whether it's okay to eat a farmed animal comes to this: eating animals of a particular kind (say, chickens in factory farms) leads to more animals like that being brought into existence. So the morality of creating demand for animals of that kind is determined by the expected moral value of that potential animal's life. Or, put more simply, do those animals have lives worth living?

By one thoughtful estimate, factory-farmed chickens have lives that are not worth living, but cows have lives that are. Pigs' lives are worth living, but only barely, and only assuming that they habituate to their miserable conditions somewhat.[95] If a chicken's life in a factory farm is horrible, then killing it sooner is better. Our ideas of long lives take on a negative value, because if each day of life would have been better to never have been lived, then a long life is bad, not good. On the other hand, if a life has an overall positive value, then allowing animals to live longer is better.

Some readers might be unimpressed by all of this, particularly if they are deeply concerned with human issues. It's easy to say that because humans can suffer more than animals, then we shouldn't concern ourselves with animal welfare. That's why the numbers are important, though, because even with a steep discount, *enough* animal suffering starts to compete with human suffering concerns when it comes to taking effective action in the world. If we're trying to optimize our goodness, when it gets right down to it, should we try to help humans or animals?

11

COMPARING HUMAN TO ANIMAL SUFFERING

So far we've reviewed what you can do to help out your fellow human beings, and also what you can do to help nonhuman animals. If you already feel strongly about one or the other, then your way forward is pretty clear, as there are excellent charities you can donate to (for example, the Against Malaria Foundation and the Humane League, respectively). If you're an extreme speciesist, for example, then just donate to the most effective human causes. The rest of this chapter is for people like me who believe that animals should get some moral consideration.

But if you're an optimizer, you'll want to know which of these causes are better. If what we really care about is increasing good feelings and decreasing bad feelings *for all entities that are capable of feeling*, then we want to know whether we should donate our money to help humans or animals. Even if humans can suffer more than animals, it might be more efficient to focus on helping animals if it's cheaper to do so.

Let's start by looking at lives saved. One of the most effective charities for reducing animal suffering is the Humane League. The website Animal Charity Evaluators has a report on this organization and estimates that a donation of $5,000 saves the lives of 35,000 animals, mostly chickens.[1]

This is a ballpark estimate, but it can help you decide where you want to put your efforts. Let's use the estimate that effective charities can save a human life for $5,000.[2] That's thirty-five thousand times more expensive than sparing an animal. Do you think that the average fish or chicken is worth more or less than one 35,000th of a human? Or, another way to think of it is: Would you rather save 35,000 animals or one human? For the most effective charities we have, the costs are about the same. If you choose one human, then you might want to invest in relieving human suffering. If you chose 35,000 animals, then you might want to invest your charitable donation in an effective animal welfare organization.

But, as we've seen, saving lives is a bit of a crude measure—what's better is looking at the number of years of life saved. Specifically, DALYs averted for humans, and the *equivalent* of years of human life saved for animals.

One of the most effective charities for saving human life is the Against Malaria Foundation. Estimates suggest that for $78, they can save one healthy year of human life.[3]

What is an animal life saved? We'll focus on chickens, because they are the main focus of animal welfare charities. Where saving a human life means *adding* years of life to a human, saving a chicken means they are not raised at all: factory-farmed chicken lives are so bad that we consider them better off to never have lived. Chickens live about 0.1151 years before slaughter (42 days).[4] Saving 7,000 chickens means preventing 805.7 years of chicken existence. This existence is worse than nonexistence; we assume that each year of chicken existence is 0.25 times as bad as going from a healthy human life to nonexistence. Then we multiply by the suffering discount rate of 0.0038, leading to the conclusion that $1000 donated to the Human League means stopping the moral equivalent of 0.765 human life years ceasing to exist. So we can estimate that the Humane League saves (the animal equivalent of) 1 year of human life for the cost of $1306.50, and the Against Malaria Foundation can do the same for $78.

Now we're talking! We have, in U.S. dollars, how much it costs to avert one human life-year's worth of suffering. Using the most effective charities we have, you reduce almost 20 times as much suffering per dollar spent fighting malaria than fighting factory farms

There is a side benefit to donating money that discourages people from eating so much meat—not only is it better for animal welfare, but it's better for the environment. Raising animals, especially cows, uses up a lot of resources. Much of the plant-based food we grow is used to feed animals, not human beings directly. So reducing meat is a double-win.

In contrast, helping more humans stay alive is great for them, but has unfortunate side effects: more humans means more environmental destruction, and more eating of meat. When you save a human life, more animals will die because that human will continue to eat meat. This reduces the good done by saving a human life. By how much, though?

A Nigerian can be expected to eat 5.91 kilograms of meat every year, or about 1.8 animals. Let's suppose that as a result of your donations, you help someone live twenty more years of life in Nigeria, where 97 percent of the population is at risk from malaria. This means that an expected thirty-six more animals will die because of your intervention.[5] This is assuming that some other person would not take their place if they died. If you're saving an American life, then after twenty years 600 more animals would be expected to die, but saving American lives is so expensive we're not considering that. So in terms of animal welfare, if you think a human life is (roughly) more than thirty-six times more valuable than animal life, then you'd want to invest in preventing malaria, rather than helping out with animal welfare. We should think that human life is more than thirty-six times more valuable than animal life if we go by the sentience discount rates.[6]

According to my calculations, in terms of doing the most good, you should help humans instead of nonhuman animals.

12

ENVIRONMENTAL MORALITY

A curious tendency in the world of philanthropy, philosophy, and economics is that human welfare, animal welfare, and environmental concerns all operate in their own silos and rarely compare results with each other. What I've been trying to do in this book is to put it all together, because what we need is advice on what to do when we've taken everything into account that we can.

Unfortunately, environmental damage covers a lot of stuff, including river pollution, extinction of species, destruction of the rainforest, and climate change. Lacking data on environmentalism as a whole, I'm going to focus on climate change, because it looms large in our minds, it's really important, and people have done studies on it. But keep in mind throughout this discussion that thinking of environmental damage only in terms of climate change is always an underestimation, as there are lots of other kinds of relevant damage.

When you want to compare things, you need to have some unit of measure that is common. The World Health Organization estimated that an additional 140,000 deaths were caused by climate change in the year 2004. This is due mostly to changes to the threat of climate-sensitive diseases, such as dengue and malaria, but also through crop failure.[1] In this book I've been using years of human life lost (DALYs). We've seen that you can roughly translate animal suffering into years of human life lost. Can we do that for climate change?

WHO IS CLIMATE CHANGE GOING TO KILL?

When you look at the relationship between temperature and human mortality, we can see that the ideal human temperature is about 23° Celsius, or 73° Fahrenheit. When the temperature goes above or below that, more people die. An extra day at 35° Celsius (95° Fahrenheit) results in 0.4 more deaths per 100,000 people annually, and an extra day at -5° Celsius (23° Fahrenheit) causes 0.3 more deaths per 100,000 people annually. (In this sidebar, all deaths reported are per 100,000 people annually.)[2] Because it's hard to understand the importance of these numbers, here are some points of comparison: in the latter half of the 20th century, around the world, about 780 out of every 100,000 die every year; 125 of those are from cancers and 11.4 are from auto accidents (in America).

But other factors are very important: normal weather conditions, wealth, age, and how adapted regions already are, or will become, to heat.

These adaptation effects are important because as the years go by, and the globe heats up, people will adapt: more places will get air conditioning, they will exercise outside early in the morning, spend more time indoors, and so on. So when you are trying to predict death caused by climate change into the future, you can't assume the world will stay the same: you need to account for how people will be expected to adapt as the heat goes up. Surface temperatures are going up at a rate of about 0.0001° Celsius per day, giving people time to make adaptations if they can afford it. Failing to account for income and adaptation makes mortality estimates to be over twice what they should be.

Seattle and Houston are similar in many ways, but temperature is not one of them. Because Houston is so much hotter than Seattle, it has already adapted to the heat—air conditioning in Houston is almost 100 percent, where in the state of Washington it's only 27 percent. This means that really hot days are rarer in Seattle, but also more deadly, because Seattle's not prepared for it. An extra day at 35° Celsius (95° Fahrenheit) will kill 1.3 more deaths (per 100,000 people annually) in Seattle than in Houston.

> Regions of the globe that are already hot and poor will be the worst affected by rising temperatures. By the end of the century, one estimate says that Accra, a city in Ghana, will have about a 19 percent increase in death compared to today (160 deaths), due to global warming. In contrast, Oslo, Norway, will enjoy a *decrease* in deaths by 28 percent (230 deaths avoided).[3]
>
> Richer areas will be better able to adapt. Regions in the top 10 percent of wealth are expected to have *no* climate-change-related deaths by 2100 because of their expected ability to adapt, and today's poorest regions will be the most affected, both because they're the hottest and because adaptation requires money they won't have.[4]
>
> Age is also important, as the elderly are particularly vulnerable to temperature changes. For people over sixty-four, a hot day will kill between 4.5 and 10.1 more. A cold day (-5° Celsius), 3.4 extra deaths.

Estimating of how many people will die due to climate change is extraordinarily complex, and the papers that even attempt to do so only started coming out in 2018.[5] These estimates look at how many people will die over the next one hundred years for every ton of greenhouse gas put into the atmosphere. Specifically, they estimate in terms of life-years lost per ton. According to these estimates, taking actions that result in putting a ton of greenhouse gases into the atmosphere results in somewhere between -0.00019 and 0.0013 human life-years lost. Put in terms of minutes, that's between 97.8 minutes gained and 668 minutes of life lost. You might be surprised to see that this range includes adding life to the world. This is because climate change causes a variety of effects, and some of them are good, like fewer people dying of cold-related deaths, and better farming in currently colder climates. But it is far more likely to cause loss of human life.[6]

Even though we can calculate the number of human life-years lost due to our consumption's release of greenhouse gases, it doesn't feel nearly as visceral as poisoning somebody's food, or strangling them to death. But deaths are just as real, and their moral implications are just as bad. If you're unconvinced, here's a compelling thought experiment by philosophers Jonathan Glover and M. J. Scott-Taggart: Suppose there's a land where there are a bunch of people who are near starvation. Each one of them has a bowl of beans. A bunch of robbers come down and each robber steals one bowl of beans, each causing one particular person to die. Then the thieves have a discussion about the ethics of what they're doing. Some of them, it turns out, are bothered by the fact that the food that they're taking from others is resulting in death. So one of the robbers suggests that rather than each robber going down and taking one bowl of beans from one person, and thereby killing that person, each robber steals one bean from each bowl. At the end of the raid, all the bowls will be equally empty, and the same number of people die, but no one will be responsible for anyone's death in particular, because they will have only taken one bean from each bowl, which is, really, an amount so small that you wouldn't even notice it. Each robber sleeps soundly, knowing that they were not responsible for anybody's death.[7]

The robber's reasoning is almost comical in its ridiculousness, but there is a direct analog to many of the harms that we can cause in the world today, including climate change. The fact that one is causing a tiny little bit of harm to lots of people, rather than a lot of harm to one person does not make the action any better, as the beans story makes clear. By now you know why it feels different, though. Human morality evolved in a world where hurting everybody a little bit, particularly people all over the world, was impossible. We have to use our principles and reasoning to see that it's bad, and not rely on our instincts.[8] This is why people don't feel bad about an employee of Disney stealing a stapler from work or pirating *The Mandalorian*. They're only taking one bean from each Disney shareholder's bowl. It just doesn't *feel* wrong.

Recall that earlier I used Monte Carlo simulations to estimate how much animals can suffer and how much it matters. I concluded that it was far worse to eat chicken than beef, primarily because chickens are treated so much worse, and because there's so little meat on them. But it turns out that raising cattle is much worse for the environment than raising chickens. So eating beef is better for animal welfare, and eating chicken is better for climate change. So which is the better meat to eat?

Now that we have estimates for the damage done to human lives for every ton of greenhouse gases emitted, and we know how many tons of greenhouse gases are emitted to raise different kinds of animals, we can put it all in a common currency and see which is better and which is worse.

We can put chicken and cow suffering in terms of its equivalent in human suffering. After 10,000 simulations, using ranges of numbers provided by experts, it turns out that chicken is still much worse to eat.

Our simulations suggest that eating a serving size of beef does a very small amount of good, because the lives of domestic cattle aren't so bad, to the tune of adding the equivalent of 0 to 0.3 days of human life over the next one hundred years.

Eating chicken, on the other hand, shortens the equivalent of human life by between -0.053 and 0.68 days of life. Put bluntly, eating one factory-farmed chicken patty (which is about one serving) does the damage to the world in the equivalent of shortening someone's life by about half a day.[9]

So far I've only talked about caring for the environment in terms of how environmental destruction will affect the lives of sentient beings. Indeed, I think that effects on sentient beings are, ultimately, the only things that do matter. However, it's hard to predict the far-future effects of environmental destruction, even if we limit ourselves to effects on sentient creatures.

Should we care about wilderness? Let's suppose we could tear down an ecosystem and replace it with another one that has less species diversity but has the same amount of good feelings. No problem? We shouldn't be too sure. Older ecosystems cannot simply be replaced, any more than the Notre-Dame cathedral can simply be rebuilt. There are potential benefits that might be gained from study or some other kind of exploitation of these resources in the future that we might not be able to imagine now, from medicines harvested from particular plants, to biological understandings from particular species that might be hard to get by studying other species. Biological diversity is a huge, untapped reservoir of future scientific discoveries that are likely to positively affect sentient beings in the future, even if we have no clear idea of what those benefits will be. As such, we should be wary of destroying old ecosystems based on simple calculations of life-years lost, for example.

For similar reasons we should care about extinction. Killing a male and female rabbit is better than killing the last male and female of a similarly sentient species. Even though they might suffer the same amount, extinction truly is forever.[10] Rendering a species extinct means we cannot study it, and future sentient beings won't be able to enjoy the benefits of what we might have learned from such study. We should preserve species and ecosystems not only because of the suffering of the now-and-future creatures in it, but for the same reasons the loss of the library of Alexandria was such a tragedy—lost knowledge that could have made the world a better place.

Local and Organic Food Antimatters

Some people think that eating organic food is better for the environment because the creation of it uses fewer pesticides, preserves soil better, and keeps water cleaner. Talking about "organic food" in general sweeps under the rug an enormous variety of foods and how they are grown. But we can talk about organic versus conventional food in general, even if there are exceptions to our conclusions.

Unfortunately the net effect of organic food seems to be mostly negative. It's not any more nutritious, and doesn't even taste any different, in most cases.[11] The pesticides on conventional food are in such small amounts that most scientists agree they are far below what is considered a risk to health.[12]

Precisely because of the lack of pesticides used in organic farming, a lot more food goes to waste. This means that more land has to be used to raise the same amount of food.[13] In fact, we would not be able to feed the world's growing population without conventional (nonorganic) agriculture.[14] Because more land and water is required, that means more habitat destruction. This means turning more wild land (which tends to be a carbon sink) into farm land (which emits carbon into the atmosphere). Organic food is, on average, worse for climate change than conventional food.[15] As policy analyst Mark Budolfson puts it, buying organic is voting with your purchasing choices—it's just voting for habitat destruction and starvation.

Local food, too, is another well-intentioned idea that is generally worse for the environment. The reason is that wherever you live, there are going to be lots of foods that would be more efficiently grown elsewhere. These foods get produced with less energy, and then shipped to you in bulk. Although they travel farther, there is so much food that the impact is lessened. In contrast, local food is often grown less efficiently, by small farmers, and then small quantities are driven in on a truck. Locally grown food has a higher carbon footprint.[16]

CHAPTER 12: ENVIRONMENTAL MORALITY

Eating organic and, in most cases, local, antimatters. As organic food is more expensive, you'd do much more good buying regular food and donating the price difference to an effective charity.

We've seen that when it comes to helping people, some people are just cheaper to help than others, and sometimes these people are far away. We get the same kind of thing with certain environmental causes.

One of the core values of the environmental movement is the preservation of species. That is, species going extinct is frowned upon. It's no secret that tropical areas have much, much more diversity than colder ones. Ecuador, for example, has more species of trees in each hectare of tropical rainforest than all of Canada and the United States combined.[17] Think about that—if you walked the length of a football field in a rain forest, you would pass within throwing distance of more species of tree than in all of Canada and the United States.

So from a species-protection point of view, owning and protecting a single hectare of Ecuadorean land would be better for the environment than protecting all of the U.S. and Canada. Of course, there are other environmental concerns than preventing species extinction. My point is that there are great inequities in terms of how much good certain actions take. As usual, some things matter much more than others.

The choices you make in your life affect the environment in many ways, mostly adversely. Much of the problem is due to the energy consumed by your purchases. If we just focus on greenhouse gases for a moment, leaving your television on all day causes some amount of CO_2e to be released into the atmosphere. How much? It turns out it greatly depends on how your electricity is generated. The most efficient, environmentally friendly way to get energy is from nuclear power, and the worst ways are burning wood or coal.[18]

Some places get most of their energy from dams, which do environmental damage to ecosystems when they are created, but after that have negligible bad effects. What this means is that in terms of electricity, if you live in a place where your energy is relatively cleanly generated, you should probably focus less on turning off the lights and more on other, more effective changes.

If you want to help the environment, there are ways you can change your behavior to do less damage. Unfortunately, doing the things you need to do to stay alive in the modern world usually requires harming the environment. But by doing things like taking fewer flights, you can do less harm. For an entertaining book describing the climate change impact of everything from drinking a glass of tap water to destroying a power plant, I recommend the book *How Bad Are Bananas?* by Mike Berners-Lee.

But if you want to maximize your help for the environment, it involves giving money to charities. Right now one of the top-rated charities is the Clean Air Task Force (CATF). The charity evaluator SoGive estimates that this is a particularly effective charity for fighting climate change, and that somebody can avert the emission of one ton of greenhouse gases for between three cents and $5.50. The range is huge because this is an area of incredible uncertainty.

We also know how much a ton of greenhouse gases emitted into the atmosphere will harm humanity in terms of human life lost over the next one hundred years: between 97.8 minutes gained and 668 minutes of life lost. Roughly, this suggests that for about $2.77 you can expect to save about 285 minutes of human life by helping the CATF fight climate change.

At the time of this writing, there are no environmental charities among the top rated charities of GiveWell.org, for example.[19] Like many other attempts to mitigate future risk, particularly catastrophic risks, such as detecting planet-destroying meteors, a malignant artificial intelligence takeover, or catastrophic climate change, there's too much uncertainty and too many unknown variables to confidently put numbers on how many lives you save by investing money now in an attempt to prevent these disasters.

13
CHOOSING CHARITIES

Let's look at blindness. Bad, right? The WHO sees living a year blind as 60 percent as bad as not living it at all. One way to help blind people is by providing them with guide dogs. Training a guide dog costs about $40,000, and the dog is of use for typically eight or nine years. So it costs about $4,444 every year.[1]

The most common cause of preventable blindness is trachoma. To keep someone from going blind from trachoma costs between $20 and $100. The difficult-to-accept truth is that for the money spent to train a guide dog for a blind person for one year means allowing about forty-four people to become blind who would otherwise have not if the money were spent on them instead.[2]

Now don't get me wrong—I love guide dogs. When I see one out and about, I get a surge of goodwill and happiness. I watch guide dog training documentaries and cry my eyes out.

In contrast, when I see a person who *isn't blind*, I don't cry with joy because they can see. Blindness doesn't even cross my mind. This is why guide dogs are so much more viscerally meaningful to donors. Preventing blindness is less heartwarming. But if I really cared about the problem of blindness, I'd donate the money to trachoma prevention instead of guide dog training.

GiveWell.org is a website that uses evidence to rank charities in terms of effectiveness.[3] GiveWell uses data to determine, for example, DALYs averted for every dollar you put in. This results in a set of highly ranked charities. By donating to one of these, you can be confident that your money is being spent in the morally best way possible.

What's missing from the rankings on GiveWell are some of the best-known charities, like United Way and CARE. This is because large organizations like this are doing so many things, and provide so little data relevant to their effectiveness, that they can't be quantitatively compared. In other words, when you donate to United Way, you have no idea if your money is doing any good. Or, for that matter, if your money is causing harm. When you can be reasonably sure that a human life will be saved with every $3,000–$7,000 donated to the Malaria Consortium, how can you justify giving anything to United Way? This doesn't stop even my own university, which seems to have an enduring, unjustified affection for United Way.

One unfortunate tendency in the effective altruism movement is to pit one charity against another. It's interesting to think about the relative amount of good you're doing for the world when you donate to a local art museum versus saving lives with bed nets, but why is it always charity versus charity?

It's because people tend to like comparing similar things. That's why there's a tendency to criticize someone's choice of charity. They tell you they donate to some relatively ineffective charity like the Make-A-Wish Foundation, and you want to say, "If you *really* cared, you'd donate to . . ." But when they tell you they spent $3,000 on a vacation to Cancún, we say, "Awesome! Did you have fun?"[4] In terms of helping the world, maybe even an inefficient charity like Make-A-Wish is better than taking a trip to Cancún.

But there is some justification for these comparisons because many people budget their charitable giving. If someone is going to donate 5 percent of their income each year to charity, then it becomes a competition for that money between charities. The money to pay for the Cancún vacation comes out of a different budget, and as such is not in competition for the same pot of money.

Furthermore, it seems that people have some internal thermostat for being good. So if they are giving to charities that unmatter, they will become morally satisfied, and thus not want to give further. We want people to direct the money they're willing to give to the best places, rather than having their sense of moral worth satiated on less-effective ones.

CHAPTER 13: CHOOSING CHARITIES

This probably isn't news to anyone, but people give to charities for more reasons than doing good for the world. I might give my friend money to help them pay for surgery for their pet. I might donate to a local thing that means something to me. These are personal reasons. Or a business might give to a charity that makes them look better. Are these reasons justifiable?

There are two ways of looking at it. On the one hand, they are less justifiable than giving the same amount of money to an effective charity. But this is true of just about all of your spending. That is, donating $20 to the school book drive is money almost as poorly spent as $20 to see a movie.

BUDGETING YOUR MONEY

Here is how I recommend you deal with your income:[5]

50 percent must-haves: rent/mortgage, insurance, food, basic phone
20 percent discretionary spending: fun stuff, less effective charities
20 percent savings and investment: emergencies, retirement, and return on investments, paying off credit card debt
10 percent charitable donation to a GiveWell-recommended charity

If you spend too much on discretionary spending, cut back, or get a higher-paying job. If you spend too much on must-haves, consider moving to a cheaper living space. Over time, you can increase the percentage you give to charity. People with higher incomes can afford to increase the percentage of money going to effective charities.

But another way to look at it is terms of your personal budget. Suppose you make $100,000 per year. You decide that you'll spend $10k per year on charity. You pick a GiveWell-recommended charity. You put away $20k for savings and long-term investment. You spend $50k on "must-haves" like food, shelter, and basic phone. ("Must-haves" are those things you'd spend money on even if you no longer had any income.) This leaves $30k per year for discretionary spending: vacations, bubble tea, Laurie Anderson concerts, and less-effective charities. You can look at suboptimal charity giving as coming from your discretionary spending budget.

This is what I do: If a friend has a fundraiser, I want to donate to it, because I want to be a good friend. I don't fool myself into thinking that I'm helping the world in any cost-effective way. I recognize it for what it is: I'm spending this money primarily for the nurturing of my relationships: my feelings, and the feelings of my friends. I can spend it because it's coming out of my bubble tea and Laurie Anderson concert budget, not my charity budget.[6]

This is important, because it's not good to spend too much energy second-guessing every single thing you do! "Huh, I'd really like to have a bubble tea. But what's more important, me having a bubble tea or giving that same money to charity? I'll have the water with no ice, please."

Questioning everything you do has cognitive load and will make you feel guilty, which is miserable for you and not sustainable for living the best life you can. You don't want to have to think about malaria or factory farming every time you pull out your wallet. By compartmentalizing you don't have to question every decision. You decide up front what percentage of your time and income is dedicated to discretionary spending, charity, and savings. Then, as long as you stay in budget, you don't need to worry about getting a bubble tea. Live it up, brothers and sisters! Tonight we feast!

Some readers might be alarmed by this reasoning about comparing charities. Some of you are thinking about, say, patronage of the arts. Does this way of thinking lead to the conclusion that donating to the arts is never a good idea? The answer is yes, if you're considering only the effectiveness of charitable giving. There are other ways to look at it, but let's talk about donation to the arts *as a charity* first.

Before we analyze it, though, think about this: Suppose there are two jars of peanut butter you could buy: some generic one from your grocery store, which costs $5, and one from your cousin's peanut butter company, which costs $10. Despite the fact that your cousin claims it tastes much better, you can't tell the difference. Physically, they are the same in all the ways that matter. But you might buy peanut butter from your cousin anyway.

In this simple example it's clear here that what you're doing is spending $5 on peanut butter and $5 on the good feeling you get from the purchase,

CHAPTER 13: CHOOSING CHARITIES

and from the protected or improved relationship you have with your cousin. In terms of food-buying, this purchase works at 50 percent efficiency. This might well be worth it to you. The point is that you compare the candidate purchase (buying from your cousin) to the most effective purchase (the no-frills peanut butter at the grocery store). This is how we're going to think about charity: by comparing some candidate charity to a really effective one.[7]

Recall that you can prevent blindness in somebody for about $50—you can dispute the exact number, but you can follow through with this reasoning with whatever numbers you think are right (just make sure it's based on data!).[8] Let's suppose that a local theater company wants to raise money to put on a production. Suppose the theater needs to raise $10,000 of donation money to produce the play, and during its run be seen by 5,000 people.[9] This means that it costs the donators $2 to have one audience member be entertained and, hopefully, deeply moved. So $50 can either allow twenty-five people to watch a play, or prevent blindness in one person. For these two charities to be morally equivalent, the benefit of those twenty-five people seeing the play would have to be as great as a person not going blind.

Philosopher Peter Singer offers this kind of thinking, which I adapted to a theater production: suppose there is an evil spirit who hates theater, and he watches people buy tickets and starts counting.[10] One, two, three . . . When he gets to twenty-five, he uses his wicked magic to inflict blindness on a random person in the world. Then he resets his counter and gleefully starts counting ticket buyers again. During the run of this production, if 5,000 people see the show, he would blind 200 people. Would you buy a ticket to this show? Even if it were really good? How good would the play have to be to be worth blinding someone for every twenty-five people who saw it? The answer: even better than Julie Taymor's *The Lion King*.

I'm not picking on theater because I don't like it—I'm a produced playwright, and helped run a theater company. Theater is close to my heart, and an important part of my life. It's certainly true that theater has the potential to really move people, and at its best, be transformative. But if you look at *actual* theater productions most of them aren't like that—for example, think about the one hundred most recent shows produced within an hour's drive of you. It's a relatively inefficient way to improve the human condition. I try not to let my love for the art fool me into believing anything different.

CAUSES AND SYMPTOMS:
AN EFFECTIVE ALTRUISM CRITIQUE

One criticism of the effective altruism movement is that it, at best, puts Band-Aids on the cuts, while ignoring the swinging knives that do the cutting. Global poverty, and the food insecurity and susceptibility to infectious disease that comes with it, are caused by aspects of the social and political systems of the world. By feeding the hungry and providing malaria nets, the argument goes, charitable donations are doing nothing to stop the ultimate causes of suffering.

At worst, charitable giving is preventing or slowing systemic change with these Band-Aids—perhaps because things need to get really bad in order for there to be a revolution or because the government doesn't do the work because the aid organization is going to.[11]

Working for systemic political change is often not conducive to the data-driven, empirical methods that effective altruists use to choose a method of philanthropy. Political advocacy often requires time to see results. It happens in a complex social environment, where any change (or lack of change) is likely the result of many uncontrolled and unobservable factors. These reasons make it more difficult to do experiment-like observations.

But even if you think political change is important, how should politics be changed? There is incredible disagreement about this. For example, some scholars believe that capitalism is actually a major *cause* of global poverty, while other thinkers believe that capitalism is one of the things that has *reduced* poverty.[12] I'm not going to venture into this debate, as it is too far outside my expertise for me to do it justice. But I will say that when the experts disagree, you might want to focus your efforts on things where there is more widespread agreement, because it minimizes the probability that your actions antimatter.

I do find it encouraging that saving lives does seem to help the economy. A 9.4 year increase in life expectancy leads to a 1 percent increase in Gross Domestic Product per capita, according to a study of 48 African countries.[13]

CHAPTER 13: CHOOSING CHARITIES

This analysis is just to evaluate donations to theater *as a charity* for someone trying to be as good as they can be. I'm not saying don't donate to the arts, I just want you to keep in mind the difference in good done between giving to the arts versus a really effective charity. That difference, the "altruistic arbitrage," tells you how much of that donation is actually for you, for your warm glow, just like a part of buying peanut butter from your cousin is for you. Does giving to the arts do some good for the world and make it better place? Sure! It's just a small amount of good compared to what you could do with the money instead, so you should recognize that most of the effect of that donation is self-serving, like buying yourself ice cream. If you give to the theater company out of your discretionary budget, and not your charity budget, you're good to go. That is, giving to the theater company is harmless as long as it doesn't prevent you from giving as much to effective charities.

But we can get a little more detailed than that, because donating to the theater does some good, and is somewhat self-serving. Just like the calculation with your cousin's peanut butter, we can calculate how much.

We'll use the assumptions we've been working with. If it costs $2 of donation money to allow someone to see the show, then a $50 donation to the theater company will allow twenty-five people to see it. We also believe that a $50 donation to an effective charity will prevent someone's blindness. In this light, the theater donation sounds like a bad deal, but seeing theater is not worthless, so, presumably, there is *some number of* people seeing the show that would be worth making someone blind. What is that number? How many people would have to enjoy the show to be worth blinding somebody? Maybe it depends on the quality of the production. Fair enough. Some shows are a waste of time, and others are incredible experiences. Ahead of time, you don't know how good the show will be, so assume the show will be of average quality. Think of all of the theater you have seen. How many of *those* shows would you need to see to make it worth blinding someone? Or, to bring it closer to home, suppose Linda has had a life in which she saw no theater but her vision was intact, and Belinda has a life where she saw some number of plays, of varying quality, and then went blind. If you had to choose between being Linda or Belinda, how many plays would Belinda have to see before she went blind to make this a tough decision for you?

Most of you are probably thinking, "There is *no* amount of theater that would make it worth going blind!" but, just to humor me, let's come up with a number and see where it goes, because this is a kind of reasoning you can apply widely, not just to theater (see the sidebar "Comparing Charities"). Suppose that number is 2,000. That is, if Belinda saw 2,000 theater productions, there would be enough really good ones in there that it would be worth going blind over. Or, if 2,000 different people see the play, the good of it balances the bad of someone going blind. Stay with me here—I'm getting to the part where I help you justify the theater donation!

SHOULD WE LET PEOPLE STARVE FOR POPULATION CONTROL?

Some people think that overpopulation is the cause of poverty, and that saving poor people now just means more suffering people in the future, so we should let them die, allowing things like famine and disease to lower the population to a sustainable level.

This idea isn't a very good one. For one thing, it is based on a few misunderstandings of how the world works. Human population seems to be a function of how many children people *want* to have. With some exceptions, people know that sex causes pregnancy and have preferences about how many children they want. Other animals don't do this. What this means is that if a whole lot of people die, say, from a famine, it might not change the underlying desire in the people of an area to have many children. Indeed, when there is a famine, the population is quickly replenished.[14]

People often want a lot of children when they know that many of the ones they have will die within the first few years of life, or if they are so poor that they can't save money to support themselves in old age.

What makes populations go down are incentives and the choice to have fewer children (that is, availability of birth control). Economic development,[15] lowered child mortality, and education are what lower populations over time, not allowing them to die of famine and disease. We can see that industrialized nations have a zero or even negative population growth rate, in cultures as different as the

United States and Japan (when controlling for immigration). Regions appear to pass through a demographic transition. As their standard of living increases from extreme poverty to more wealth, population increases. But as living standards increase further, population drops as people get more access to birth control and have less of an incentive to have lots of children.[16]

This means that a $50 donation would have to allow 2,000 people, not 25, to make it a competitive charity with those preventing blindness at $50 a shot. So 2,000/25 is 80, and 80 is 4 percent of 2,000. By this reasoning, your donation to the theater is 4 percent as *effective* as the donation to prevent blindness. Put another way, if you donate $50 to the theater production, $2 of it is for the good of the world, and $48 of it is for your own warm glow, just like $5 of buying peanut butter from your cousin is for the peanut butter, and $5 is for your positive feelings about doing it.

Again, this is not to say that it's not worth doing. Buying $50 worth of ice cream for yourself probably means $50 of warm glow (cold glow?) while doing just about *nothing* good for the world. The theater donation works at 4 percent of charitable efficiency, and buying ice cream works at 0 percent efficiency. But you can spend discretionary money on ice cream, theater donations, or whatever you like.

Suppose you want to give 10 percent of your income to charity. You choose something really effective, like the Against Malaria Foundation, for the majority of your giving. But then your friend is in a show, and she takes part in a funding drive for the theater company, and asks you for a donation. She does this in person and you'd feel like a tool if you don't give something. You donate $50. Then you go back to your budget. How much of this $50 expenditure counts as charitable giving, and how much counts as discretionary spending?

You can do a calculation like the one I did above. If you use the same numbers, for budgeting purposes you would conclude that this $50 donation resulted in $2 coming out of the charity budget and $48 coming out of the discretionary spending budget, where it competes with ice cream, vacations, and other self-serving sources of happiness. It's not that giving to the arts is bad, it's just not very efficient as a charity. You should be clear about what good it's actually doing, and how much of what it is for *you*, and budget accordingly.

COMPARING CHARITIES

Here is a calculation you can do to compare a candidate charity to a really effective one.

Donation =	how much you would donate (default: $100)
Helped =	how many people helped
Prevent =	cost of helping someone with an effective charity (default: $50)
Count =	number of people who would have to be helped with this charity to be equivalent goodness of one person having the thing that would be prevented (default: going blind)
Cost =	Donation divided by Helped. This is how much it costs to help one person with the candidate charity.
Efficiency =	Count Divided by Helped
ActualHelp =	Efficiency Times Donation. This is the dollars' worth of good you're doing with your Donation, counting as your doing good for the world.
Glow =	Donation minus ActualHelp. How much of this donation is for your own good feelings.

So, given some Donation amount, you can calculate how much of that money is actually doing good for the world, and thus comes out of your charity budget, with this:

*ActualHelp = (Count/Helped) * Donation*

And you can calculate how much of that Donation is self-serving, and should come out of your discretionary budget, with this:

*Glow = Donation − ((Count / Helped) * Donation)*

CHAPTER 13: CHOOSING CHARITIES

This isn't so bad. Lots of people don't think twice about spending $48 for a bottle of wine at a restaurant for a warm glow. But nobody *needs* wine, so it's discretionary spending. So go ahead and donate to ineffective charities for the warm glow. Just calculate how much of it was actual charity and how much was just making yourself feel good.

You have to spend money on you, because giving too much can lead to misery and burnout. Even from an effective altruism perspective, giving away *everything* you have will inhibit your ability to do good in the future for these reasons.

You've got to look presentable for work, stay healthy, etc.[17] What about close personal relationships? Utilitarianism is often criticized on the grounds that you have a moral obligation to more needy people before even your own children. But if you don't take care of your own children, what will that do to you? Being maximally good means maximizing good for the long haul, which means acting in the moment in a way that will maximize your good over the course of your whole life. You can burn out of charity just like you can burn out of anything. To avoid burnout, you need to be happy, and studies show that happier people are more likely to help others.[18] That's the only way you'll be able to be a good person for your whole life.[19] That means nurturing close relationships, and nurturing close relationships sometimes means throwing a surprise party for someone you care about, even if that money could have been spent on a bed net. When you are realistic with yourself, and the limits of what you can do, you see that maximizing goodness *requires* some amount of prioritizing yourself over the most needy in the short term, for the long-term good of the world.

Most people, of course, get nowhere near this ideal. Lots of people could give a lot more and be just as happy as they would be otherwise, if not happier. And in practice, there are no effective altruists who live anywhere near the global poverty level. Again, it's not about being perfectly good, it's about being as good as you realistically have the capacity to be. Just like me, you probably don't have the mettle to be a perfectly good. Being a practicing utilitarian is about being better, but being perfect is practically impossible.

Now we have looked at three of the biggest problems facing the world: animal welfare, human health, and climate change. We also know how much it costs to reduce these problems using charity, using the most effective charities, in terms of the equivalent of human life years gained.

If you want to help through climate change, you can donate to the Clean Air Task Force, which can save a year of human life for about $5,108. If you want to increase animal welfare, you can donate to the Humane League and save the equivalent of one year of human life for $1306.50. If you want to help human health directly, you can donate to the Against Malaria Foundation and save a year of life for only $78.[20]

Even though the difference between $78 and $5,108 seems large, keep in mind that all of these numbers have a great deal of uncertainty behind them, and to give you a sense of perspective, the average American regulatory agency requires about $27,000 to save one year of life.[21] Donating generously to *any* of these charities is a far more effective way of helping the world than what most people do.

Given current data about animal welfare, climate change, and human health, my recommendation for the best way to help the world, and as such be the most good person you can be, is to get as much money as you can to the Against Malaria Foundation.

14
HOW TO MOBILIZE PEOPLE TO BE GOOD

What about getting other people on board with being good? One obvious way to try to motivate people to help with some of the big problems the world is facing is to convince them of how bad these problems are. Unfortunately, this can backfire. It is much more effective to offer them solutions to the problems: what they can do to help.[1]

How People Give, How People Donate

Although there are vast differences between the effectiveness of different charities, only six percent of individual donors do any comparison at all. Thirty-three percent do some research, more often for humanitarian and environmental causes, and rarely for arts and religion causes.[2] Congratulations, you're one of the better third because you're reading this book!

An experiment wanted to test what kinds of pitches people would respond to when asked to donate money to a charitable cause. One group was told a single child needed a drug that cost $300,000 to make. This group saw a picture of a child and was told her name and age. Another group was told that $300,000 was needed to help eight children, and they got names, ages, and pictures, too. Who gave more? The people who thought they were only helping one child.

Here we have an example where $37,500 to save someone was less attractive than spending $300,000.³ Why do people think this way?

The accepted answer is that stories of single individual people trigger our empathy more than groups of people. In fact, the more people a charity reports it will help, the less money people will give. The sad truth is that sometimes doing the most good is at odds with what will warm our hearts the most.

It gets more interesting: another study sent out scientific information about the effectiveness of the charity to some of its regular donors, and a description of an individual who was helped to the others. It turns out that for people who regularly gave more than $100, the scientific information increased the size of donations, but for small-amount donors, it decreased it. This suggests that the big donors care about effectiveness and evidence, and people who give to get a warm glow in their hearts tend to give little and care about effectiveness less.⁴

If evidence, magnitudes, numbers, and statistics don't generate empathy, and the things that generate empathy result in decreased donations, then we have reason to doubt the idea that significant, effective change in the world will be driven by empathy. If the world will be saved, it will be saved by people capable of abstract, scientific reasoning. If you've made it this far in the book, you are probably one of those people.

Many people think that when you're doing good in this world, it doesn't count if you take credit for it, that giving anonymously is the only legitimate means of being good. If you take credit for it, show it off, or if, God forbid, it also causes some good to come to you, in terms of reputation or even something financial, then it negates the good that you've done, or maybe the goodness of the person.

This is a viewpoint I have a lot of trouble relating to. Suppose Chris saves three people's lives through charity. The most extreme version of this view would say that if Chris takes credit for it, then the goodness of what he did was nullified, making the situation morally equivalent to her having done nothing. But objectively speaking, the giving version of Chris saves three lives that the non-giving version did not. Is taking credit so bad that it's worth three people dying over?

This might have to do with the doctrine of double-effect. Sometimes, when billionaires do charitable giving, they are criticized because what they *really* wanted was some benefit to themselves and their companies. People

CHAPTER 14: HOW TO MOBILIZE PEOPLE TO BE GOOD 355

are wont to think of rich people as being selfish, so if they can see some self-serving motive for the charity, it plays on people's cynicism, and they assume that the self-serving motivation was the primary one.

But there's a good argument for why people *should* take credit for their charity: social proof. When people perceive that some behavior is something a lot of other people are doing, it becomes more socially acceptable for them to do it. If the people donating to charity are being quiet about it, it makes it seem less normal.

Suppose someone makes $150,000 per year, and gives away $100,000 of it to charity. This strikes some people are alarming, and possibly freakish. But if you already know about say, ten or twenty people in your social circle behaving similarly, it would not seem so weird.

When trying to make people act well, by far the strongest intervention is by changing people's ideas of what is socially acceptable. This social proof also can make people act badly. People tend to act in a way that feels normal, whether that behavior is objectively good or bad.

This was demonstrated in a brilliant experiment run in the real world. There was this park in Arizona with a petrified forest, and one of the problems that the park was facing was that people were walking out with petrified pieces of wood. Over time there was less and less petrified wood there. Robert Cialdini placed signs on different trails in the park, encouraging people to leave the wood there, but worded in different ways. Some signs read, "Many past visitors have removed the petrified wood from the park, changing the state of the Petrified Forest." Other signs read, "The vast majority of past visitors have left the petrified wood in the park, preserving the natural state of the Petrified Forest." Later, he placed pieces of petrified wood on the trail and measured how many of them got stolen. People stole far less wood when they were told that most people were being good, and more wood when they were told that lots of people were breaking the rules.[5]

This is profound, because we often will try to get people to act better by emphasizing how often people act badly. For example, we might try to discourage teen pregnancy or suicide by giving alarming statistics on how prevalent these problems are. This approach backfires. That study found that telling park-goers that lots of other people stole wood actually *increased* theft!

This same kind of intervention was replicated with not having your towels washed every day in hotel rooms, promoting recycling, and other domains.⁶ It's particularly effective because telling people what to do or not do can feel heavy-handed. People know they are being manipulated. But merely describing how others act goes under the radar, so to speak, and people don't put up the same resistance and skepticism because they don't know they are being manipulated.⁷ I use a lot of manipulation in this book, but I figure the reader is reading the book in the first place because they *want* to be manipulated to be more productive, happy, and good. But when you're approaching people out of the blue, or sticking a flyer in their mailbox, your audience is certainly not asking to be told what to do!

Bringing it back to charity, if you are generous, and don't tell anybody about it, you're being humble, but also making charity look less normal, and thus contributing to others' reluctance to give. Similarly, we should all be celebrating people who are generous, rather than seeing their announcement of giving as a reason for questioning their motives and being scornful of them.

Furthermore, studies show that allowing people to make their generosity public makes them more generous, even though the reason might be to increase prestige or to prevent damage to their reputation. People are more likely to vote when they are given the opportunity to wear a sticker that says, I Voted, or even to put a virtual sticker on their social media account. When companies violate environmental regulations and laws, publicly putting them on a list of offenders causes more change than fines and penalties.⁸ People care about their reputations, and this can be used for good, by rewarding helpers and shaming hurters.

By emphasizing that others are giving, it makes the practice seem more acceptable, even expected. People tend to underestimate how giving the people around them really are, and this reduces the amount of giving they want to do.⁹

Fighting Feelings of Futility

One thing that holds a lot of people back is a feeling of futility. It might seem that whatever you do is useless because there is so much suffering. You can try to reduce your carbon footprint, or donate what you can, but in the grand scheme of things, sometimes it can feel like it doesn't matter, because the problems you're fighting are so overwhelming.

CHAPTER 14: HOW TO MOBILIZE PEOPLE TO BE GOOD

Let's take climate change as an example. People sometimes think of climate change as something that will or won't happen. Looking at it like this can engender a feeling of hopelessness, because what you can personally do will probably not affect whether it will happen or not. Since it appears to be happening, and gigantic organizations and millions of individual people are making choices that make climate change worse, it overwhelms what you can do.

But a better way of thinking about climate change isn't that it will or won't happen—it *is happening*, and the outcomes can be better or worse, and the longer we can delay the bad outcomes, the better. This means that every choice you make regarding your carbon footprint matters, even if it only matters a little bit. Every time you choose not to take a transcontinental flight, for example, it slows global warming by some small amount, and makes the world a bit better.

For animal welfare, the situation is even clearer. When you find yourself thinking that factory farms are going to continue hurting animals, and that people will continue to support them, your individual choice not to eat chicken isn't going to solve the problem. But it will contribute to fewer chickens being raised in terrible conditions, and that is a win for that chicken who might have been.

When it comes to helping human beings survive disease, you're actually saving lives. Your donations are unlikely to change the global poverty or disease statistics in any measurable way, but you are actually saving human lives, and each one you do is very meaningful.

Let's make an analogy with murder. If someone murders, they are not really affecting the worldwide murder rate in any appreciable way. That is, even if someone kills five people, the numbers for world homicide rates will not budge. But is this mean those murders didn't matter? That they were just fine, because the global murder rate stayed the same? Of course not. Each murder leaves misery in its wake. Loved ones suffer, and the murdered person feels the pain of death, and, more importantly, loses years of future life.

The situation is the same with saving people from disease. Each person you save from disease is similar to the misery avoided by preventing a murder. There's a real person out there whose family will not suffer so much and who will get to live years of life more because of what you did. And that's valuable, even if it doesn't appreciably change the rate of malaria death in the world.

Not every situation is like this. Let's suppose someone is dying from a bacterial infection, and you do something to kill some bacteria. There is some amount of bacteria that would need to be killed to save the person, and if that threshold is not met, the person will die anyway. So if you remove just a little bacteria, it doesn't matter. But fighting climate change is not like this. Saving human lives is not like this. Unlike the infection example, every little bit actually does matter.

A LOVE LETTER TO PEANUT BUTTER

My readers and students notice that I talk a lot about jars of peanut butter. I don't know of a perfect food, but I think peanut butter comes close. How do I love thee? Let me count the ways!

Peanut butter is vegetarian. The creation of peanut butter causes very little harm to animal welfare.

Peanut butter can be eaten raw. You don't need to cook it, you can eat it on a spoon, or you can use it as a condiment on bread, apples, or celery.

Peanut butter is calorie- and protein-dense. For 100 grams of peanut butter, you get 25 grams of protein—the same as steak. This isn't great for weight loss, but in terms of nutrition per unit weight, peanut butter packs a punch.

It is the cheapest food there is, in terms of calories per dollar. It costs five cents for 100 calories, comparable to pinto beans. Just for some contrast, sirloin steak is one dollar for 100 calories—twenty times as expensive.[10]

Peanut butter is delicious. Kids love it. It's incredibly versatile: It can be eaten by itself, in desserts (it's great with chocolate) or in savory dishes (it's part of many Asian sauces), in a smoothie, and on sandwiches. I eat it mixed with steel-cut oats for breakfast.

Peanut butter lasts a long time and doesn't require refrigeration. There are very few nonperishable, high-protein foods.

Per unit calorie and protein, peanuts have the lowest carbon footprint of any common food, and are a sustainable crop that requires little water, helps fix nitrogen, and can be planted in a way that reduces soil erosion (but it's worse if it's made with palm oil).[11]

For many of these reasons, peanut butter is an ideal disaster-relief and food bank food: it's nutritious and calorie-dense, and doesn't go bad. You can save nations with peanut butter!

The downsides? If you don't eat it slowly and with something to drink, you can get an ache in your upper chest due to your esophagus flexing uncomfortably, trying to get it down. It's not great for weight loss, being high calorie.

Oh yeah, and it can kill people with allergies to it. No food is perfect. But what comes closer than peanut butter? Maybe mussels?

Encouraging Other People to Reduce Animal Suffering

When people eat unethical meat, it contributes to animal suffering. One way to get people to reduce their harm to animals is to encourage them to change their diets. But the evangelical vegan is an oft-mocked stereotype.

It's probably better to get people to reduce the amount of animal products they consume, rather than trying to convince them to be completely vegan. Convincing two people to cut their meat consumption in half does the same amount of good as convincing someone to be vegan, and it's probably a hell of a lot easier. Not eating any meat at all is hard. It's so hard that even most self-identified American vegetarians regularly eat meat.[12] That is, most people you know who you classify as vegetarians probably eat meat sometimes.

Getting back to the lessons of social proof, this is because veganism is viewed as a radical, extreme position, and it's tough to convince someone to live by something they view as radical and extreme. Being completely vegan has social costs that being a reducetarian does not. Nobody mocks people who try to eat less meat.

It's the same with any restrictive diet. I have a friend who is a devout Jew. He told me that he would be completely kosher, except that it would interfere too much with his social life. He wouldn't be able to eat at restaurants, and would only be able to attend dinner parties of people who kept a kosher kitchen. So he compromised and keeps kosher at home, and is more lax outside the house (he still avoids pork). Similarly, unless you live in India, maintaining a vegan diet isn't easy. There are many restaurants you either can't go to or just need to order french fries wherever you go. People report that family concerns are a serious barrier to becoming vegetarian.[13] Telling your mother you can't eat her roasted chicken anymore has a high social cost.

Many people are resistant to reducing their meat consumption because they perceive vegetarian dishes as less appetizing. One study showed omnivores images of vegan dishes, and asked them which ones looked the most appetizing. They found that familiar vegetarian foods, such as roasted potatoes or pasta with tomato sauce, looked much tastier than vegan replacement foods, such as veggie burgers or a tofu scramble.[14] So we shouldn't try to convince people that tofu is going to be just as delicious as turkey on Christmas night.

What we don't want is to try to convince someone to be vegan, and then later they reject it (perhaps after trying it for a while), and then abandon the whole project of trying to eat better, and just eat whatever they want. Instead, by encouraging people to reduce the amount of animal product they consume has the potential to influence someone a little, and as they habituate to their new diet, their commitment to reducing can grow over time. A gradual change reduces the chances of burnout.

This is what is happening to me. My happiness and energy are better when I have meat in my diet. I tried to be vegetarian and failed at it. But the reasons I wanted to avoid eating meat are still there, and I slowly have worked those ideas into my diet. I try to eat lentils or mussels for lunch, and steel-cut oatmeal for breakfast, instead of my usual bacon and eggs.

Even when people do manage to resolve to be vegetarian, it often doesn't last the rest of their lives. The average vegetarian stays that way for about five years.[15] There are five times as many former vegetarians as current vegetarians.[16] This suggests that even convincing someone to be vegetarian once cannot be expected to last. You have to look at it in terms of years of vegetarianism you get per dollar invested. This makes sense also because, even if someone does stay vegetarian for their whole lives, the number of years of vegetarianism gained depends on how old they were when they decided.

We can use scientific methods to come up with estimates of the effectiveness of trying to get people to reduce meat consumption. I want to

CHAPTER 14: HOW TO MOBILIZE PEOPLE TO BE GOOD

emphasize that these studies are estimates. Measuring these things is very difficult, and the people who conduct them stress that they are ballpark estimates, and that the exact numbers reported might change given better data and methods. Nonetheless, I'll give you a sample of the cost-effectiveness of these kinds of interventions.

One study looked at $1,500 worth of investment to try to get people to have meatless Mondays. According to this study, it for every dollar spent, 0.082 animals were spared (the range is 0.065 to 0.099). This means that one dollar saves 1 percent of an animal, or, it costs about $100 to spare a single animal. In this sense, "sparing" an animal means it was a land animal (probably a chicken) never raised, or a fish never caught.[17]

Other studies are less encouraging. One way animal advocacy groups try to get people to consume less meat is by passing out leaflets. A big study of the short-term effect of doing this found that it resulted in *more* animals consumed. The people doing this study emphasize that their results are only estimates, but why might this be happening?

We're not sure, but one reason might be that leaflets tend to focus on cows and pigs. This might have made people opt for eating chicken instead. But because each chicken produces so much less meat, even if people reduced their meat consumption (by the pound) they still might have ended up with more animal death by eating more chicken.[18] Here's an example of how an attempt to help ends up antimattering. Getting people to eat less pork and beef makes them turn to another animal, chicken, that's worse for animal welfare.

Protests are another form of activism, but are difficult to study. Many studies of protests are poorly done, or have a lack of sufficient data to conclude on the effects of protest for animal welfare.[19]

Estimates have been made for the effectiveness of online advertisements, too. One estimate suggests that you can convince someone to be a vegetarian for one year for $100 donated to an effective charity, using online ads. Let's talk about this number.[20]

Although being a vegetarian is generally good for the environment, some people find it very difficult to do. For these people, forcing themselves to be vegetarian would take enormous mental resources and willpower, and that comes at a cost. There's only so much you can do in your life, and you should spend your willpower wisely on the most effective things. Now, if it costs only $100 to convince someone *else* to be vegetarian for one year, you can look at your finances and make a decision: would you rather put in the effort to be vegetarian or spend $100 a year? Because the effect on animal welfare and the environment is about the same.

MARKETING REDUCETARIANISM

Getting people to eat less meat is, in part, a marketing issue. People find dairy the hardest thing to give up, followed by eggs and fish. Asking someone to give up their cheese is a hard sell. So we start with reducing meat.[21] A large survey found that some marketing materials work better than others.[22]

People respond better to descriptions of cruel animal conditions on farms than on the fact that animals die, or that animals have a complex psychology.

Advocacy should promote high-protein foods like beans, lentils, and grains. This is because almost half of reducetarians, vegans, and vegetarians eat more vegetables in place of meat, which can cause an unhealthy reduction in protein and caloric intake.

Women are more likely to be convinced to reduce meat consumption. This is probably because men crave meat (and other savory foods) more than women do, and women crave sweets more than men.[23] Marketing might be more effective when targeted at women.

Maybe you have $200 you could spend a year on effective animal welfare charities, like the Humane League. This means that a meat-eater who gives $200 per year to the Humane League is doing twice as much good for animal welfare as a vegetarian who doesn't donate anything.

I'm one of those people who couldn't muster the will be to vegetarian. But I donate about $600 per year to the Humane League, which means I'm doing *six times* as much for animal welfare as a vegetarian![24]

Some caveats are in order: this $100 is a very rough estimate, but is probably within an order of magnitude of the true cost. Also, you should be spending your "meat offsets" from your discretionary spending, not your charitable spending. For example, I donate a portion of my income every

CHAPTER 14: HOW TO MOBILIZE PEOPLE TO BE GOOD

year to charity. Of the $600 I give to the Humane League, $500 comes from the charity budget, and $100 of it comes from my discretionary spending money. Otherwise you're not really offsetting anything, you're just taking money away from your other charitable spending.

ARE HUMANS GOOD FOR THE WORLD?

Some of the things in this book look like they suggest that humanity's very existence is bad for the world. If insects feel pain and suffering, it just might be true, in the short term.

But there are reasons to suspect that even if it is true, we should soldier on.

First, if humans were to vanish somehow, the ecosystem would rapidly replace us with other living things. To the extent that the existence of humans is bad because of the suffering of wild animals needed to support our species, removing us, while leaving the rest of the world intact, likely won't make a dent in wild animal suffering.

But even if we are causing more harm than the animals that would replace us, it could be that we're passing through a stage, and when we come out of it, everything will be much better. But only some form of *intelligent* life will be able to get through this stage. So even if humans are doing things that right now, on net, make the world worse, we might reach a utopian stage for all beings at some point in the future. If humans were to go extinct then the wild animal suffering, for example, would continue for a very long time, and might only change if some other animal evolved to become intelligent, like raccoons or something, and they might go through the same phases that we are going through right now. So at least we have a head start.

Are humans the right ones to do this? Well, there's a window of time for life on Earth. We're about three-quarters of the way through it, and eventually the Sun is going to fry the Earth and it will be no more life here. There very well might not be enough time in that window for another intelligent species to arise and develop the capacity to substantially reduce suffering. Given that humans have enjoyed some amount of moral progress over time, we have a head start on morality over any potential new species and might be the best bet for goodness in the future.

15
WHEN GIVING GIVES BACK

If you've gotten this far, I think it's safe to say you're the kind of person I really admire: someone who really cares about being a good person, and the kind of person who strives to be better. If you're one of these awesome people, and if you identify with the idea that you are one of these people, great!

But maybe you're not. Maybe you *want* to be that kind of person, but you just don't feel it. Recall the section on moral psychology, and think about how your evolved and cultural morality is not particularly in line with how you can maximize your goodness in our modern world. There are many parts of your moral psychology that are barriers to maximizing your goodness. Here are some examples.

The best way to be good is to help people we don't know: the poorest people in distant countries. But it feels less rewarding to help those we don't know. It feels less rewarding to give money and not know exactly what it's doing, or the individual it's helping. People you don't know can't reciprocate the kindness, and giving has been found to be more rewarding when it helps form bonds between people.[1] Helping people by giving them bed nets is one of the most effective things you can do to be good, but you don't know the people getting the nets, and you don't know if it will save them—it's all about probabilities, and our minds aren't good about thinking of those things.

People are more satisfied with their charitable giving if they know what the money is doing. A study found that giving to UNICEF provided no happiness benefit, but giving to a charity that bought bed nets did, presumably because it was more specific—with UNICEF, they're doing so many things you don't know what they'll use "your" donation for.[2] When you give money for bed nets, you don't know who will get the net, so you have to try to find satisfaction in the *knowledge* that it's the most effective thing you can do, but it doesn't have the same positive visceral punch as helping the homeless person you see freezing on the street every day.

We also believe that spending money on ourselves makes us happier than spending it on other people, but this has been shown to be incorrect.[3] A study of 136 countries found there was a correlation between giving and happiness, even when controlling for income and other demographic variables. This was true for 120 of the countries. It's even been found to be true for toddlers.[4] It's one of the many ways we're wrong about what is going to make us happy.

Even being reminded of money makes us less concerned about other people's problems![5]

So there are lots of bits of legacy software installed in your brain that are trying to keep you from being as good a person as you can be, just like there are bits of legacy code in there trying to keep you from being maximally productive and happy.

Here's what you do: right now, put the book down and donate $8 to the Against Malaria Foundation or the Humane League, depending on whether you feel like being good to humans or animals right now. It's just eight bucks; anybody can afford that. Once you do, though, there's a foot in the door.

You are now someone who gives to an effective charity.

You rock.

This small act of giving is something you can leverage in the future for more generous acts, helping you be the effective, generous person you want to be. Over time, your commitment to this identity can grow. You will feel pride, and this feeling of identity will encourage you to do more and more things to make the world a better and better place.

Studies also show that making a pledge, or a commitment, or even a prediction that you will give in the future makes you more likely to do so. You can pledge on websites like givingwhatwecan.org and thelifeyoucansave.org.

That's thinking about your personal identity. But there is also a growing social identity you can feel a part of. Local and online communities of charitable people are popping up all over. There are effective altruism communities on Facebook. It feels great to be in a group of like-minded people, who reward and encourage each other for the good they do. Particularly generous people are featured on websites like boldergiving.org, but you don't have to be a high-flyer like these people to get a lot of gratification out of giving.

People coming of age in the new millennium are more interested in having an impact on the world than they are for earning money for its own sake. When you're old, wouldn't you be proud that you were at the vanguard of this movement, rather than a Johnny-come-lately to the generosity bandwagon?

WHERE DOES IT ALL END? HOW MUCH IS TOO MUCH?

It's hard to accept that how nice you are to other people really doesn't matter that much. I have some very smart friends who strenuously object to this idea. Their argument is this: when you're mean to somebody it has a ripple effect. This means that you might lower their mood, and make them worse to other people.

Events in networks start locally, but then usually only have effects at a certain distance. The same is almost certainly true for how you treat the people around you. If you're in a position to be nice or mean to the president of the United States, for example, then the point where your influence peters out might be quite distant. Because if you put the president in a good or bad mood, or you make the president feel slighted, or grateful, it might have enormous effect on lots of people. But opening the door for someone at the grocery store? It's unlikely that this will have serious impact in the world, good or bad. These ripple effects can spread, but experimental evidence suggests it only goes out about three people.[6] This is good, but unlikely to make a huge difference on the world or those in it who need the most help.

But there are ripple effects, and people think: who knows, the nice or mean thing you say to someone might end up having a big effect down the line. And it doesn't cost you much to be nice. All of that is true. And I try to be nice to people. But what I don't do is let that be the full extent of my goodness.

CHAPTER 15: WHEN GIVING GIVES BACK

Let's think of it in terms of numbers. Suppose you talk to someone who's feeling miserable, and you do or say something so nice that it turns into a great day for them. Let's further suppose that they improve the lives of two additional people, as the research suggests. Let's suppose that you do this every day. This is a rough estimate: if you're generally nice most of the time, and sometimes short with people, there will be additional small effects that might cancel out. For now we will assume that every single day you completely brighten someone's day (and in turn, two other people's), which I think is probably an upper boundary of niceness that in truth would be difficult to maintain. But let's go with it.

That's three days brightened 365 days a year. Another way to think of it is that you're transforming three miserable years into three good years every year you do this throughout your life, for people in rich countries whose lives are already pretty good, relative to most people on earth. Now ask this: how many years of miserable years being turned into good ones is the same amount of good as extending someone's life for one year? Earlier we estimated that generic misery is about 7 percent as bad as being dead (as mild depression is 14 percent as bad according to disability weights). We can think of getting three people out of this misery for one year as 7 percent of three years, which is about seventy-seven days, or, about two and a half months. In other words, working *very* hard at being nice to everyone you meet every single day is like adding 2.5 months of healthy life each year. That's great, but keep in mind that you can save one year of healthy life for $78, which means you can save 2.5 months of life for a little less than six bucks.

If these numbers are anywhere near accurate, it means that by donating a mere *$12 per year* to an effective charity, you're making the world twice as good as someone who is so nice to people that they turn three miserable days into good days every single day of their lives.

Nice is not the same thing as good. If your being nice is completely satisfying your desire to be a good person, then you are probably being much less effective than you could be. Rather than being nice and hoping somebody else does something really good due to a ripple effect, you are better off just doing that good thing yourself. Why turf it out to the next guy?

To be the best person you can be, you need to overcome your instincts in strategic ways.

There is something about this line of thinking that doesn't sit right with people. They want to hold on to their ideas of morality because they have a remarkable benefit: they allow you to be a perfectly good person. When you think of being good as not being bad, and, specifically, not breaking a handful of rules, like not stealing, not cheating, not lying, and so on, then it's possible to be a perfectly good person. It might be difficult over the course of your lifetime, of course, but you might be able to be proud that, say, yesterday, you didn't do anything morally wrong, therefore you were good.

I'll admit this is attractive. The alternative presented in this book turns this on its head a bit. Being good isn't just not being bad. There are ever-increasing levels of good. There is no perfectly good, only better and better. If you're concerned with being perfectly good, this system simply doesn't allow it. If you can't be optimally good, then it puts you in a position of always being a little bad, and that's unacceptable to many people. It tempts them to jettison the entire thing.

A reframing can help make it more acceptable. Let's take nutrition as an example. Although science seems to change its mind about what is nutritionally good diet, let's assume for the moment that it's most rational to believe the state of the art regarding what makes for a healthy diet. At the time of this writing, we're pretty sure that eating too much sugar isn't good for you. I'm confident that most of my readers already know this, and that most of my readers acknowledge that they probably eat more sugar than they should. We try to eat less sugar, but often fail, and often think it's just worth it to indulge once in a while, because eating sugary things is so pleasurable.

I think it's helpful to think of morality as more like nutrition than as the Ten Commandments. You can have an idea of what it means to eat an optimally nutritious diet that changes over time. At any given moment, it's something to strive for, even though you may never fully achieve it. But just as the idea of eating a perfectly nutritional diet for years is practically unattainable, this does not mean that you abandon the entire idea of a nutritious diet. You just do the best you can. This is how I think about being good. Just strive to be as good as you can, but accept that you can always improve and that nobody's perfect. Rather than simply thinking of a diet as nutritious or nonnutritious, or a person being rich or not rich, we think about it on a scale. There are better and worse diets. There are richer and poorer people. Any line we draw to make it into a binary distinction will be, to some extent, arbitrary. It reflects reality more realistically if we think of it as continuous.

CHAPTER 15: WHEN GIVING GIVES BACK

Sometimes you skip your exercise class. Sometimes you eat too much junk food. Sometimes you say hurtful things, Or don't get enough sleep, or speed on the highway. That doesn't mean you should give up trying to be better.

That's how you should think about moral behaviors and lifestyles. Practicing effective altruism probably does have an end state, a place to stop, where you can confidently say that you're now *good enough*. What is that state?[7]

Keep in mind that living as a pauper will affect your ability to do good in the future. You won't get promoted at your job as a real estate agent if you dress like a pauper and are emaciated. You will also have less influence on other people without some level of decent appearance. Brian Tomasik puts it this way: you should only buy luxuries "to earn a higher income or to engage in social networking with powerful people."[8] This might mean taking flights and dressing well. This is why it might be okay for Al Gore to have flown around the world, talking to world leaders and giving talks about climate change. The expectation is that the good influence he has makes up for the carbon put into the atmosphere by buying a laptop and flying all over the place.

But I don't think this ideal is one that we should even keep in mind. I was hesitant to even write about it. My advice is to not think about this extreme state. Comparing yourself to this ideal can be depressing—you feel guilty for not meeting it, and at the same time dread the idea of meeting it, and risk abandoning getting better altogether.

But I want people to be honest about the trade-off they make in their lives between their own pleasure and moral obligation. Be aware that many of the things you do are self-serving, and don't do much good for the world. Some of the things you do might be bad for the world (like eating factory-farmed meat) but you do it anyway.

Sometimes people give too much of themselves, and it's worse for them. If you give to the extent that you're not looking after your own well-being, it's called pathological altruism, and can even lead to health problems.[9] Students who give too much, for example, end up with lower grades.[10] Adam Grant calls givers who also look after themselves successfully "otherish."[11]

Compassion fatigue is another problem—when you give too much too fast, you can burn out, and give up the giving enterprise entirely. This is a particular problem for care professions, such as nursing, therapy, and social work, where caregiving can increase depression and stress. This also happens when being a caregiver for a loved one who needs a lot of help, such as someone with dementia.[12] However, this kind of caregiving is detrimental to the caregiver only in situations where the health of the recipient is severely impaired, or if the caregiver is putting in an excessive amount of time. Controlling for these factors, caregiving makes one happier.[13] In any case, intensive caregiving to one particularly needy individual is not the kind of doing good that this book is about. You don't get compassion fatigue from donating money every year.

Let's look at some people who give in the extreme.

Zell Kravinsky had two kidneys. He knew that he could live with just one. As long as there are people with no kidneys, he thought, how could anyone justify having two? He looked up the danger of donating a kidney to himself. The chance of his death was about 1 in 4,000. So, by *not* donating a kidney, he was effectively saying that his life was four thousand times more valuable than somebody else's. That thought, to Dr. Kravinsky, was "obscene."[14]

He set about trying to donate his kidney to a stranger. This proved somewhat difficult, not because nobody would take the kidney, but because this kind of generosity was so unusual that the doctors figured there must be something mentally wrong with him. He had to pass psychiatric tests. His family strongly urged him not to do it. But ultimately he managed to do it. In spite of all of the efforts of the people around him trying to keep him from being a better person, he managed to donate one of his kidneys to a stranger and save her life.

That's not all for Zell, though, who happened to have amassed millions of dollars by investing in real estate around the University of Pennsylvania. He vowed to give most of it away, and has donated at least $45 million to charitable causes. Hearing Zell's story might make you feel that he's foolish, or he went too far, or that he's a hero we should all emulate.

I feel all of these things.

People like Zell Kravinsky are rare. They represent one extreme of what we can think of as a spectrum of generosity. There is a temptation to look at these stories and use them to justify why you don't, or won't, be generous with what you have. But it's important to understand that there are many more people, and dollars, at the other extreme: people who hoard enormous amounts of money and don't give any of it away. These are the people we should really look down on.

There's a big middle road between selfish millionaires and Zell Kravinsky.

MORAL HEROES: JULIA WISE AND JEFF KAUFMAN

Jeff works at Google and makes a good salary. Julia went to school for social work—a degree a lot of people choose if they want to do good in the world.[15]

These two keep track of their donations online for everybody to see.[16] For the past several years (from 2012 to 2018) they gave away about half of their pretax income to charity. In 2018 they made a combined household income of $407,094 and gave away $209,585. Not only is this a huge amount of money, but because they're giving to the Against Malaria Foundation, they are ensuring that they do the most good.

They live on 6 percent of their income ($24,425), and the rest goes to savings and taxes.

They're happy. And I bet the eighty-one people whose lives they save every year are, too.

Let's suppose a person gives some, then a little more, then a little more every year. Let's say they give 1 percent of their income, then 5 percent, then 10 percent. At what percentage is the person doing enough? Utilitarianism's answer is pretty unpalatable, so let's just not think about it.[17]

Instead, let's ask at what percentage the person should even start considering that it's maybe too much. There's no percentage here for everybody, because 10 percent of a poor person's wealth or income is much more important to them than 10 percent of a rich person's. What people are concerned about is that if they give their money away, there won't be enough for them left, and that will make them unhappy.

Following this, we can say that giving more and more is just fine for you *if it doesn't noticeably lower your happiness*. Rather than trying to be perfect, the way to think about it instead is: how can I become better than I was? Even if you don't sacrifice all of your happiness for the welfare of others, for most people they have never tried to give so much that their happiness was even in danger of being slightly strained. For example, if you gave 5 percent of your income to effective charities, would you even notice it? Would you be noticeably less happy? If the answer is no, then you should be giving away that 5 percent. Then go ahead and try to give away 7 percent next year, and keep increasing it every year until you at least notice some change to happiness. When it is, then you can decide whether you want to continue giving at that rate, cut back, or even give a little more. If most people even tried to approach this standard, the world would be a lot better off.

Who's Going to Listen to You?

Some of you might be thinking, at this point, that having me as a friend would be insufferable—that my constant moralizing and condemnation of everything everybody does would make me a whole lot of annoying to hang out with.

It speaks to a real tension for people who have fundamental disagreements with how people conduct themselves, or the political and social systems we live in. On the one hand, you want to be true to yourself and your beliefs, but on the other, you don't want to be an insufferable bore.

This tension can exist for many causes, of which effective altruism is just one example. People who are radical left-wingers or reactionary right-wingers, and extreme environmentalists or animal welfare activists can suffer it, too. The tension manifests itself in effectiveness: you can go off-grid and try to reject society completely, but if you do, you drastically reduce your ability to influence other people's behavior or the system you're trying to reject. On the other hand, if you integrate with society to have that influence, you end up participating in, and perpetuating, the very system you are there to oppose.[18] That's why you have to purchase books on communist philosophy.

I realize that many of the views I endorse in this book are off-putting. They can make people feel guilty or ashamed. If introduced in a casual way, they can produce cognitive dissonance, which will make the listener less willing to entertain the same ideas in the future, perhaps when introduced in a more persuasive, nuanced way. So I basically try to shut up about it unless asked.

For example, recently I was talking to friends who told me that they rescued a baby squirrel whose mother had died. They nursed it back to health, then drove an hour and half round trip to a wildlife refuge to pass it on for further treatment. Then we proceeded to eat a steak and chicken dinner. I kept my mouth shut.

And why? I care deeply about these friends, and my relationship with them, and bringing up the moral inconsistency would have little positive effect on their behavior, and a potentially large, negative effect on the friendship. Haranguing your friends about their moral behavior is not effective altruism. It just makes you unlikable and have a worse social life. If you put uncomfortable admonitions on social media for your friends to read, you'll get animated GIFs of eye-rolls in the comments.

Instead, I recommend you lead by example and express your views and arguments in appropriate venues.

Maintaining Compliance

They say that the best diet that you can be on is the one that you can stay on, and the best exercise regimen that you can have is one that you will stick to. If you want to maximize your good in the world, thinking of activities that will do the most good is an important aspect of it, but just as important, or perhaps even more important, is making sure that you live your life in a way that you can sustain being the kind of person that you want to be.

Recall the earlier discussion on adaptation, and how you adapt more quickly to routine things that you stop thinking about. This has interesting implications for philanthropy. On the one hand, if you get really good feelings by donating to something, then there are few things you can do to keep yourself from adapting to it. One is that you can give just once a year. This way it feels special, and not routine. Also, it is minorly better for the charity as well, because a single large payment has the same overhead as a small individual payment, so giving a lot at once reduces their administrative costs. You can also give to a different charity every time. Given our uncertainty about what's the best area to focus on—animal welfare, human welfare, or environmental protection—you might as well use the uncertainty to your advantage and give to a different one each year.

However if the idea of philanthropy, of giving something in a particular instance, smarts a little bit because you feel the loss of money, then what you should do is try to habituate to it. You can sign up for a monthly automatic donation to a good charity, and eventually you'll get used to it and never even think about it, just like my eating cabbage chips while I work.

Compliance is also related to happiness. If what you're trying to do makes you miserable, you're going to have a hard time sticking to it. One finding is that setting attainable goals for yourself leads to more happiness.[19] This is another reason to have a healthy, realistic attitude toward just how morally good you're going to try to be.

Being less happy because you're giving away money probably won't be a problem for you, for two reasons. First, you probably won't be able to give away enough for that to happen. We strongly cling to what we perceive to be rightfully ours. But the other reason is that giving away money, within reason, actually makes you happier. You get a feeling of warm glow or "helper's high" when you do it. Numerous studies across many cultures have found that spending money on yourself doesn't make you as happy as spending it on others.[20]

Moving beyond spending money, giving behaviors also make you generally happier, whether you are giving to the people around you or specifically trying to help the world in general, compared to doing good things just for yourself. In fact, acts of "self-kindness" sometimes show no improvement at all in psychological well-being—acts of self-kindness are different from self-compassion, however, which is more of a mindset than an action that one can take. Self-compassion is good for your well-being.[21]

What seems to really matter is where you put your donation money. That is, there's very little you can do in the real world that has a real strong positive effect compared to donating money to an effective charity. Even local activism tends to have only small effects, with one exception.

THE ONLY LOCAL ACTIVISM THAT MAKES SENSE

A popular slogan is "Think Globally, Act Locally." You can probably guess that I'm not a fan of this advice. Acting locally almost always means that

CHAPTER 15: WHEN GIVING GIVES BACK

the good your doing is very inefficient. It's great that today we can "Think Globally, Act Globally."

Optimizing the good you do often ends up meaning that you are focused on helping people across the world. This is because it's very unlikely that the cheapest people to help live right next to you.

The same goes for volunteering your time. Granted, there are fewer opportunities to help those most in need when they are a world away. But there is one way you can act locally that is really effective: convincing other rich people to give their money to good causes.

It's particularly important because people are doing bad things globally, like cheating on car emissions or overfishing the ocean.

When encouraging people to give, studies show that you can encourage giving by emphasizing that even a little bit helps. For example, putting "even a penny helps" on charity advertisements increases the number of donors but does not reduce the amount each donor gives.[22]

We saw that having a giving lifestyle is good for your career. But we also saw that being good to the people around you isn't the best way to maximize your positive impact on the world.

Giving to charity doesn't have the same rewards that being good to your friends and coworkers does. There is no opportunity for reciprocity, and rarely an increase in reputation.

But how much of a sacrifice is giving away money? Money is good because of what it can buy. And you buy things for self-gratification, security, and lots of other reasons that all eventually have to do with good feelings. Money, spent or not, that doesn't help foster good feelings, at least at some point, is wasted. Because the good of having money is only the good feelings it can get for you, then if giving away money doesn't make you any less happy, then *it's not a sacrifice at all.*

For many people, giving away 2 percent, 5 percent, or even 20 percent of their income wouldn't make them any less happy. Remember that any extra income over $105k provides only a negligible increase in happiness in the industrialized world. If you make $120k, and give away $15k per year, you will very likely be just as happy as if you kept the money or spent it on yourself.

If you are skeptical of this, you're not alone. A pair of studies asked people how much of a year was spent in a bad mood when one is only making $20k per year, or $55k. They also asked people actually making those incomes how happy they were. Turns out, people vastly underestimate how happy people at these income levels are. Happiness is indeed related to income, as we've discussed, but not nearly as much as people think.[23]

It's no secret that many people in the industrialized world have an unhealthy relationship with their material possessions, and their constant craving for more of them. Many people have storage facilities rented out to house the stuff their house can't house, and can't fit their cars in their garages because they're so full of junk. Having lots of clutter is associated with stress, which makes the accumulation look like an even worse idea.[24] But when people think about what they might want to spend their money on, lots of times they are thinking of more stuff.

Another surprising finding is that giving away money actually makes you happier.

There are several studies that give people some money and give them a chance to spend it on themselves or on others. Giving it to others makes them happier than spending it on themselves.[25] A massive study, looking at 234,917 people in 136 countries, found that generous spenders are happier. People are happier when thinking about past instances of generosity than when thinking about past instances of purchasing things for themselves.[26] The difference in happiness between those who reported giving to charity and those who didn't had a happiness boost equivalent to that caused by a *doubling* of income.[27] Not only that, the link appears to be causal: people get happier when they are *assigned* to spend the money on others than when they spend it on themselves.[28]

Of course, the recipients of these gifts are happier too, particularly if they perceive good intentions in the giver.[29]

Not only does giving make the giver happy, and the recipient happy, but people watching this transaction get happy, too: Jonathan Haidt talks about a feeling he calls "elevation," which is a warm, moving feeling caused by perceptions of others' giving.[30] If you spend money on a video game, you do so because you expect owning the game will make you happier. Knowing that charitable donation can make you happier, perhaps you can save people's lives with a similar mindset. That is, don't think of it as giving money away, think of it as spending money, just like you always do, to save people's lives. It might even make you happier than a video game.

Not only that, it also appears to make you more money. Economic analyses show richer people give away more money. For every dollar more

CHAPTER 15: WHEN GIVING GIVES BACK

they make in income, they give away an additional 14 cents. But what's surprising is that for every dollar of charitable giving done, there is an subsequent increase in income to the tune of $3.75.[31]

What the what? How can giving away money make you more money?

One reason might be that giving makes you happy, causing a "helper's high" or a "warm glow of giving," activating the reward centers of your brain.[32] Volunteering your time also makes you happier, more satisfied with your life, and increases your self-esteem.[33] When you are happier, you work harder,[34] so what's probably happening is that you give, you get happier, you work harder, and you end up making more money as a result.[35]

A chief reason that being kind to others makes you happier is when you recognize how much the recipient appreciated the kind act.[36] This explanation of the warm glow feeling helps explain why it sometimes feels less rewarding to donate to charities, where you don't ever meet the people that you're helping. Unfortunately, the most effective charities, and doing the most good in this world, doesn't always get associated with the strongest warm glow. This is where you have the opportunity to be better than the warm-glow rabble out there.

When generous giving makes you happier and even richer, it doesn't look so much like a sacrifice.

ACKNOWLEDGMENTS

In spite of what I wrote about social media in this book, I used Facebook to great effect for advice and information. A big thank-you to Facebook for making the service, and to everybody on it, "friends" or otherwise, who commented on my posts and questions.

Thank you for the loving support of my parents, James and Janet Davies, and my sister, novelist J. D. Spero, and her family.

Thanks to Leona Tucker, Dana Peters, Brian Tomasik, Mary Berard, Jane Ransom, microbiologist Alex Gill, computer scientist Charles Isbell, my beloved wife Vanessa Davies, and my former student and now tech entrepreneur Jon Gagné.

Thanks to philosophers Jeanette Bicknell, Heidi Maibom, Myrto Mylopolous, Bob Fischer, John Broome, Mark Budolfson, Gordon Davis, Gabriele Contessa, and Joshua Shepherd.

Thanks to neuroscientists Kent Berridge, and my podcasting partner and friend Kim Hellemans.

Thanks to economists Nicholas Rivers, Catherine Deri Armstrong, and especially Richard Bruns, who spent many hours helping me on this book.

Thanks to psychologists Cheryl Harasymchuk, John Zelenski, Nadiya Slobodenyuk, John Opfer, and Robert West.

Darren McKee, cohost of the *Reality Check* podcast, gets a great big thank-you, for introducing me to Effective Altruism, pointing me to lots of information that helped with this book, and for being a great friend. Darren has donated over $100,000 of his own money to effective charities. He's a hero and an inspiration.

Thank you to my agent, Don Fehr of Trident Media, who inspired the writing of this book when I sent him some book ideas, and he didn't like any of them because he thought I should be "thinking bigger." This book is me trying to think bigger. The team at Pegasus Books, especially my editor Jessica Case, gets a big thank-you for taking on another one of my books, and being such a pleasure to work with.

The biggest thank-yous go to my academic advisors over the years, who trained me to think like a scholar, including philosophers Rebecca Holson, Paul Blunt, and especially Nancy J. Nersessian, psychologist Dorrit O. Billman, and computer scientist Ashok K. Goel.

Although many people have helped me, the weird ideas in this book are my own.

ENDNOTES

Front Unmatters and Front Antimatters
1. D. Hagmann, E. H. Ho, and G. Loewenstein, "Nudging Out Support for a Carbon Tax," *Nature Climate Change*, 9, 2019, 484–489.

Chapter 1. Productivity
1. I'm using the word "optimizer" here because "maximizer" is already a term in scientific psychology with a slightly different meaning: one who does an extensive search of options, looking for the best one. Maximizers have goals to have the best options (that is, having high standards) and a way of trying to find them—searching through options. But people who search through alternatives and who have difficulty making decisions are less happy. This is in contrast with people who are satisfied with good-enough alternatives. (See J. Hughes and A. A. Scholer, "When Wanting the Best Goes Right or Wrong: Distinguishing Between Adaptive and Maladaptive Maximization," *Personality and Social Psychology Bulletin* 43(4), 2017, 570–583.) But other studies find that maximizers are just as happy as others. (See S. E. Highhouse, D. L. Diab, and M. A. Gillespie, "Are Maximizers Really Unhappy? The Measurement of Maximizing Tendency," *Judgment and Decision Making Journal* 3(5), 2008, 364.) I'm using the term "optimizer" to refer to someone who delights in finding better ways to do things, doesn't mind being told about alternatives, and who is willing to try things out.
2. This is related to the explore-exploit trade-off well-known in computer science. B. Christian and T. Griffiths, *Algorithms to Live by: The Computer Science of Human Decisions* (New York: Macmillan, 2016), chapter 2.
3. C. Newport, *Deep Work: Rules for Focused Success in a Distracted World* (New York: Hachette, 2016), 4.
4. S. Stephens-Davidowitz, *Everybody Lies: Big Data, New Data, and What the Internet Can Tell Us About Who We Really Are* (New York: HarperCollins, 2017).
5. M. Chui, J. Manyika, J. Bughin, R. Dobbs, C. Roxburgh, H. Sarrazin, and M. Westergren, "The Social Economy: Unlocking Value and Productivity Through Social Technologies," *McKinsey Global Institute* 4, 2012, 35–58.
6. S. Kessler, "38% of College Students Can't Go 10 Minutes Without Tech," *Mashable Tech*, 2011, http://mashable.com/2011/05/31/college-tech-device-stats; C. Marci, "A (Biometric) Day in the Life: Engaging Across Media." Paper presented at Re:Think (March 28, 2012), New York.
7. C. Bailey, *The Productivity Project: Accomplishing More by Managing Your Time, Attention, and Energy* (Toronto: Random House Canada, 2016), 79.
8. L. M. Carrier, L. D. Rosen, N. A. Cheever, and A. F. Lim, "Causes, Effects, and Practicalities of Everyday Multitasking," *Developmental Review* 35, 2015, 64–78.
9. I'd quote the actual line, which is much more colorful, but getting permission to put verbatim lyrics in a book is an expensive, maddening process. If you want to look it up, it's the lyric with the word "Billboard" in it.

10. M. R. Dickey, "FREAKY: Your Breathing Patterns Change When You Read Email," *Business Insider*, December 5, 2012, retrieved June 18, 2019, from https://www.business insider.com/email-apnea-how-email-change-breathing-2012-12; 2012.
11. L. D. Rosen, L. M. Carrier, and N. A. Cheever, "Facebook and Texting Made Me Do It: Media-induced Task-switching While Studying," *Computers in Human Behavior* 29(3), 2013, 948–958.
12. W. J. Horrey and C. D. Wickens, "Examining the Impact of Cell Phone Conversations on Driving Using Meta-analytic Techniques," *Human Factors* 48(1), 2006, 196–205.
13. D. J. Levitin, *The Organized Mind: Thinking Straight in the Age of Information Overload* (New York: Penguin, 2014), 96.
14. Z. Wang, P. David, J. Srivastava, S. Powers, C. Brady, J. D'Angelo, et al., "Behavioral Performance and Visual Attention in Communication Multitasking: A Comparison Between Instant Messaging and Online Voice Chat," *Computers in Human Behavior* 28, 2012, 968–975.
15. T. Lesiuk, "The Effect of Music Listening on Work Performance," *Psychology of Music* 33(2), 2005, 173–191. People with a lot of musical skill are more likely to find music distracting.
16. J. Andrade, "What Does Doodling Do?" *Applied Cognitive Psychology* 24, 2010, 100–106. J. Davies and M. Fortney, "The Menton Theory of Engagement and Boredom," in *1st Annual Conference on Advances in Cognitive Systems, Palo Alto, CA.* (December 6–8, 2012); http://www.cogsys.org/2012.
17. C. I. Karageorghis and D. L. Priest, "Music in the Exercise Domain: A Review and Synthesis (Part I)," *International Review of Sport and Exercise Psychology* 5(1), 2012, 44–66.
18. A. Gazzaley and L. D. Rosen, *The Distracted Mind: Ancient Brains in a High-tech World* (Cambridge, MA: MIT Press, 2016), 223.
19. S. Reimers and E. A. Maylor, "Task Switching Across the Life Span: Effects of Age on General and Specific Switch Costs," *Developmental Psychology* 41(4), 2005, 661–671.
20. M. Richtel, "Lost in E-mail, Tech Firms Face Self-made Beast," *New York Times*, June 14, 2008.
21. G. Mark, V. M. Gonzalez, and J. Harris, "No Task Left Behind?: Examining the Nature of Fragmented Work," *Proceedings of the SIGCHI conference on Human factors in computing systems*, ACM, April 2005, 321–330.
22. L. Marulanda-Carter and T. W. Jackson, "Effects of E-mail Addiction and Interruptions on Employees," *Journal of Systems and Information Technology* 14, 2012, 82–94.
23. R. Alzahabi, M. W. Becker, and D. Z. Hambrick, "Investigating the Relationship Between Media Multitasking and Processes Involved in Task-switching," *Journal of Experimental Psychology: Human Perception and Performance* 43(11), 2017, 1872.
24. L. M. Carrier, L. D. Rosen, N. A. Cheever, and A. F. Lim, "Causes, Effects, and Practicalities of Everyday Multitasking," *Developmental Review* 35, 2015, 64–78.
25. Gazzaley and Rosen, *The Distracted Mind*, 196.
26. K. L. Fonner and M. E. Roloff, "Testing the Connectivity Paradox: Linking Teleworkers' Communication Media Use to Social Presence, Stress from Interruptions, and Organizational Identification," *Communication Monographs* 79(2), 2012, 205–231.
27. M. Czerwinski, G. Smith, T. Regan, B. Meyers, G. G. Robertson, and G. K. Starkweather, "Toward Characterizing the Productivity Benefits of Very Large Displays," *Interact* 3, August 2003, 9–16; R. Ball and C. North, "Analysis of User Behavior on High-resolution Tiled Displays," *IFIP Conference on Human-Computer Interaction*, Springer, Berlin, Heidelberg, September 2005, 350–363.
28. S. M. Siha and R. W. Monroe, "Telecommuting's Past and Future: A Literature Review and Research Agenda," *Business Process Management Journal* 12(4), 2006, 455–482. This overview emphasizes that much of the work on telework is poorly done, and better research is needed.

29 W. O'Brien and F. Y. Aliabadi, "Does Telecommuting Save Energy? A Critical Review of Qualitative Studies and Their Research Methods," *Energy and Buildings* 225(15), 2020, 110–298.
30 S. Wuchty, B. F. Jones, and B. Uzzi, "The Increasing Dominance of Teams in Production of Knowledge," *Science* 316(5827), 2007, 1036–1039; K. Lee, J. S. Brownstein, R. G. Mills, and I. S. Kohane, "Does Collocation Inform the Impact of Collaboration?" *PloS one* 5(12), 2010, e14279.
31 Fonner and Roloff, "Testing the Connectivity Paradox," 205–231.
32 A. Gazzaley and L. D. Rosen, "Are You a Self-Interrupter?" *Nautilus* May/June 2017, 16–18.
33 J. Wajcman and E. Rose, "Constant Connectivity: Rethinking Interruptions at Work," *Organizational Studies* 32, 2011, 941–961.
34 Gazzaley and Rosen, *The Distracted Mind; Ancient Brains in a High-tech World*, 11.
35 E. Ophir, C. Nass, and A. D. Wagner, "Cognitive Control in Media Multitaskers," *Proceedings of the National Academy of Sciences* 106(37), 2009, 15583–15587.
36 Carrier, et al., "Causes, Effects, and Practicalities of Everyday Multitasking," 64–78; C. Bailey, *The Productivity Project: Accomplishing More by Managing Your Time, Attention, and Energy* (Toronto: Random House Canada, 2016), 191.
37 C. Duhigg, *The Power of Habit: Why We Do What We Do in Life and Business* (New York: Random House, 2012).
38 Gazzaley and Rosen, *The Distracted Mind*, 12.
39 J. W. Pennebaker, "Writing About Emotional Experiences as a Therapeutic Process," *Psychological Science* 8(3), 1997, 162–166. See also K. A. Baikie and K. Wilhelm, "Emotional and Physical Health Benefits of Expressive Writing," *Advances in Psychiatric Treatment* 11(5), 2005, 338–346.
40 W. Hofmann, K. D. Vohs, and R. F. Baumeister, "What People Desire, Feel Conflicted About, and Try to Resist in Everyday Life," in *Self-Regulation and Self-Control* (London: Routledge, 2018), 256–266.
41 Z. Wang and J. M. Tchernev, "The 'Myth' of Media Multitasking: Reciprocal Dynamics of Media Multitasking, Personal Needs, and Gratifications," *Journal of Communication* 62(3), 2012, 493–513.
42 This information foraging theory is well-articulated in Gazzaley and Rosen, *The Distracted Mind*.
43 H. Jahncke, S. Hygge, N. Halin, A. M. Green, and K. Dimberg, "Open-plan Office Noise: Cognitive Performance and Restoration," *Journal of Environmental Psychology* 31(4), 2011, 373–382.
44 S. A. Samani, S. Z. Rasid, and S. Sofian, "A Workplace to Support Creativity," *Industrial Engineering & Management Systems* 13(4), 2014, 414–420.
45 J. Kim and R. De Dear, "Workspace Satisfaction: The Privacy-Communication Trade-off in Open-plan Offices," *Journal of Environmental Psychology* 36, 2013, 18–26.
46 C. Knight and S. A. Haslam, "The Relative Merits of Lean, Enriched, and Empowered Offices: An Experimental Examination of the Impact of Workspace Management Strategies on Well-being and Productivity," *Journal of Experimental Psychology: Applied* 16(2), 2010, 158.
47 C. Bailey, *The Productivity Project*.
48 Levitin, *The Organized Mind*, 199.
49 P. J. Bieling, A. L. Israeli, and M. M. Antony, "Is Perfectionism Good, Bad, or Both? Examining Models of the Perfectionism Construct," *Personality and Individual Differences*, 36(6), 2004, 1373–1385.

Chapter 2. Hacking Your Brain, Hacking Your Life

1. T. Schroeder, A. L. Roskies, and S. Nichols, "Moral Motivation," in J. M. Doris (ed.), *The Moral Psychology Handbook* (Oxford, UK: Oxford University Press, 2010), 72–110.
2. K. A. Amaya and K. S. Smith, "Neurobiology of Habit Formation," *Current Opinion in Behavioral Sciences* 20, 2018, 145–152.
3. B. Verplanken and W. Wood, "Interventions to Break and Create Consumer Habits," *Journal of Public Policy & Marketing* 25(1), 2006, 90–103.
4. The "central executive" system also involves the cingulate. Levitin, *The Organized Mind*, 46, 167.
5. I talk about mind-wandering in greater detail in J. Davies, *Imagination: The Science of Your Mind's Greatest Power* (New York: Pegasus, 2019). For the trade-off between the cognitive and mind-wandering system activation levels, see Levitin, *The Organized Mind*, 41.
6. Levitin, *The Organized Mind*, 43–48.
7. The supplementary motor area has to do with self-initiated actions, which we're most interested in here, and the premotor cortex has to do with actions guided by the outside world, such as moving in sync with someone else. P. Haggard, "The Neurocognitive Bases of Human Volition," *Annual Review of Psychology* 70, 2019, 9–28; T. Schroeder, A. L. Roskies, and S. Nichols, "Moral Motivation," in J. M. Doris (ed.), *The Moral Psychology Handbook* (Oxford, UK: Oxford University Press, 2010), 72–110.
8. A. L. Duckworth, T. S. Gendler, and J. J. Gross, "Situational Strategies for Self-Control," *Perspectives on Psychological Science* 11(1), 2016, 35–55.
9. The existence of a "lizard brain" in the human brain is widely held, but not universally believed. For a counter-opinion, see J. Cesario, D. J. Johnson, and H. L. Eisthen, "Your Brain Is Not an Onion With a Tiny Reptile Inside." *Current Directions in Psychological Science,* 2020, 255–260.
10. S. M. McClure, D. I. Laibson, G. Loewenstein, and J. D. Cohen, "Separate Neural Systems Value Immediate and Delayed Monetary Rewards," *Science* 306(5695), 2014, 503–507; S. M. McClure, K. M. Ericson, D. I. Laibson, G. Loewenstein, and J. D. Cohen, "Time Discounting for Primary Rewards," *Journal of Neuroscience* 27(21), 2007, 5796–5804.
11. Duckworth, et al., "Situational Strategies for Self-Control," 35–55.
12. K. McGonigal, *The Willpower Instinct: How Self-Control Works, Why It Matters, and What You Can Do to Get More of It* (New York: Penguin, 2012), 12.
13. M. R. Leary, E. B. Tate, C. E. Adams, A. Batts Allen, and J. Hancock, "Self-compassion and Reactions to Unpleasant Self-relevant Events: The Implications of Treating Oneself Kindly," *Journal of Personality and Social Psychology* 92(5), 2007, 887.
14. J. A. Brefczynski-Lewis, A. Lutz, H. S. Schaefer, D. B. Levinson, and R. J. Davidson, "Neural Correlates of Attentional Expertise in Long-term Meditation Practitioners." *Proceedings of the National Academy of Sciences* 104(27), 2007, 11483–11488.
15. R. J. Davidson and A. W. Kaszniak, "Conceptual and Methodological Issues in Research on Mindfulness and Meditation," *American Psychologist* 70(7), 2015, 581. However, it is very easy to meditate badly. It's actually hard to meditate well. One could have your control group engage in rumination, or self-analysis. This idea comes from economist Richard Bruns (personal communication).
16. B. Verplanken and W. Wood, "Interventions to Break and Create Consumer Habits," *Journal of Public Policy & Marketing* 25(1), 2006, 90–103.
17. Gazzaley and Rosen, *The Distracted Mind*, 204.
18. J. E. Gillham, A. J. Shatté, K. J. Reivich, and M. E. Seligman, "Optimism, Pessimism, and Explanatory Style," in E. C. Chang (ed.), *Optimism and Pessimism: Implications for Theory, Research, and Practice* (Washington, DC: American Psychological Association, 2001), 53–75.

19 L. S. Nes and S. C. Segerstrom, "Dispositional Optimism and Coping: A Meta-Analytic Review," *Personality and Social Psychology Review* 10(3), 2006, 235–251; T. D. Wilson, *Redirect: Changing the Stories We Live By* (London, UK: Hachette, 2011).

20 For a full description of this, and specific exercises you can do, see Wilson, *Redirect: Changing the Stories We Live By*.

21 R. F. Baumeister, "Suicide As Escape from Self," *Psychological Review* 97(1), 1990, 90.

22 Lest you think I'm exaggerating about Buddhist antagonism to ambition, here is a quotation: "We tend to always be going in a certain direction, and having a particular aim in mind. [Zen] Buddhism, on the other hand, has a certain respect for enlightened 'aimlessness' . . . Don't sit to attain a goal . . . Whatever you are doing, whether it is watering the garden, brushing your teeth, or doing the dishes, see if you can do it in a way that is 'aimless.' Sitting in silence can be wonderfully aimless." T. N. Hanh, *Silence: The Power of Quiet in a World Full of Noise* (New York: Random House, 2015), 187–188.

23 T. Kasser and R. M. Ryan, "Further Examining the American Dream: Differential Correlates of Intrinsic and Extrinsic Goals," *Personality and Social Psychology Bulletin* 22(3), 1996, 280–287.

24 T. A. Judge and J. D. Kammeyer-Mueller, "On the Value of Aiming High: The Causes and Consequences of Ambition," *Journal of Applied Psychology* 97(4), 2012, 758–775.

25 Duckworth, et al., "Situational Strategies for Self-control," 35–55.

26 J. Pencavel, "The Productivity of Working Hours," *The Economic Journal* 125(589), 2014, 2052–2076.

27 Levitin, *The Organized Mind*, 98; K. D. Vohs, R.. F. Baumeister, B. J. Schmeichel, J. M. Twenge, N. M. Nelson, and D. M. Tice, "Making Choices Impairs Subsequent Self-control: A Limited-resource Account of Decision Making, Self-regulation, and Active Initiative," *Journal of Personality and Social Psychology* 94(5), 2014, 883–898.

28 M. S. Hagger, et al., "A Multilab Preregistered Replication of the Ego-depletion Effect," *Perspectives on Psychological Science* 11(4), 2016, 546–573. But see also R. F. Baumeister, D. M. Tice, and K. D. Vohs, "The Strength Model of Self-regulation: Conclusions from the Second Decade of Willpower Research," *Perspectives on Psychological Science* 13(2), 2018, 141–145.

29 M. Friese, J. Frankenbach, V. Job, and D. D. Loschelder, "Does Self-control Training Improve Self-control? A Meta-analysis," *Perspectives on Psychological Science* 12(6), 2017, 1077–1099.

30 Duckworth, et al., "Situational Strategies for Self-control," 35–55.

31 Families who eat at the dining room table weigh less than families who eat in front of the television. B. Wansink and E. van Kleef, "Dinner Rituals that Correlate with Child and Adult BMI," *Obesity* 22(5), 2014, E91–E95.

32 J. J. Arch, K. W. Brown, R. J. Goodman, M. D. Della Porta, L. G. Kiken, and S. Tillman, "Enjoying Food Without Caloric Cost: The Impact of Brief Mindfulness on Laboratory Eating Outcomes," *Behaviour Research and Therapy* 79, 2016, 23–34.

33 K. A. Amaya and K. S. Smith, "Neurobiology of Habit Formation," *Current Opinion in Behavioral Sciences* 20, 2018, 145–152.

34 Duckworth, et al., "Situational Strategies for Self-control," 35–55.

35 Ibid.

36 B. Wansink and K. Van Ittersum, "Portion Size Me: Downsizing Our Consumption Norms," *Journal of the Academy of Nutrition and Dietetics*, 107(7), 2007, 1103–1106. This is also known as "unit bias." A. B. Geier, P. Rozin, and G. Doros, "Unit Bias: A New Heuristic that Helps Explain the Effect of Portion Size on Food Intake," *Psychological Science* 17(6), 2006, 521–525.

37. Duckworth, et al., "Situational Strategies for Self-control," 35–55.
38. P. Rozin, K. Kabnick, E. Pete, C. Fischler, and C. Shields, "The Ecology of Eating: Smaller Portion Sizes in France than in the United States Help Explain the French Paradox," *Psychological Science* 14(5), 2003, 450–454.
39. C. Duhigg, *The Power of Habit: Why We Do What We Do in Life and Business* (New York: Random House, 2012).
40. Amaya and Smith, "Neurobiology of Habit Formation," 145–152. You can remember using these using the mnemonic CLEAT.
41. B. Verplanken and W. Wood, "Interventions to Break and Create Consumer Habits," *Journal of Public Policy & Marketing* 25(1), 2006, 90–103.
42. Amaya and Smith, "Neurobiology of Habit Formation."
43. Verplanken and Wood, "Interventions to Break and Create Consumer Habits," 90–103.
44. C. Duhigg, *The Power of Habit*.
45. H. Aarts, T. Paulussen, and H. Schaalma, "Physical Exercise Habit: On the Conceptualization and Formation of Habitual Health Behaviours," *Health Education Research* 12(3), 1997, 363–374.
46. K. A. Finlay, D. Trafimow, and A. Villarreal, "Predicting Exercise and Health Behavioral Intentions: Attitudes, Subjective Norms, and Other Behavioral Determinants 1," *Journal of Applied Social Psychology* 32(2), 2012, 342–356.
47. This effect is stronger if the change in environment is relevant to the habit in question. Changing jobs but staying in the same building is less likely to help you change your habit of going to the donut shop for a break. Verplanken and Wood, "Interventions to Break and Create Consumer Habits," 90–103.
48. W. Wood and D. T. Neal, "A New Look at Habits and the Habit-goal Interface," *Psychological Review* 114(4), 2007, 843.
49. P. M. Gollwitzer, "Implementation Intentions: Strong Effects of Simple Plans," *American Psychologist* 54(7), 1999, 493.
50. Verplanken and Wood, "Interventions to Break and Create Consumer Habits," 90–103.
51. Ibid.
52. K. C. Berridge, T. E. Robinson, and J. W. Aldridge, "Dissecting Components of Reward: 'Liking,' 'Wanting,' and 'Learning'," *Current Opinion in Pharmacology* 9(1), 2009, 65–73.
53. K. C. Berridge and M. L. Kringelbach, "Affective Neuroscience of Pleasure: Reward in Humans and Animals," *Psychopharmacology* 199(3), 2008, 457–480.
54. Ibid.
55. Ibid.
56. J. Olds, "Pleasure Centers in the Brain," *Scientific American* 195(4), 1956, 105–117.
57. Berridge and Kringelbach, "Affective Neuroscience of Pleasure," 457–480.
58. K. C. Berridge, C. Y. Ho, J. M. Richard, and A. G. DiFeliceantonio, "The Tempted Brain Eats: Pleasure and Desire Circuits in Obesity and Eating Disorders," *Brain Research*, 1350, 2010, 43–64, 19.
59. Ibid., 3, 19.
60. Kent Berridge calls the mesolimbic wanting "incentive salience." The wanting system seems to be associated with the lateral hypothalamus, parts of the nucleus accumbens, the amygdala, the dorsal striatum (especially the ventrolateral portion), and the ventral pallidum (following GABA neuron disinhibition). Berridge and Kringelbach, "Affective Neuroscience of Pleasure," 457–480; Berridge, et al., "The Tempted Brain Eats," 43–64.
61. N. Dieker, "Can an Electronic Shock Bracelet Kill Your Bad Habits?" *Medium* August 2, 2019, retrieved August 18, 2019, from https://forge.medium.com/can-an-electronic-shock-bracelet-kill-your-bad-habits-35681ddb62cc.

62 This study found an effect of unconscious conditioning: A. G. Greenwald and J. De Houwer, "Unconscious Conditioning: Demonstration of Existence and Difference from Conscious Conditioning," *Journal of Experimental Psychology: General* 146(12), 2017, 1705. This study, in contrast, found no effect for non-conscious stimuli: T. Heycke and C. Stahl, "No Evaluative Conditioning Effects with Briefly Presented Stimuli," *Psychological Research*, 2018, 1–8.
63 Masochists enjoy pain, but only in some contexts. They might enjoy getting their back turned to hamburger by someone with a whip, but then get a lot of negative feeling when they stub their toe afterward.
64 J. Davis, "The Perfect Human," *Wired*, January 1, 2007, retrieved June 13, 2019, from https://www.wired.com/2007/01/ultraman/.
65 K. L. Milkman, J. A. Minson, and K. G. Volpp, "Holding the Hunger Games Hostage at the Gym: An Evaluation of Temptation Bundling," *Management Science* 60(2), 2013, 283–299.
66 Duckworth, et al., "Situational Strategies for Self-control," 35–55.
67 Z. Chance, R. Dhar, M. Hatzis, and M. Bakker, "How Google Optimized Healthy Office Snacks," *Harvard Business Review* [online], March 3, 2016, retrieved May 14, 2019, from https://hbr.org/2016/03/how-google-uses-behavioral-economics-to-make-its-employees-healthier.
68 L. S. Levitz, "The Susceptibility of Human Feeding Behavior to External Controls," in G. Bray (ed.), *Obesity in Perspective* (vol. NIH: 75–708, 53–60; 1976), Washington, DC: Department of Health, Education, and Welfare.
69 S. Achor, *The Happiness Advantage: The Seven Principles of Positive Psychology That Fuel Success and Performance at Work* (New York: Random House, 2011), 162.
70 Duckworth, et al., "Situational Strategies for Self-control," 35–55.
71 Ibid.
72 C. Newport, *Deep Work: Rules for Focused Success in a Distracted World* (New York: Hachette, 2016), 33.
73 Ibid., 70–79.
74 S. Misra, L. Cheng, J. Genevie, and M. Yuan, "The iPhone Effect: The Quality of In-person Social Interactions in the Presence of Mobile Devices," *Environment and Behavior* 48(2), 2016, 275–298.
75 Levitin, *The Organized Mind*, 68.
76 See study 2 of E. J. Masicampo and R. F. Baumeister, "Consider It Done! Plan Making Can Eliminate the Cognitive Effects of Unfulfilled Goals," *Journal of Personality and Social Psychology* 101(4), 2011, 667–685.
77 Levitin, *The Organized Mind*. People are good at prioritization: V. Bellotti, B. Dalal, N. Good, P. Flynn, D. G. Bobrow, and N. Ducheneaut, "What a To-do: Studies of Task Management Towards the Design of a Personal Task List Manager," in *Proceedings of the SIGCHI conference on Human factors in computing systems* (735–742).
78 D. K. Simonton, *Greatness: Who Makes History and Why* (New York: Guilford Press, 1994), 189.
79 R. Boice, "Contingency Management in Writing and the Appearance of Creative Ideas: Implications for the Treatment of Writing Blocks," *Behaviour Research and Therapy* 21(5), 537–543. 1983; R. Boice, "Procrastination, Busyness and Bingeing," *Behaviour Research and Therapy* 27(6), 1989, 605–611. For an interesting reinterpretation of the results, and a clear description of the original study, see S. Krashen, "Optimal Levels of Writing Management: A Reanalysis of Boice," *Education* 122(3), 2002, 605–608.
80 H. Sword, "'Write Every Day!': A Mantra Dismantled," *International Journal for Academic Development*, 21(4), 2016, 312–322.
81 Now she's got a master's degree in legal anthropology from the London School of Economics. #ProudSpouse

82 U. N. Sio and T. C. Ormerod, "Does Incubation Enhance Problem Solving? A Meta-analytic Review," *Psychological Bulletin*, 135(1), 2009, 94.
83 Levitin, *The Organized Mind*, 210.
84 C. K. Hsee, A. X. Yang, and L. Wang, "Idleness Aversion and the Need for Justifiable Busyness," *Psychological Science* 21(7), 2010, 926–930; W. C. Wang, C. H. Kao, T. C. Huan, and C. C. Wu, "Free Time Management Contributes to Better Quality of Life: A Study of Undergraduate Students in Taiwan," *Journal of Happiness Studies* 12(4), 2011, 561–573.
85 J. McGregor, "What the Most Productive Workers Have in Common," *Washington Post* August 5, 2014, retrieved August 18, 2019, from https://www.washingtonpost.com/news/on-leadership/wp/2014/08/05/what-the-most-productive-workers-have-in-common/; C. Newport *Deep Work: Rules for Focused Success in a Distracted World* (New York: Hachette, 2016), 70–80.
86 I got the term "strategic quitting" from E. Barker, *Barking Up the Wrong Tree: The Surprising Science Behind Why Everything You Know About Success Is (Mostly) Wrong* (New York: HarperOne, 2017), 96.
87 A. V. Whillans, E. W. Dunn, P. Smeets, R. Bekkers, and M. I. Norton, "Buying Time Promotes Happiness," *Proceedings of the National Academy of Sciences*, 114(32), 2017, 8523–8527.
88 A. G. LeBlanc, J. D. Barnes, T. J. Saunders, M. S. Tremblay, and J. P. Chaput, "Scientific Sinkhole: The Pernicious Price of Formatting," *PloS one* 14(9), 2019, e0223116.
89 C. Bailey, *The Productivity Project: Accomplishing More by Managing Your Time, Attention, and Energy* (Toronto: Random House Canada, 2016), 140.
90 I should mention that this book also has lots of advice that I think is quite good. B. Tracy, *Eat That Frog! 21 Great Ways to Stop Procrastinating and Get More Done in Less Time.* (San Francisco: Berrett-Koehler Publishers, Inc., 2007), 11.
91 See the Pomodoro Technique Blog for an introduction: https://francescocirillo.com/pages/pomodoro-technique.
92 Alexandra Michel, "Burnout and the Brain," Association for Psychological Science, January 29, 2016, https://www.psychologicalscience.org/observer/burnout-and-the-brain.
93 "Job burnout: How to spot it and take action," Mayo Clinic, November 21, 2018, https://www.mayoclinic.org/healthy-lifestyle/adult-health/in-depth/art-20046642.
94 B. N. Waber, D. Olguin Olguin, T. Kim, and A. Pentland, "Productivity Through Coffee Breaks: Changing Social Networks by Changing Break Structure," January 11, 2010. Available at SSRN: https://ssrn.com/abstract=1586375.
95 D. H. Pink, *When: The Scientific Secrets of Perfect Timing* (New York: Penguin, 2019), 54.
96 J. P. Trougakos, D. J. Beal, S. G. Green, and H. M. Weiss, "Making the Break Count: An Episodic Examination of Recovery Activities, Emotional Experiences, and Positive Affective Displays," *Academy of Management Journal* 51(1), 2008, 131–146.
97 A. Ariga and A. Lleras, "Brief and Rare Mental 'Breaks' Keep You Focused: Deactivation and Reactivation of Task Goals Preempt Vigilance Decrements," *Cognition* 118(3), 2011, 439–443; E. M. Hunter and C. Wu, "Give Me a Better Break: Choosing Workday Break Activities to Maximize Resource Recovery," *Journal of Applied Psychology* 101(2), 2016, 302; H. Zacher, H. A. Brailsford, and S. L. Parker, "Micro-breaks Matter: A Diary Study on the Effects of Energy Management Strategies on Occupational Well-being," *Journal of Vocational Behavior* 85(3), 2014, 287–297.
98 S. Kim, Y. Park, and Q. Niu, "Micro-break Activities at Work to Recover from Daily Work Demands," *Journal of Organizational Behavior* 38(5), 2017, 28–44.
99 M. Sianoja, U. Kinnunen, J. D. Bloom, K. Korpela, and S. A. E. Geurts, "Recovery During Lunch Breaks: Testing Long-term Relations with Energy Levels at Work," *Scandinavian Journal of Work and Organizational Psychology* 1(1), 2016, 1–12; J. P. Trougakos, I. Hideg, B. H. Cheng, and D. J. Beal, "Lunch Breaks Unpacked: The Role of Autonomy as a Moderator of Recovery During Lunch," *Academy of Management Journal* 57(2), 2014, 405–421.

100 I wrote a whole book about what feels important. J. Davies, *Riveted: The Science of Why Jokes Make Us Laugh, Movies Make Us Cry, and Religion Makes Us Feel One with the Universe* (New York: St. Martin's Press, 2014).
101 R. Beuhler and C. McFarland, "Intensity Bias in Affective Forecasting: The Role of Temporal Focus," *Psychological Science* 16, 2001, 626–630.
102 S. Pinker, *Enlightenment Now: The Case for Reason, Science, Humanism, and Progress* (New York: Viking, 2018), 4.
103 Stephens-Davidowitz, *Everybody Lies*.
104 I'm talking, of course, about typical news focused on politics and human-interest stories, not science news, which is a little better about this stuff.
105 R. M. Mar, K. Oatley, J. Hirsh, J. de la Paz, and J. B. Peterson, "Bookworms Versus Nerds: Exposure to Fiction Versus Non-fiction, Divergent Associations with Social Ability, and the Stimulation of Fictional Social Worlds," *Journal of Research in Personality* 40(5), 2006, 694–712; B. Johnson, "Religion and Philanthropy." Unpublished manuscript, University of North Carolina at Chapel Hill, 2012; P. A. Katz and S. R. Zalk, "Modification of Children's Racial Attitudes," *Developmental Psychology* 14(5), 1978, 447–461.
106 The main issues I have relate to indexing things using keywords different from what I wrote them down as, e.g., kids, children, offspring, etc.
107 I discuss memory palaces more extensively in my previous book: J. Davies, *Imagination: The Science of Your Mind's Greatest Power* (New York: Pegasus Books, 2019).
108 This idea comes from my aunt, Mary Berard.
109 Walker, *Why We Sleep*.
110 L. M. Juliano and R. R. Griffiths, "A Critical Review of Caffeine Withdrawal: Empirical Validation of Symptoms and Signs, Incidence, Severity, and Associated Features," *Psychopharmacology* 176(1), 2004, 1–29.
111 S. Stranges, W. Tigbe, F. X. Gómez-Olivé, M. Thorogood, and N. B. Kandala, "Sleep Problems: An Emerging Global Epidemic? Findings from the INDEPTH WHO-SAGE Study Among More than 40,000 Older Adults from 8 Countries Across Africa and Asia," *Sleep* 35(8), 2012, 1173–1181; M. Walker, *Why We Sleep: Unlocking the Power of Sleep and Dreams* (New York: Simon and Schuster, 2017), chapter 7.
112 J. J. Iliff, M. Wang, Y. Liao, B. A. Plogg, W. Peng, G. A. Gundersen, and E. A. Nagelhus, "A Paravascular Pathway Facilitates CSF Flow through the Brain Parenchyma and the Clearance of Interstitial Solutes, Including Amyloid β," *Science Translational Medicine* 4(147), 2012, 147ra111–147ra111.
113 M. A. Grandner, "Sleep Deprivation: Societal Impact and Long-Term Consequences," in *Sleep Medicine*, eds. Sudhansu Chokroverty and Michel Billiard (New York: Springer, 2015), 495–509.
114 M. Walker, *Why We Sleep: Unlocking the Power of Sleep and Dreams* (New York: Simon and Schuster, 2017).
115 Phones often have a function where you can set it to be in a "do not disturb" mode during your normal sleeping hours. If you're worried about emergency calls, most phones will let a call through if it's from someone in your contacts and they call twice. Walker, *Why We Sleep*, appendix.
116 This is due to a variety of factors, including increased sleep fragmentation due to a weakened bladder, and a weakening melatonin and circadian rhythm cycle. Walker, *Why We Sleep*, 85.
117 Walker, *Why We Sleep*, 64.
118 D. H. Pink, *When: The Scientific Secrets of Perfect Timing* (New York: Penguin, 2019), 63–68.
119 A. Brooks and L. Lack, "A Brief Afternoon Nap Following Nocturnal Sleep Restriction: Which Nap Duration Is Most Recuperative?" *Sleep* 29(6), 2016, 831–840.
120 Ibid.

121 The effects of orally consumed caffeine peaks at 30 minutes after consumption. Walker, *Why We Sleep*.
122 L. A. Reyner and J. A. Horne, "Suppression of Sleepiness in Drivers: Combination of Caffeine with a Short Nap," *Psychophysiology* 34(6), 1997, 721–725.
123 Levitin, *The Organized Mind*, 192.
124 A very useful feature I'm shocked is not in phones is the ability to put a phone on airplane mode for a specific length of time. If you're taking a flight, watching a movie, or napping, you know how long you will need the phone to be in this mode, and it would be nice to not have to remember to take it out of this mode.
125 I. N. Lee, "Nap Pods Making a Return to UBC after Decades of Absence," *The Thunderbird*, November 21, 2018, retrieved April 29, 2019, from https://thethunderbird.ca/2018/11/21/nap-pods-making-a-return-to-ubc-after-decades-of-absence/.
126 Much of the findings about what to do when I got from Daniel Pink's excellent book *When*.
127 Pink Daniel, H. "When: The Scientific Secrets of Perfect Timing." (2018). New York: Riverhead books. 26-31. In medicine, 47–53; Morality, 53.
128 Ibid., 29–45.
129 N. G. Pope, "How the Time of Day Affects Productivity: Evidence from School Schedules," *Review of Economics and Statistics* 98(1), 2016, 1–11.
130 Pink, *When*, 26–31; In medicine, 47–53; Morality, 53.
131 C. Vetter, D. Fischer, J. L. Matera, and T. Roenneberg, "Aligning Work and Circadian Time in Shift Workers Improves Sleep and Reduces Circadian Disruption," *Current Biology* 25(7), 2015, 907–911.
132 M. S. Clark and J. Mils, "The Difference Between Communal and Exchange Relationships: What It Is and Is Not," *Personality and Social Psychology Bulletin* 19(6), 1993, 684–691.
133 A. M. Grant, *Give and Take: A Revolutionary Approach to Success* (New York: Penguin, 2013), 16.
134 This is "pathological altruism." B. Oakley, A. Knafo, G. Madhavan, and D. S. Wilson (eds.), *Pathological Altruism* (Oxford, UK: Oxford University Press, 2011); Health problems: V. S. Helgeson, "Relation of Agency and Communion to Well-being: Evidence and Potential Explanations," *Psychological Bulletin* 116(3), 1994, 412; Crime victimization: R. J. Homant, "Risky Altruism as a Predictor of Criminal Victimization," *Criminal Justice and Behavior* 37(11), 2010, 1195–1216.
135 J. A. Frimer, L. J. Walker, W. L. Dunlop, B. H. Lee, and A. Riches, "The Integration of Agency and Communion in Moral Personality: Evidence of Enlightened Self-interest," *Journal of Personality and Social Psychology* 101(1), 2011, 149.
136 M. Feinberg, J. T. Cheng, and R. Willer, "Gossip As an Effective and Low-cost Form of Punishment," *Behavioral and Brain Sciences* 35(1), 2012, 25.
137 https://www.adamgrant.net/give-and-take-assessment.
138 Clark and Mils, "The Difference Between Communal and Exchange," 684–691.
139 This book is so good I try to reread it every few years. Grant, *Give and Take*.
140 P. A. Thoits and L. N. Hewitt, "Volunteer Work and Well-being," *Journal of Health and Social Behavior* 42(2), 2001, 115–131.
141 N. Weinstein and R. M. Ryan, "When Helping Helps: Autonomous Motivation for Prosocial Behavior and its Influence on Well-being for the Helper and Recipient," *Journal of Personality and Social Psychology*, 98(2), 2010, 222.
142 Thoits and Hewitt, "Volunteer Work," 115–131.
143 S. L. Brown, R. M. Nesse, A. D. Vinokur, and D. M. Smith, "Providing Social Support May Be More Beneficial than Receiving It: Results from a Prospective Study of Mortality," *Psychological Science* 14(4), 2003, 320–327.
144 100 hours: M. C. Luoh and A. R. Herzog, "Individual Consequences of Volunteer and Paid Work in Old Age: Health and Mortality," *Journal of Health and Social Behavior* 43(4),

2012, 490–509; Chunking volunteering: S. Lyubomirsky, K. M. Sheldon, and D. Schkade, "Pursuing Happiness: The Architecture of Sustainable Change," *Review of General Psychology* 9(2), 2005, 111–131.
145 J. H. Fowler and N. A. Christakis, "Cooperative Behavior Cascades in Human Social Networks," *Proceedings of the National Academy of Sciences* 107(12), 2010, 5334–5338.

Chapter 3. Happiness

1 P. Van Cappellen, M. Toth-Gauthier, V. Saroglou, and B. L. Fredrickson, "Religion and Well-being: The Mediating Role of Positive Emotions," *Journal of Happiness Studies* 17(2), 2016, 485–505.
2 Happiness makes you work harder: S. G. Barsade and D. E. Gibson, "Why Does Affect Matter in Organizations?" *Academy of Management Perspectives* 21(1), 2007, 36–59; Happiness improves your health: M. Olfson, C. Blanco, and S. C. Marcus, "Treatment of Adult Depression in the United States," *JAMA Internal Medicine* 176(10), 2016, 1482–1491.
3 S. Kagan, *How to Count Animals, More or Less* (Oxford, UK: Oxford University Press, 2019).
4 G. MacKerron, "Happiness Economics from 35,000 Feet," *Journal of Economic Surveys* 26(4), 2012, 705–735.
5 S. Pinker, *Enlightenment Now: The Case for Reason, Science, Humanism, and Progress* (New York: Viking, 2018), 266; MacKerron, "Happiness Economics," 705–735.
6 J. Haidt, *The Happiness Hypothesis: Finding Modern Truth in Ancient Wisdom* (New York; Basic Books, 2006), 220.
7 M. Csikszentmihalyi, *Flow: The Psychology of Optimal Experience* (New York: Harper Perennial Modern Classics, 2008), 158; M. Csikszentmihalyi and J. LeFevre, "Optimal Experience in Work and Leisure," *Journal of Personality and Social Psychology* 56(5), 1989, 815; C. K. Hsee, A. X. Yang, and L. Wang, "Idleness Aversion and the Need for Justifiable Busyness," *Psychological Science* 21(7), 2010, 926–930; Flow states activate the left prefrontal cortex, especially areas 44, 45, and the basal ganglia. The self-critical part of the prefrontal cortex and the amygdala are deactivated; Levitin, *The Organized Mind*, 203.
8 A. E. Clark, S. Flèche, R. Layard, N. Powdthavee, and G. Ward, *The Origins of Happiness: The Science of Well-Being Over the Life Course* (Princeton, N.J.: Princeton University Press, 2018), 74.
9 Work ranks close to last (commuting to and from work is even worse) whether you look at reflections on the day, or ask people's emotional states at random times during the day (experience sampling method). A. B. Krueger, D. Kahneman, D. Schkade, N. Schwarz, and A. A. Stone, "National Time Accounting: The Currency of Life," in *Measuring the Subjective Well-being of Nations: National Accounts of Time Use and Well-being*, ed. Alan B. Krueger (Chicago: University of Chicago Press, 2009), 9–86; Unemployment is worse: Clark, et al., *The Origins of Happiness*, 61.
10 J. F. Helliwell and S. Wang, "Weekends and Subjective Well-being," *Social Indicators Research* 116(2), 389–407.
11 S. Steiner, "Top Five Regrets of the Dying," *Guardian*, February 1, 2012.
12 L. Golden and B. Wiens-Tuers, "To Your Happiness? Extra Hours of Labor Supply and Worker Well-being," *The Journal of Socio-Economics* 35(2), 2006, 382–397.
13 R. M. Ryan, J. H. Bernstein, and K. W. Brown, "Weekends, Work, and Well-being: Psychological Need Satisfactions and Day of the Week Effects on Mood, Vitality, and Physical Symptoms," *Journal of Social and Clinical Psychology* 29(1), 2010, 95–122.
14 I. Granic, A. Lobel, and R. C. Engels, "The Benefits of Playing Video Games," *American Psychologist* 69(1), 2014, 66.

15. N. Ravaja, T. Saari, J. Laarni, K. Kallinen, M. Salminen, and J. Holopainen, "The Psychophysiology of Video Gaming: Phasic Emotional Responses to Game," in *DiGRA 2005 conference changing views–worlds in play* 2005, 1–13.
16. This study found that they use the game as a break, for recovery after strain. L. Reinecke, "Games at Work: The Recreational Use of Computer Games During Working Hours," *CyberPsychology & Behavior* 12(4), 2009, 461–465.
17. J. E. Kim and P. Moen, "Retirement Transitions, Gender, and Psychological Well-being: A Life-course, Ecological Model," *The Journals of Gerontology Series B: Psychological Sciences and Social Sciences*, 57(3), 2002, P212–P222.
18. R. F. Baumeister and K. D. Vohs, "The Pursuit of Meaningfulness in Life," *Handbook of Positive Psychology* 1, 2002, 608–618.
19. Pinker, *Enlightenment Now*, 267.
20. J. McGonigal, *Reality Is Broken: Why Games Make Us Better and How They Can Change the World* (New York: Penguin, 2011), 29–30.
21. I. Granic, A. Lobel, and R. C. Engels, "The Benefits of Playing Video Games," *American Psychologist* 69(1), 2014, 66.
22. Ibid., 74.
23. L. A. Jackson, E. A. Witt, A. I. Games, H. E. Fitzgerald, A. Von Eye, and Y. Zhao, "Information Technology Use and Creativity: Findings from the Children and Technology Project," *Computers in Human Behavior* 28(2), 2012, 370–376.
24. Granic, et al., "The Benefits of Playing Video Games," 66; L. Denworth, "Brain-changing Games," *Scientific American Mind* 23(6), 2013, 28–35; R. Stephens and C. Allsop, "Effect of Manipulated State Aggression on Pain Tolerance," *Psychological Reports: Disability and Trauma* 111(1), 2012, 311–321; M. Konnikova, "Why Gamers Can't Stop Playing First-person Shooters," *The New Yorker*, November 25, 2013, retrieved April 18, 2019, from https://www.newyorker.com/tech/annals-of-technology/why-gamers-cant-stop-playing-first-person-shooters.
25. Granic, et al., 73.
26. McGonigal, *Reality Is Broken*, 67.
27. Ibid., 4.
28. Jane McGonigal says, "Good games *are* productive. They're producing a higher quality of life." Though, to her credit, she talks about how to use wisdom from game design to increase actual productivity later in her book. McGonigal, *Reality Is Broken*, 51.
29. R. Hunicke, M. LeBlanc, and R. Zubek, "MDA: A Formal Approach to Game Design and Game Research," in *Proceedings of the AAAI Workshop on Challenges in Game AI* (Vol. 4, No. 1, 2004) 1722.
30. For a readable account, with links to studies, see W. Boot, "Evidence Behind 'Brain Training' Games Remains Lacking," *Undark*, June 11, 2019, retrieved June 18, 2019, from https://undark.org/2019/06/11/brain-training-lacks-evidence/.
31. V. J. Shute, M. Ventura, and F. Ke, "The Power of Play: The Effects of Portal 2 and Lumosity on Cognitive and Noncognitive Skills," *Computers & Education* 80, 2015, 58–67.
32. K. Bainbridge and R. E. Mayer, "Shining the Light of Research on Lumosity," *Journal of Cognitive Enhancement* 2(1), 2018, 43–62.
33. Granic, et al., "The Benefits of Playing Video Games," 66.
34. This is the "peak end rule." V. Tiberius and A. Plakias, "Well-being," in J. M. Doris (ed.), *The Moral Psychology Handbook* (Oxford, UK: Oxford University Press, 2010), 402–432, 405.
35. Note that this is because people *perceive* that the disabled are less happy, which might not be true. Tiberius and Plakias, "Well-being," e402–432, 418.
36. J. M. Zelenski (in press), *Positive Psychology: The Science of Well-Being*, chapter 2.

37 S. Oishi, "The Experiencing and Remembering of Well-being: A Cross-cultural Analysis," *Personality and Social Psychology Bulletin* 28(10), 2002, 1398–1406.
38 Tiberius and Plakias, "Well-being," 402–432, 412.
39 B. S. Frey, C. Benesch, and A. Stutzer, "Does Watching TV Make Us Happy?" *Journal of Economic Psychology* 28(3), 2007, 283–313.
40 Levitin, *The Organized Mind*.
41 Zelenski, *Positive Psychology*, chapter 3. People differ a lot in how much they suffer after the loss of a spouse. G. A. Bonanno, C. B. Wortman, D. R. Lehman, R. G. Tweed, M. Haring, J. Sonnega, and R. M. Nesse, "Resilience to Loss and Chronic Grief: A Prospective Study from Preloss to 18-months Postloss," *Journal of Personality and Social Psychology* 83(5), 2002, 1150.
42 A. E. Clark, E. Diener, Y. Georgellis, and R. E. Lucas, "Lags and Leads in Life Satisfaction: A Test of the Baseline Hypothesis," *The Economic Journal* 118, 2008, 222–243.
43 S. Lyubomirsky, "Hedonic Adaptation to Positive and Negative Experiences," in S. Folkman (ed.), *The Oxford Handbook of Stress, Health, and Coping* (Oxford, UK: Oxford University Press, 2011), 200–224, 203.
44 Lyubomirsky, *The Oxford Handbook of Stress, Health, and Coping*, 200–224. I explored adaptation about confusing things that you can't make sense of in my previous book *Riveted*.
45 Zelenski, *Positive Psychology*, chapter 3.
46 Pinker, *Enlightenment Now*, 268.
47 Haidt, *The Happiness Hypothesis*, 89.
48 Zelenski, *Positive Psychology*, chapter 3.
49 Ibid.
50 D. G. Blanchflower and A. J. Oswald, "Is Well-being U-shaped Over the Life Cycle?" *Social Science & Medicine* 66(8), 2008, 1733–1749.
51 A. T. Jebb, L. Tay, E. Diener, and S. Oishi, "Happiness, Income Satiation and Turning Points Around the World," *Nature Human Behaviour* 2(1), 2018, 33. You can see a chart at https://qz.com/1211957/how-much-money-do-people-need-to-be-happy/.
52 T. Kasser and R. M. Ryan, "Further Examining the American Dream: Differential Correlates of Intrinsic and Extrinsic Goals," *Personality and Social Psychology Bulletin* 22(3), 1996, 280–287.
53 Levitin, *The Organized Mind*, 78.
54 Lyubomirsky, "Hedonic Adaptation," 200–224.
55 Ibid.
56 Ibid.; A. R. Hariri, S. Y. Bookheimer, and J. C. Mazziotta, "Modulating Emotional Responses: Effects of a Neocortical Network on the Limbic System," *Neuroreport* 11(1), 2000, 43–48.
57 T. D. Wilson, D. J. Lisle, J. W. Schooler, S. D. Hodges, K. J. Klaaren, and S. J. LaFleur, "Introspecting About Reasons Can Reduce Post-choice Satisfaction," *Personality and Social Psychology Bulletin* 19(3), 1993, 331–339.
58 T. D. Wilson, D. B. Centerbar, D. A. Kermer, and D. T. Gilbert, "The Pleasures of Uncertainty: Prolonging Positive Moods in Ways People Do Not Anticipate," *Journal of Personality and Social Psychology* 88(1), 2005, 5.
59 N. Paumgarten, "There and Back Again," *The New Yorker*, April 16, 2007; Clark, et al., *The Origins of Happiness*, 63, 70.
60 A. Stutzer and B. S. Frey, "Stress That Doesn't Pay: The Commuting Paradox," *Scandinavian Journal of Economics* 110(2), 2008, 339–366.
61 Exercise makes people happy: K. Weir, "The Exercise Effect," *American Psychological Association* website cover story, 42(11), 2011, 48, retrieved October 24, 2019, from https://www

.apa.org/monitor/2011/12/exercise; People are happier walking outside: E. K. Nisbet and J. M. Zelenski, "Underestimating Nearby Nature: Affective Forecasting Errors Obscure the Happy Path to Sustainability," *Psychological Science* 22(9), 2011, 1101–1106.
62. J. Jachimowicz, J. J. Lee, B. R. Staats, J. Menges, and F. Gino, "Between Home and Work: Commuting as an Opportunity for Role Transitions," *Harvard Business School NOM Unit Working Paper* (16–077), 2018, 16–17.
63. Paumgarten, "There and Back Again," 16.
64. Of course, it's species-specific, too. Mice would probably think we spend an inordinate amount of time in the daylight. https://en.wikipedia.org/wiki/List_of_countries _by_meat_consumption_per_capita
65. Haidt, *The Happiness Hypothesis*, 83.
66. Lyubomirsky, "Hedonic Adaptation," 200–224.
67. Haidt, *The Happiness Hypothesis*, 88.
68. The original meaning of the word "happy" favored the "cortical lottery" theory of happiness: it meant "favored by fortune" or "lucky." The "hap" being the same one as in "happenstance" or "haphazard."
69. Zelenski, *Positive Psychology*, chapter 3.
70. Lyubomirsky, "Hedonic Adaptation," 200–224.
71. "Prozac" is Haidt's shorthand for whatever SSRI works for you—there are many on the market. Haidt, *The Happiness Hypothesis*, 35.
72. M. E. Seligman, "The Effectiveness of Psychotherapy: The Consumer Reports Study," *American Psychologist* 50(12), 1995, 965.
73. Paul Reps (trans.), *Zen Flesh, Zen Bones* (New York: Anchor/Doubleday, 1958), 22–23.
74. There is some evidence that meditation can help you focus. H. A. Slagter, A. Lutz, L. L. Greischar, A. D. Francis, S. Nieuwenhuis, J. M. Davis, and R. J. Davidson, "Mental Training Affects Distribution of Limited Brain Resources," *PLoS Biology* 5(6), 2007, e138; Y. Y. Tang and M. I. Posner, "Attention Training and Attention State Training," *Trends in Cognitive Sciences* 13(5), 2009, 222–227.
75. Haidt, *The Happiness Hypothesis*, Kindle location about 115.
76. Ibid., 26.
77. M. A. Killingsworth and D. Gilbert, "A Wandering Mind Is an Unhappy Mind," *Science* 330, 2010, 932.
78. Pink, *When*, 191; F. B. Bryant, C. M. Smart, and S. P. King, "Using the Past to Enhance the Present: Boosting Happiness through Positive Reminiscence," *Journal of Happiness Studies* 6(3), 2005, 227–260.
79. M. Baldwin, M. Biernat, and M. J. Landau, "Remembering the Real Me: Nostalgia Offers a Window to the Intrinsic Self," *Journal of Personality and Social Psychology* 108(1), 2015, 128.
80. J. Nawijn, M. A. Marchand, R. Veenhoven, and A. J. Vingerhoets, "Vacationers Happier, but Most not Happier after a Holiday," *Applied Research in Quality of Life* 5(1), 2010, 35–47. I discuss more extensively about how imagination can make you happier or more miserable in my previous book *Imagination*.
81. Haidt, *The Happiness Hypothesis*, 95.
82. Zelenski, *Positive Psychology*, chapter 3.
83. D. C. Glass and J. E. Singer, *Urban Stress: Experiments on Noise and Social Stressors* (New York: Academic Press, 1972).
84. Haidt, *The Happiness Hypothesis*, 94.
85. R. M. Sapolsky, *Why Zebras Don't Get Ulcers: The Acclaimed Guide to Stress, Stress-related Diseases, and Coping—Now Revised and Updated* (New York: Holt, 2004).

ENDNOTES (CHAPTER 3)

86 "Manage stress: Strengthen your support network," American Psychology Organization, lc, retrieved October 25, 2019, from https://www.apa.org/helpcenter/manage-stress.
87 2015 Stress in America Snapshot. www.apa.org.
88 J. A. Updegraff and S. E. Taylor, "From Vulnerability to Growth: Positive and Negative Effects of Stressful Life Events," *Loss and Trauma: General and Close Relationship Perspectives* 25, 2000, 3–28.
89 R. A. Emmons, "Personal Goals, Life Meaning, and Virtue: Wellsprings of a Positive Life," in C. L. M. Keyes and J. Haidt (eds.), *Flourishing: Positive Psychology and the Life Well-lived*, American Psychological Association, 2003, 105–128; For a review of this "adversity hypothesis," see Haidt, *The Happiness Hypothesis*, 143–154.
90 J. Henrich, *The Secret of Our Success: How Culture is Driving Human Evolution, Domesticating our Species, and Making Us Smarter* (Princeton, N.J.: Princeton University Press, 2017), 209.
91 D. T. Cox, D. F. Shanahan, H. L. Hudson, K. E. Plummer, G. M. Siriwardena, R. A. Fuller, and K. J. Gaston, "Doses of Neighborhood Nature: The Benefits for Mental Health of Living with Nature," *BioScience*, 67(2), 2017, 147–155.
92 Zelenski, *Positive Psychology*, chapter 7.
93 D. Peters, "What Drives Quality of Life in Iowa Small Towns?" *Extension and Outreach Publications* 297, 2017.
94 M. Beck, "City vs. Country: Who is Healthier?" *Wall Street Journal*, July 12, 2011, retrieved June 9, 2020, from https://www.wsj.com/articles/SB10001424052702304793504576434442652581806.
95 J. P. Suraci, M. Clinchy, L. Y. Zanette, and C. C. Wilmers, "Fear of Humans as Apex Predators Has Landscape-scale Impacts from Mountain Lions to Mice," *Ecology Letters*, 2019, doi: 10.1111/ele.13344.
96 S. Arbesman, *The Half-life of Facts: Why Everything We Know Has an Expiration Date* (New York: Penguin, 2012), 135.
97 M. Kremer, "Population Growth and Technological Change: One Million BC to 1990," *The Quarterly Journal of Economics*, 108(3), 1993, 681–716; L. M. Bettencourt, J. Lobo, D. Helbing, C. Kühnert, and G. B. West, "Growth, Innovation, Scaling, and the Pace of Life in Cities," *Proceedings of the National Academy of Sciences* 104(17), 2007, 7301–7306; Arbesman, *The Half-life of Facts*, 60.
98 Stephens-Davidowitz, *Everybody Lies*.
99 R. A. Emmons and R. Stern, "Gratitude as a Psychotherapeutic Intervention," *Journal of Clinical Psychology* 69(8), 2013, 846–855.
100 The statistics are a little different depending on whether you count safety by distance, by hour, or by trip. That is, if you look at a per mile chance of and severity of injury, cars are a bit safer. But if you look at it by the trip, or by the hour, they are the same. I think it's more rational to look at it by the trip, because the fact is you take shorter trips if you're on a bike. You might drive forty minutes to a restaurant, but you'd be unlikely to bike that same distance. Using a bike means you choose different destinations.
101 "Bicycling: the SAFEST Form of Transportation," *Mr. Money Mustache* (blog), June 13, 2013, http://www.mrmoneymustache.com/2013/06/13/bicycling-the-safest-form-of-transportation/.
102 This data is from 1993, compiled by the Failure Analysis Associates, Inc. I cannot find the original reference, but I got it from https://www.helmets.org/stats.htm.
103 "The Health Benefits of Cycling," Bicycle Helmet Research Foundation, https://www.cyclehelmets.org/1015.html; for an overview.
104 Zelenski, *Positive Psychology*, chapter 9; M. Koo, S. B. Algoe, T. D. Wilson, and D. T. Gilbert, "It's a Wonderful Life: Mentally Subtracting Positive Events Improves People's Affective States, Contrary to their Affective Forecasts," *Journal of Personality and Social Psychology* 95(5), 2008, 1217–1224.

105 Koo, et al., "It's a Wonderful Life," 1217–1224.
106 For a good review of the literature on the benefits of sleep and exercise, see G. Wells, *The Ripple Effect* (Toronto: HarperCollins, 2017), chapters 2 and 3.
107 World Health Organization, *Physical Activity*. Fact sheet 385. June 2016.
108 G. Cooney, D. T. Gilbert, and T. D. Wilson, "The Unforeseen Costs of Extraordinary Experience," *Psychological Science* 25(12), 2014, 2259–2265.
109 S. J. Heintzelman, J. Trent, and L. A. King, "Encounters With Objective Coherence and the Experience of Meaning in Life," *Psychological Science* 24(6), 2013, 991–998.
110 S. A. Turner and P. J. Silvia, "Must Interesting Things Be Pleasant? A Test of Competing Appraisal Structures," *Emotion* 6(4), 2006, 670–674.
111 Zelenski, *Positive Psychology*, chapter 3.
112 Haidt, *The Happiness Hypothesis*, 88.
113 P. Van Cappellen, M. Toth-Gauthier, V. Saroglou, and B. L. Fredrickson, "Religion and Well-being: The Mediating Role of Positive Emotions," *Journal of Happiness Studies* 17(2), 2016, 485–505; Haidt, *The Happiness Hypothesis*, 199.
114 S. Folkman, "Positive Psychological States and Coping with Severe Stress," *Social Science & Medicine*, 45(8), 1997, 1207–1221; Zelenski, *Positive Psychology*, chapter 3.
115 M. Joshanloo, "Cultural Religiosity as the Moderator of the Relationship Between Affective Experience and Life Satisfaction: A Study in 147 Countries," *Emotion* 19(4), 2019, 629–638.
116 Pinker, *Enlightenment Now*, 438.
117 Ibid., 438–439.
118 S. Crabtree, "Religiosity Highest in World's Poorest Nations," Gallup.com, August 31, 2010, retrieved September 5, 2019, from https://news.gallup.com/poll/142727/religiosity-highest-world-poorest-nations.aspx.
119 Clark, et al., *The Origins of Happiness*, 125.
120 "People Who Take All Their Vacation Get Better Performance Reviews," *The Huffington Post*, August 27, 2014, retrieved August 15, 2019, from https://www.huffpost.com/entry/vacation-days-performance-review_n_5723548.
121 Interestingly, it also appears that anticipation of the trip boosts happiness. People look forward to their trips, and this seems to make them mildly happier before they go (this effect is very small: on a five-point scale, vacationers rated an average happiness of 2.25, and everybody else rated 2.07).; Nawijn, et. al., "Vacationers Happier, but Most not Happier after a Holiday," 35–47.
122 J. De Bloom, S. A. Geurts, and M. A. Kompier, "Vacation (After-) Effects on Employee Health and Well-being, and the Role of Vacation Activities, Experiences and Sleep," *Journal of Happiness Studies* 14(2), 2013, 613–633.
123 J. T. Cacioppo, J. H. Fowler, and N. A. Christakis, "Alone in the Crowd: The Structure and Spread of Loneliness in a Large Social Network," *Journal of Personality and Social Psychology* 97(6), 2009, 977.
124 Levitin, *The Organized Mind*, 12.
125 Zelenski, *Positive Psychology*,. Chapter 3.
126 R. H. Frank, *Luxury Fever: Why Money Fails to Satisfy in an Era of Excess* (New York: Simon and Schuster, 2001).
127 This amount is in 2007 dollars. N. Powdthavee, "Putting a Price Tag on Friends, Relatives, and Neighbours: Using Surveys of Life Satisfaction to Value Social Relationships," *The Journal of Socio-Economics* 37(4), 2008, 1459–1480.
128 A meta-analysis showed that extraverts tend to act moderately extraverted 5–10 percent more than introverts. W. Fleeson and P. Gallagher, "The Implications of Big Five Standing for the Distribution of Trait Manifestation in Behavior: Fifteen Experience-sampling Studies and a Meta-analysis," *Journal of Personality and Social Psychology* 97(6), 2009, 1097.

129 S. Pinker, *The Village Effect: How Face-to-Face Contact Can Make Us Happier and Healthier* (Toronto: Random House Canada, 2014); The best personality predictors of happiness are being extraverted and having a low neuroticism. Agreeable and more conscientious people are also a bit happier. Zelenski, *Positive Psychology*, chapter 3.

130 W. Fleeson, A. B. Malanos, and N. M. Achille, "An Intraindividual Process Approach to the Relationship between Extraversion and Positive Affect: Is Acting Extraverted as 'Good' as Being Extraverted?" *Journal of Personality and Social Psychology* 83(6), 2002, 1409.

131 L. Henderson, P. Zimbardo, and B. Carducci, "Shyness," *The Corsini Encyclopedia of Psychology* (Hoboken, NJ: Wiley & Sons, 2010), 1–3.

132 W. Isaacson, *Einstein: His Life and Universe* (New York: Simon and Schuster, 2008).

133 J. Hambrick, *"Mozart Minute: Wolfgang's Labor, Constanze's Birth Pangs,"* WOSU Radio, 2015, retrieved May 10, 2019, from https://radio.wosu.org/post/mozart-minute-wolfgangs-labor-constanzes-birth-pangs#stream/0.

134 D. H. Pink, *Drive: The Surprising Truth about What Motivates Us* (New York: Penguin, 2001), 143.

135 Pinker, *The Village Effect*, 30.

136 Ibid., 179.

137 J. Davies, "Why Facebook Is the Junk Food of Socializing," *Nautilus* (blog), June 1, 2015, retrieved October 28, 2019, from http://nautil.us/blog/why-facebook-is-the-junk-food-of-socializing.

138 S. Turkle, *Alone Together* (New York: Basic Books, 2011).

139 Haidt, *The Happiness Hypothesis*, 132.

140 G. Miller, "Why Loneliness Is Hazardous to Your Health," *Science* 331, 2011, 138–140.

141 A. C. Zimmermann and R. A. Easterlin, "Happily Ever After? Cohabitation, Marriage, Divorce, and Happiness in Germany," *Population and Development Review* 32(3), 2006, 511–528.

142 R. E. Lucas and A. E. Clark, "Do People Really Adapt to Marriage?" *Journal of Happiness Studies* 7(4), 405–426.

143 Pinker, *The Village Effect*, 221.

144 C. C. Mann, "The Coming Death Shortage," *Atlantic Monthly* 295(4), 2005, 92–102.

145 Clark, et al., *The Origins of Happiness*, 27.

146 Information from Statistics Canada, retrieved April 25, 2019, from https://www150.statcan.gc.ca/n1/pub/89-503-x/2010001/article/11546-eng.htm#a8.

147 Zimmermann and Easterlin, "Happily Ever After?" 511–528.

148 Stephens-Davidowitz, *Everybody Lies*.

149 B. Tomasik, "The Cost of Kids," August 4, 2012, retrieved July 5, 2019, from https://reducing-suffering.org/the-cost-of-kids/.

150 Clark, et al., *The Origins of Happiness*, 83–84.

151 See the graph on page 8; G. L. Brase and S. L. Brase, "Emotional Regulation of Fertility Decision Making: What Is the Nature and Structure of 'Baby Fever?'" *Emotion* 12(5), 2012, 1141.

152 Tomasik, "The Cost of Kids."

153 Haidt, *The Happiness Hypothesis*, 124.

154 Romantic love is a cross-cultural phenomenon, and not a new, historically recent thing, as some would have you believe. W. R. Jankowiak, and E. F. Fischer, "A Cross-Cultural Perspective on Romantic Love," *Ethnology* 31(2), 1992, 149–155.

155 G. Sinha, "You Dirty Vole," in O. Sacks (ed.), *The Best American Science Writing 2003* (New York: HarperCollins, 2003), 132–137. There's a great podcast that explains all of this, called, "This Is Your Brain On Love" from the excellent series "Radiolab." I highly recommend it. Very entertaining and interesting. http://www.wnyc.org/shows/radiolab/episodes/2007/08/28.

156 A. Cloutier and J. Peetz, "Relationships' Best Friend: Links Between Pet Ownership, Empathy, and Romantic Relationship Outcomes," *Anthrozoös* 29(3), 2016, 395–408.
157 S. Kanazawa, "Why Productivity Fades with Age: The Crime–Genius Connection," *Journal of Research in Personality* 37(4), 2003, 257–272.
158 V. J. Knox, "Cognitive Strategies for Coping with Pain: Ignoring vs. Acknowledging," unpublished doctoral dissertation (Waterloo, Ontario: University of Waterloo, 1972).
159 R. Nozick, *Anarchy, State, and Utopia* vol. 5038 (New York: Basic Books, 1974).
160 I'm not suggesting that computer programs cannot have subjective experience, just that, in this example, this one doesn't.
161 F. Hindriks and I. Douven, "Nozick's Experience Machine: An Empirical Study," *Philosophical Psychology* 31(2), 2018, 278–298.
162 F. De Brigard, "If You Like It, Does It Matter if It's Real?" *Philosophical Psychology* 23(1), 2010, 43–57.
163 T. Hurka, M. D. Adler, and M. Fleurbay, "Objective Goods," *The Oxford Handbook of Well-Being and Public Policy* 2014, 379–402.
164 S. Lyubomirsky, K. M. Sheldon, and D. Schkade, "Pursuing Happiness: The Architecture of Sustainable Change," *Review of General Psychology* 9(2), 2005, 111–131.
165 A. M. Isen and P. F. Levin, "Effect of Feeling Good on Helping: Cookies and Kindness," *Journal of Personality and Social Psychology* 21(3), 1972, 384.
166 Clark, et al., *The Origins of Happiness*, 222.
167 Haidt, *The Happiness Hypothesis*, 101.

Chapter 4. What We Think Is Right and Wrong

1 This is by income, not wealth. See www.globalrichlist.com. This information was retrieved April 2, 2019.
2 W. MacAskill, *Doing Good Better: Effective Altruism and a Radical New Way to Make a Difference* (New York: Gotham Books, 2015).
3 J. Greene, *Moral Tribes: Emotion, Reason, and the Gap Between Us and Them* (New York: Penguin, 2013), 126.
4 F. Harinck, C. K. De Dreu, and A. E. Van Vianen, "The Impact of Conflict Issues on Fixed-Pie Perceptions, Problem Solving, and Integrative Outcomes in Negotiation," *Organizational Behavior and Human Decision Processes* 81(2), 2000, 329–358.
5 I. Pyysiäinen and M. Hauser, "The Origins of Religion: Evolved Adaptation or By-Product?" *Trends in Cognitive Sciences* 14(3), 2010, 104–109.
6 People today tend to read the Bible with the moral background of humanism. Pinker, *Enlightenment Now*, 429.
7 M. Godoy, "Lust, Lies, and Empire: The Fishy Tale Behind Eating Fish On Friday," *NPR*, April 6, 2012, retrieved August 18, 2019, from https://www.npr.org/sections/thesalt/2012/04/05/150061991/lust-lies-and-empire-the-fishy-tale-behind-eating-fish-on-friday.
8 N. Epley, B. A. Converse, A. Delbosc, G. A. Monteleone, and J. T. Cacioppo, "Believers' Estimates of God's Beliefs Are More Egocentric than Estimates of Other People's Beliefs," *Proceedings of the National Academy of Sciences* 106(51), 2009, 21533–21538; J. Davies, "Religion Does Not Determine Your Morality," *The Conversation*, published online July 24, 2018.
9 Greene, *Moral Tribes*, 83.
10 K. A. Wade-Benzoni, A. E. Tenbrunsel, and M. H. Bazerman, "Egocentric Interpretations of Fairness in Asymmetric, Environmental Social Dilemmas: Explaining Harvesting Behavior and the Role of Communication," *Organizational Behavior and Human Decision Processes* 67(2): 1996, 111–126.
11 Greene, *Moral Tribes*, 99.

12 D. B. Krupp, L. A. Sewall, M. L. Lalumière, C. Sheriff, and G. T. Harris, "Nepotistic Patterns of Violent Psychopathy: Evidence for Adaptation?" *Frontiers in Psychology* 3, 2012, 305.

13 This is excluding the 90 percent of what we assume to be "junk" DNA that's just along for the ride and doesn't code for useful proteins. "Why Mouse Matters," National Human Genome Research Institute, July 23, 2010, https://www.genome.gov/10001345/importance-of-mouse-genome.

14 Haidt, *The Happiness Hypothesis*, 48.

15 P. Singer, *The Expanding Circle* (Oxford, UK: Clarendon Press, 1981); M. Shermer, *The Moral Arc: How Science and Reason Lead Humanity Toward Truth, Justice, and Freedom* (New York: Macmillan, 2015).

16 M. Xue and J. B. Silk, "The Role of Tracking and Tolerance in Relationship Among Friends," *Evolution and Human Behavior* 33, 2012, 17–25.

17 J. J. Massen and S. E. Koski, "Chimps of a Feather Sit Together: Chimpanzee Friendships Are Based on Homophily in Personality," *Evolution and Human Behavior* 35(1), 2014, 1–8; J. Henrich, *The Secret of Our Success: How Culture is Driving Human Evolution, Domesticating Our Species, and Making Us Smarter* (Princeton, N.J.: Princeton University Press, 2017), 206.

18 Greene, *Moral Tribes*, 35.

19 This is assuming that the proportion of close relatives was similar to that of contemporary hunter-gatherer societies. Henrich, *The Secret of Our Success*, 154.

20 Greene, *Moral Tribes*, 23.

21 J. Urist, "Which Deaths Matter?" *The Atlantic* September 29, 2004, retrieved July 22, 2019, from https://www.theatlantic.com/international/archive/2014/09/which-deaths-matter-media-statistics/380898/.

22 This in-group bias is also called tribalism, intergroup bias, or parochial altruism. Greene, *Moral Tribes*, 49.

23 Davies, *Riveted*.

24 C. K. Lai, et al. "Reducing Implicit Racial Preferences: I. A Comparative Investigation of 17 Interventions," *Journal of Experimental Psychology: General* 143(4), 2014, 1765.

25 B. L. Hughes, N. Ambady, and J. Zaki, "Trusting Outgroup, but Not Ingroup Members, Requires Control: Neural and Behavioral Evidence," *Social Cognitive and Affective Neuroscience* 12(3), 2017, 372–381; J. A. Everett, Z. Ingbretsen, F. Cushman, and M. Cikara, "Deliberation Erodes Cooperative Behavior—Even Toward Competitive Out-groups, Even When Using a Control Condition, and Even When Eliminating Selection Bias," *Journal of Experimental Social Psychology* 73, 2017, 76–81.

26 Y. N. Harari, *Homo Deus: A Brief History of Tomorrow* (New York: Random House, 2016).

27 J. Haidt, *The Righteous Mind: Why Good People Are Divided by Politics and Religion* (New York: Pantheon Books, 2012), 164.

28 Levitin, *The Organized Mind*, 28.

29 G. Fiorito and P. Scotto, "Observational Learning in Octopus Vulgaris," *Science* 256(5056), 1992, 545–547.

30 D. R. Boyd, *The Rights of Nature: A Legal Revolution that Could Save the World* (Toronto: ECW Press, 2017), 34.

31 E. Brooke-Hitching, *Fox Tossing, Octopus Wrestling and Other Forgotten Sports* (New York: Simon and Schuster, 2015).

32 J. Reese, "Survey of US Attitudes Towards Animal Farming and Animal-Free Food October 2017," *Sentience Institute*, November 2017, retrieved February 12, 2020, from https://www.sentienceinstitute.org/animal-farming-attitudes-survey-2017.

33 A. Shriver, "Knocking Out Pain in Livestock: Can Technology Succeed Where Morality Has Stalled?" *Neuroethics* 2(3), 2009, 115–124.

34. Boyd, *The Rights of Nature*, 48.
35. Ibid., 56.
36. Ibid., 119.
37. Ibid., 131–134.
38. Ibid., 220.
39. J. J. Prinz, and S. Nichols, "Moral Emotions," in J. M. Doris (ed.), *The Moral Psychology Handbook* (Oxford, UK: Oxford University Press, 2010), 111–146.
40. J. Greene, "The Cognitive Neuroscience of Moral Judgment," in M. Gazzaniga (ed.), *The Cognitive Neurosciences* 4th ed., (Cambridge, Mass.: MIT Press, 2009) 987–1002.
41. Prinz and Nichols, "Moral Emotions," 111–146.
42. Ibid., 111–146.
43. Ibid.
44. Ibid.
45. M. W. Merritt, J. M. Doris, G. Harman, "Character," in J. M. Doris (ed.), *The Moral Psychology Handbook* (Oxford, UK: Oxford University Press, 2010), 355–401.
46. D. Lieberman, "Objection! The Evolution of Pathogen, Sexual, and Moral Disgust," talk at *Disgust, Morality, and Society. A Conference Addressing the Emotion of Disgust.* Duke Institute for Brain Sciences, April 6–7, 2017.
47. A. R. Andrews, T. Crone, C. B. Cholka, T. V. Cooper, and A. J. Bridges, "Correlational and Experimental Analyses of the Relation Between Disgust and Sexual Arousal," *Motivation and Emotion* 39(5), 2015, 766–779.
48. Haidt, *The Righteous Mind*, 171. For an opposing view, that disgust does not spill over, see J. F. Landy and G. P. Goodwin, "Does Incidental Disgust Amplify Moral Judgment? A Meta-analytic Review of Experimental Evidence," *Perspectives on Psychological Science* 10(4), 2015, 518–536.
49. Haidt, *The Happiness Hypothesis*, 186.
50. Lieberman, "Objection!"
51. W. Sinnot-Armstrong, L. Young, and F. Cuchman, "Moral Intuitions," in J. M. Doris (ed.), *The Moral Psychology Handbook* (Oxford, UK: Oxford University Press, 2010), 246–272.
52. J. M. Tybur, Y. Inbar, L. Aarøe, P. Barclay, F. K. Barlow, M. De Barra, and N. S. Consedine, "Parasite Stress and Pathogen Avoidance Relate to Distinct Dimensions of Political Ideology Across 30 Nations," *Proceedings of the National Academy of Sciences*, 113(44), 2016, 12408–12413.
53. E. J. Masicampo, M. Barth, and N. Ambady, "Group-based Discrimination in Judgments of Moral Purity-related Behaviors: Experimental and Archival Evidence," *Journal of Experimental Psychology: General* 143(6), 2014, 2135.
54. F. Cushman, L. Young, and J. D. Greene, "Multi-system Moral Psychology," in J. M. Doris (ed.), *The Moral Psychology Handbook* (Oxford, UK: Oxford University Press, 2010), 47–71.
55. Ibid.
56. Ibid.
57. Ibid., 47–71, 265.
58. Sinnot-Armstrong, et al., "Moral Intuitions," 246–272.
59. That's a real thing, as I learned when I studied for my Canadian citizenship exam. It's not from Tolkien, as I suggested when I first learned about it.
60. E. Machery and R. Mallon, "Evolution of Morality," in J. M. Doris (ed.), *The Moral Psychology Handbook* (Oxford, UK: Oxford University Press, 2010), 3–46.
61. For an interesting discussion of how complicated this is, see https://slatestarcodex.com/2019/05/07/5-httlpr-a-pointed-review/.
62. Sinnot-Armstrong, et al., "Moral Intuitions," 246–272.

63. T. Hayden, "Bug Splat," Council on Foreign Relations, Rep. Keith Ellison Call for Drone Reform, *The Nation*, January 18, 2013, retrieved July 3, 2019, from https://www.thenation.com/article/bug-splat-council-foreign-relations-rep-keith-ellison-call-drone-reform/.
64. Sinnot-Armstrong, et al., "Moral Intuitions," 246–272.
65. Greene, *Moral Tribes*, 113.
66. E. W. Dunn and C. Ashton-James, "On Emotional Innumeracy: Predicted and Actual Affective Responses to Grand-scale Tragedies," *Journal of Experimental Social Psychology* 44, 2008, 692–698.
67. Described in Haidt, *The Happiness Hypothesis*, 62.
68. S. Sachdeva, R. Iliev, and D. L. Medin, "Sinning Saints and Saintly Sinners: The Paradox of Moral Self-regulation," *Psychological Science* 20(4), 2009, 523–528; P. M. Gollwitzer, P. Sheeran, V. Michalski, and A. E. Seifert, "When Intentions Go Public: Does Social Reality Widen the Intention-Behavior Gap?" *Psychological Science* 20(5), 2009, 612–618.
69. P. Singer, *The Most Good You Can Do: How Effective Altruism Is Changing Ideas About Living Ethically* (New Haven, CT: Yale University Press, 2015), 51.
70. Some argue that his contribution to defeating Germany in World War II counts for something, but this is also debated. See the chapter "A Statue for Stalin?" in P. Singer, *Ethics in the Real World: 87 Brief Essays on Things that Matter* (Princeton, N.J.: Princeton University Press, 2017), 239.
71. Singer, *The Most Good You Can Do*, 57.
72. Haidt, *The Happiness Hypothesis*, 29.
73. A. Tversky and D. Kahneman, "Loss Aversion in Riskless Choice: A Reference-Dependent Model," *The Quarterly Journal of Economics* 106(4), 1991, 1039–1061.
74. P. Rozin and E. B. Royzman, "Negativity Bias, Negativity Dominance, and Contagion," *Personality and Social Psychology Review* 5(4), 2001, 296–320.
75. J. Gottman, J. M. Gottman, and N. Silver, *"Why Marriages Succeed or Fail: And How You Can Make Yours Last* (New York: Simon and Schuster, 1995).
76. Haidt, *The Happiness Hypothesis*, 31.
77. R. Dawkins, *The Universe Is Queerer than We Can Suppose*, TED talk, 2005.
78. The idea that our natural morality is reflective of what is *actually* right and wrong is presented as "preservationism" and torn down very persuasively in P. K. Unger, *Living High and Letting Die: Our Illusion of Innocence* (Oxford, UK: Oxford University Press, 1996), 10.
79. All moral theories rely on appealing to our moral intuitions in some way, and when ethicists criticize different moral theories, they often do it by showing that the theory leads to some conclusion that doesn't match with our moral intuitions. Sinnot-Armstrong, et al., "Moral Intuitions," 246–272.
80. For example, the moral rules that deontologists endorse are often considered to be reasonable heuristics to a utilitarian, in that the utilitarian thinks that the rule, if followed, will often lead to doing the most good. So a utilitarian might think that murder is wrong because it often leads to bad outcomes, where the deontologist might think that murder's wrongness does not need to be justified by anything more fundamental. Kagan, *How to Count Animals, More or Less*, 173.
81. Greene, *Moral Tribes*, 26.
82. For philosophers: as a total hedonic utilitarian, I've had to accept that utility monsters and tiling the universe with hedonium are good things, as foreign as those ideas feel to me.
83. For one thing, if people thought that hospitals were places where they might be killed, they would be very loath to go to them. Hospitals need to protect the people that go there, even if a short-term cost-benefit analysis says they should not. This is one of the justifications for "rule utilitarianism," which says that we should follow some rules for utilitarian reasons. It might be good in the moment to sacrifice one person to save five, but the longer-term

84 Merritt, et al., "Character," 355–401.
85 Haidt, *The Happiness Hypothesis*, 64.
86 Y. N. Harari, *Homo Deus: A Brief History of Tomorrow* (New York: Random House, 2016), 239.
87 Haidt, *The Happiness Hypothesis*, 160–166.
88 This argument is summarized in Greene, *Moral Tribes*, 178.

Chapter 5. What's Actually Right and Wrong

1 "Where keeping a promise will harm someone, for example, what we ought to do will depend on how serious the promise is, to whom it was made, the size of the potential harm, and so on. To weigh all these factors requires discernment and judgment." This is a description of David Ross's deontology theory, even though deontology normally tries to avoid numerical thinking. D. McNaughton and P. Rawling, "Deontology," in H. LaFollette (ed.), *Ethics in Practice: An Anthology* (Chichester, UK: John Wiley & Sons, 2014), 37–48.
2 Doctors tend to have a more deontological moral outlook, but public health professionals are more utilitarian, and it's easy to see why. Greene, *Moral Tribes*, 128.
3 Thank you to my friend and skeptical podcaster Darren McKee for this example.
4 People will disagree on what helping and hurting entails, and who counts as people, and other details, but in general the idea that hurting people is bad is universal. Haidt, *The Righteous Mind*.
5 In philosophy, we would say that hitting people is bad because it harms another. This means that hitting people is "instrumentally" bad, meaning that it's bad because of something else that is bad. If something needs no justification, it is self-evident. If something is good or bad by itself, without needing to appeal to some more fundamental moral, it is "inherently" good or bad.
6 In philosophy, the biggest debate in ethics is between utilitarianism and deontology, which have differing answers to this question, among others. The third major competing philosophy is "virtue ethics." But this theory is more about developing one's moral character, not providing a calculus for determining the moral value of particular actions. In the time I took to research this book, I didn't find enough virtue ethics theory that was relevant to the questions I wanted to answer in this book, so I'm leaving it out.
7 E. Roedder and G. Harman, "Linguistics and Moral Theory," in J. M. Doris (ed.), *The Moral Psychology Handbook* (Oxford, UK: Oxford University Press, 2010), 273–296.
8 McNaughton and Rawling, "Deontology," 42.
9 Pinker, *Enlightenment Now*, 417.
10 Greene, *Moral Tribes*.
11 I can't do justice to the hundreds of years of ethical theorizing, but we might look to a respected source for a reasonable summary: *The Stanford Encyclopedia of Philosophy*. In its article on deontological ethics, it lists the generally accepted advantages. One is that it "leaves space for agents to give special concern to their families, friends, and projects. At least that is so if the deontological morality contains no strong duty of general beneficence, or, if it does, it places a cap on that duty's demands." I read this as saying that an advantage to this theory is that it allows me to benefit me, my kin, the people I care about, and my preferred social group more than other people, and to value my own projects over those of others. That sounds a lot like selfishness to me.
12 Greene, *Moral Tribes*, 125–127.

Chapter 6. A Work in Progress

1. Each field's facts have a different half-life, in years.
   Physics: 13.07
   Economics: 9.38
   Math: 9.17
   Psychology: 7.15
   History: 7.13
   Religion: 8.76
   S. Arbesman, *The Half-Life of Facts*.
2. 9,441 is the estimate, but the optimistic estimate is 562 metric tons, and the pessimistic estimate is 12,730 metric tons. This paper also has data estimated for other countries, which don't consume as much per capita carbon as the U.S. (page 18, table 2). P. A. Murtaugh and M. G. Schlax, "Reproduction and the Carbon Legacies of Individuals," *Global Environmental Change* 19(1), 2009, 14–20.
3. D. Notz and J. Stroeve, "Observed Arctic Sea-ice Loss Directly Follows Anthropogenic CO_2 Emission," *Science* 354(6313), 2016, 747–750.
4. Murtaugh and Schlax, "Reproduction and the Ccarbon Legacies of Individuals," 14–20.
5. D. Benatar, *Better Never to Have Been: The Harm of Coming into Existence* (Oxford, UK: Oxford University Press, 2008).
6. P. Singer, *Ethics in the Real World*, 31.
7. D. Parfit, *Reasons and Persons* (Oxford, UK: Oxford University Press, 1984).
8. M. A. Kline and R. Boyd, "Population Size Predicts Technological Complexity in Oceania," *Proceedings of the Royal Society B: Biological Sciences* 277(1693), 2010, 2559–2564.
9. J. Diamond, *Guns, Germs, and Steel: The Fates of Human Societies* (New York: W.W. Norton and Company, 1999), chapter 10.
10. Parents send children to work out of desperation, not greed. Pinker, *Enlightenment Now*, 93, 232; N. D. Kristof, "Where Sweatshops Are a Dream," *New York Times*, January 14, 2009, retrieved August 29, 2019, from https://www.nytimes.com/2009/01/15/opinion/15kristof.html.
11. MacAskill, *Doing Good Better*, chapter 8.
12. C. Bellamy, *The State of the World's Children 1998: Summary*, UNICEF, 1997.
13. MacAskill, *Doing Good Better*, chapter 8.
14. C. Cramer, D. Johnston, B. Mueller, C. Oya, and J. Sender, "Fairtrade and Labour Markets in Ethiopia and Uganda," *The Journal of Development Studies* 53(6), 2017, 841–856.
15. Harari, *Homo Deus*, chapters 6 and 7.
16. War has been disappearing, too, possibly because wealth today is less about material resources and more about knowledge and skill, which can't be stolen. There's more money to be made with cooperation. Harari, *Homo Deus*, 17.
17. It was defined as $1.90 as measured in 2011 dollars. You can use online calculators to find out what that means at the time of your reading. See the Wikipedia page on "extreme poverty" for updated information.
18. United States poverty guidelines: https://aspe.hhs.gov/poverty-guidelines.
19. P. Singer, *Practical Ethics*, third edition (Cambridge, UK: Cambridge University Press, 2011), 191.
20. MacAskill, *Doing Good Better*.

Chapter 7. The Numerical Value of Human Life

1. C. M. Farmer, "The Effects of Higher Speed Limits on Traffic Fatalities in the United States, 1993–2017," August 2019, retrieved August 30, 2019, from https://trid.trb.org/view/1607583.
2. A. Van Benthem, "What Is the Optimal Speed Limit on Freeways?" *Journal of Public Economics* 124, 2015, 44–62.

3. You can divide by about 30 to see what one year of life is worth to people. Updated numbers can be found on the Wikipedia page for "Value of Life." Also Office of the Assistant Secretary for Planning and Evaluation, 2016. Guidelines for Regulatory Impact Analysis. *US Department of Health and Human Services.* Retrieved November 2, 2020, https://aspe.hhs.gov/system/files/pdf/242926/HHS_RIAGuidance.pdf; For a readable account see S. Ben-Achour, "How to Value a Life, Statistically Speaking," *Marketplace,* March 20, 2019, retrieved February 12, 2020, from https://www.marketplace.org/2019/03/20/how-value-life/.

4. These numbers are in 2003 dollars. J. F. Morrall III, "Saving Lives: A Review of the Record" *Journal of Risk and Uncertainty* 27(3), 2003, 221–237; Roy Gamse, "How Much Are We Willing to Spend to Save a Life?" *The Life You Can Save* (blog), June 30, 2014, https://www.thelifeyoucansave.org/blog/id/93/how-much-are-we-willing-to-spend-to-save-a-life.

5. Unfortunately, there is no data on the world's most effective theater charities.

6. Philosophically minded readers might note that not everybody is utilitarian. Fair enough. Many deontologists believe that we have special obligations to people close to us. But they are often vague on how strong this obligation is. Maybe you have an obligation to be nice to your brother, so you might justify spending $78 on a meal to cheer him up. This is the cost of extending an African person's life for one year with the most effective charities. So the obligation would have to be very strong for the meal to be the better action than donating the same money to the Against Malaria Foundation.

7. A. D. Lopez, C. D. Mathers, M. Ezzati, D. T. Jamison, and C. J. Murray (eds.), *Global Burden of Disease and Risk Factors*, The World Bank, 2006, 402.

8. Some economists try to put everything in terms of dollars, but this has some serious problems: it makes changes in happiness less valuable to poor people, and asking people about their willingness to pay to not get this or that disease ends up with nonsensical data. Clark, et al., *The Origins of Happiness*, 203.

9. A. D. Lopez, et al., *Global Burden*.

10. "Disability-adjusted life year," Wikipedia, retrieved November 28, 2019, from https://en.wikipedia.org/wiki/Disability-adjusted_life_year.

11. C. Hjorthøj, A. E. Stürup, J. J. McGrath, et al., "Years of Potential Life Lost and Life Expectancy in Schizophrenia: A Systematic Review and Meta-analysis," *Lancet Psychiatry*, 4(4), 2017, 295–301.

12. There are several ways to estimate these weightings, and scholars are not in agreement over which one makes the most sense. Although there are significant differences between the results of these different methods, in one study of twelve diseases, all of the different methods actually used agreed, at least, on the rank ordering of how bad the diseases were. B. Robberstad, "QALYs vs DALYs vs LYs Gained: What Are the Differences, and What Difference Do They Make for Health Care Priority Setting?" *Norsk Epidemiologi* 15(2), 2005, 183–191.

13. MacAskill, *Doing Good Better*.

14. See the chapter on imagining the future in J. Davies, *Imagination: The Science of Your Mind's Greatest Power* (New York: Pegasus Books, 2019).

15. G. L Albrecht and P. J. Devlieger, "The Disability Paradox: High Quality of Life Against All Odds," *Social Science & Medicine* 48(8), 1999, 977–988.

16. Clark, et al., *The Origins of Happiness*, 96–97.

17. "Disability and Health Care Rationing," *Stanford Encyclopedia of Philosophy*, January 29, 2016, retrieved December 1, 2019, from https://plato.stanford.edu/entries/disability-care-rationing/.

18. Robberstad, "QALYs vs DALYs," 183–191.

19. Roedder, "Linguistics and Moral Theory," 273–296. The debate on age weighting is reviewed here: Robberstad, B. "QALYs vs DALYs," 2005, 183–191.

20. L. Green, J. Myerson, and E. McFadden, "Rate of Temporal Discounting Decreases with Amount of Reward," *Memory & Cognition* 25(5), 1997, 715–723.

21. Technically, a lower marginal utility of income. Clark, et al., *The Origins of Happiness*, 204.
22. Although discounting future money is not controversial, discounting health benefits is. Robberstad, "QALYs vs DALYs," 183–191; W. K. Viscusi, "Discounting Health Effects for Medical Decisions," *Valuing Health Care* 7(828), 1995, 125–47.
23. The Against Malaria Foundation is consistently ranked as one of the world's most effective charities. Scholars estimate that the AMF avoids a DALY for approximately $78 USD through giving people mosquito bed nets. M. Capriati and H. Hillebrandt, Against Malaria Foundation, 2018, retrieved July 18, 2019, from https://www.givingwhatwecan.org/report/against-malaria-foundation/.
24. C. F. Bowen and V. Skirbekk, "Old Age Expectations Are Related to How Long People Want to Live," *Ageing & Society* 37(9), 2017, 1898–1923.
25. Harari, *Homo Deus*, 33, 48.

Chapter 8. Choosing a Career

1. R. J. Vallerand, et al., "Les Passions de l'ame: On Obsessive and Harmonious Passion," *Journal of Personality and Social Psychology* 85(4), 2003, 756.
2. Wayne Gretzky biography from the *Encyclopedia of World Biography*, retrieved July 5, 2019, from https://www.notablebiographies.com/Gi-He/Gretzky-Wayne.html.
3. Y. Harari, *Sapiens: A Brief History of Humankind* (New York: Random House, 2014).
4. T. A. Judge and R. Klinger, "Promote Job Satisfaction through Mental Challenge," in E. Locke (ed.), *Handbook of Principles of Organizational Behavior* (Chichester, UK: John Wiley, 2000), 75–89.
5. R. N. Bellah, R. Madsen, W. M. Sullivan, A. Swidler, and S. M. Tipton, *Habits of the Heart: Individualism and Commitment in American Life* (Berkeley, CA: University of California Press, 2007).
6. A. Wrzesniewski, J. E. Dutton, and G. Debebe, "Interpersonal Sensemaking and the Meaning of Work," *Research in Organizational Behavior* 25, 2003, 93–135.
7. J. Nakamura and M. Csikszentmihalyi, "The Construction of Meaning Through Vital Engagement," in C. L. Keyes and J. E. Haidt (eds.), *Flourishing: Positive Psychology and the Life Well-lived* (Worcester, MA: American Psychological Association, 2003), 83–104.
8. Except for environmental protection, this list includes the main categories on a major effective altruism website. Their list is "global health and development," "animal welfare," "long-term future," and "effective altruism meta." Retrieved June 20, 2019, from https://app.effectivealtruism.org/funds.
9. D. Matthews and B. Pinkerton, "How to Pick a Career that Counts," *Vox*, November 28, 2018, retrieved June 20, 2019, from https://www.vox.com/future-perfect/2018/11/28/18114601/future-perfect-podcast-career-choice.
10. Matthews and Pinkerton, "How to Pick a Career that Counts."
11. MacAskill, *Doing Good Better*.
12. This concept originated independently in several places. A notable one is William MacAskill in a 2011 talk called "Want an ethical career? Become a banker." For a published analysis, see W. MacAskill, "Replaceability, Career Choice, and Making a Difference," *Ethical Theory and Moral Practice* 17(2), 2014, 269–283. The term "earning to give" appears to have been invented by Brian Tomasik. B. Tomasik, "Why Activists Should Consider Making Lots of Money," 2006, retrieved July 4, 2019, from https://reducing-suffering.org/why-activists-should-consider-making-lots-of-money/.
13. MacAskill, "Replaceability, Career Choice, and Making a Difference," 269–283.
14. This is a bit of a simplification. For an in-depth analysis, see Tomasik, "Why Activists Should Consider Making Lots of Money."
15. MacAskill, *Doing Good Better*, chapter nine.

16 Of course, Schindler also deliberately sabotaged some of the manufacturing so the munitions wouldn't work. Likewise, someone in a morally problematic career might take action to mitigate harm more than their replacement would. But I'm not focusing on this aspect of it because I don't think this possibility is necessary to justify such a career when donations are high enough—it can be ethically justifiable even if the worker causes *more* harm than their replacement would because of higher competence. You can see Schindler's story in Steven Spielberg's film *Schindler's List*. MacAskill, "Replaceability, Career Choice, and Making a Difference," 269–283.

17 "List of Serial Killers by Number of Victims," Wikipedia, https://en.wikipedia.org/wiki/List_of_serial_killers_by_number_of_victims.

18 B. Tomasik, "Employers with Huge Matching-Donation Limits," 2017, retrieved July 4, 2019, from https://reducing-suffering.org/employers-with-huge-matching-donations-limits/.

19 H. Rolston III, "Feeding People versus Saving Nature," in H. LaFollette (ed.), *Ethics in Practice: An Anthology* (Chichester, UK: John Wiley & Sons, 2014), 583–591.

20 S. Mukherjee, *Fighting Chance*, in O. Sacks (ed.), *The Best American Science Writing 2003* (New York: HarperCollins, 2002).

Chapter 9. Measuring Good Done

1 MacAskill, *Doing Good Better*.
2 Ibid.
3 S. N. Zane, B. C. Welsh, and G. M. Zimmerman, "Examining the Iatrogenic Effects of the Cambridge-Somerville Youth Study: Existing Explanations and New Appraisals," *British Journal of Criminology* 56(1), 2016, 141–160.
4 MacAskill, *Doing Good Better*.
5 Zane, et al., "Examining the Iatrogenic Effects of the Cambridge-Somerville Youth Study," 141–160.
6 MacAskill, *Doing Good Better*.
7 This charity, "Deworm the World," is ranked as very effective and you can donate to them.
8 MacAskill, *Doing Good Better*.

Chapter 10. Animals

1 S. T. Weathers, L. Caviola, L. Scherer, S. Pfister, B. Fischer, J. B. Bump, and L. M. Jaacks, "Quantifying the Valuation of Animal Welfare Among Americans," *Journal of Agricultural and Environmental Ethics* 33, 2020, 1–22.
2 Although most agree that consciousness of pleasant and unpleasant states is sufficient, people disagree on whether or not it is necessary. Some think that trees or ecosystems, and sometimes even mountains, should have moral standing on their own—that is, independently of how harming them harms other conscious beings, such as humans who can appreciate them. Kagan, *How to Count Animals*, 12–15.
3 Some disagree with this, and hold that preference violation without consciousness is morally relevant. Kagan, *How to Count Animals*.
4 P. Singer, "All Animals Are Equal," in H. LaFollette (ed.), *Ethics in Practice: An Anthology* (Chichester, UK: John Wiley & Sons, 2014), 172–180.
5 This analogy is from neuroscientist Susan Greenfield.
6 What I'm calling the muted animal theory has been called the "decreased suffering approach," and what I am calling the "Tinker Bell theory" has been called the "increased suffering" approach by S. S. Rakover, "Animals Suffer Too—A Response to Akhtar's 'Animal Pain and Welfare: Can Pain Sometimes Be Worse for Them than for Us?'" *The Journal of Mind and Behavior* 40(3/4), 2019, 195–204.
7 P. Harrison, "Do Animals Feel Pain?" *Philosophy* 66(255), 1991, 25–40.

8   This view is endorsed by Judith Jarvis Thompson: J. J. Thomson, *The Realm of Rights* (Cambridge, MA: Harvard University Press, 1990), 292.
9   S. Akhtar, "Animal Pain and Welfare: Can Pain Sometimes be Worse for Them than for Us," *The Oxford Handbook of Animal Ethics* (Oxford, UK: Oxford University Press, 2011), 495–518.
10 Many thinkers hold that it is an animal's lack of self-awareness, particularly in terms of being able to think of itself as a being that has a past and future, that makes a big difference—in fact, that it defines "personhood." The idea is that only with a conception of yourself in the future can you have frustrated goals. Singer, *Practical Ethics*, 52, 65–75.
11 D. Chamovitz, *What a Plant Knows: A Field Guide to the Senses* (New York: Scientific American, 2012). When this excellent book was published, there was no evidence yet that plants had any sense of hearing.
12 M. Gagliano, M. Grimonprez, M. Depczynski, and M. Renton, "Tuned In: Plant Roots Use Sound to Locate Water," *Oecologia* 184(1), 2017, 151–160; H. M. Appel and R. B. Cocroft, "Plants Respond to Leaf Vibrations Caused by Insect Herbivore Chewing," *Oecologia* 175(4), 2014, 1257–1266.
13 C. R. Jain, "The Practical Dharma of The Practical Path," (Allahabad: The Indian Press Ltd., 1929), 49, 55.
14 B. Tomasik, *Is Brain Size Morally Relevant?* June 19, 2013, retrieved July 2, 2019, from https://reducing-suffering.org/is-brain-size-morally-relevant/#Small_brains_matter_more_per_neuron.
15 A. Shriver, "Knocking Out Pain in Livestock: Can Technology Succeed Where Morality Has Stalled?" *Neuroethics* 2(3), 2009, 115–124.
16 A. Lutz, D. R. McFarlin, D. M. Perlman, T. V. Salomons, and R. J. Davidson, "Altered Anterior Insula Activation During Anticipation and Experience of Painful Stimuli in Expert Meditators," *Neuroimage* 64, 2013, 538–546. Interestingly, another study had people study meditation for only a few months and found a similar reduction in the suffering associated with pain. F. Zeidan, K. T. Martucci, R. A. Kraft, N. S. Gordon, J. G. McHaffie, and R. C. Coghill, "Brain Mechanisms Supporting the Modulation of Pain by Mindfulness Meditation," *Journal of Neuroscience* 31(14), 2011, 5540–5548.
17 R. A. Band, R. A. Salhi, D. N. Holena, E. Powell, C. C. Branas, and B. G. Carr, "Severity-adjusted Mortality in Trauma Patients Transported by Police," *Annals of Emergency Medicine* 63(5), 2014, 608–614.
18 Of course, giving a human with no money $1,000 gives more pleasure than giving a mouse with no money $1,000 (if it were cash, the mouse might be able to use it to make a nest). So we'll restrict our discussion to things that have similar effects, though the effects might vary in intensity.
19 Shelly Kagan believes that if two beings have toothaches, and the felt pain is equal for both beings, and you only have resources to relieve one of them, and one of the beings is a mouse and the other is a human, then you have a moral obligation to help the mouse rather than the human. Because I think mice have muted pain (relative to humans), for this thought experiment to be plausible the toothache would have to be much more severe, medically speaking, for the experienced pain to be actually the same. Given that, I disagree with Kagan, and believe that it would be equally morally virtuous to help either the mouse or the human in this situation. Kagan, *How to Count Animals, More or Less*, 125.
20 Boström admits to speciesism with respect to artificial intelligence pleasure and pain. N. Boström, *Superintelligence: Paths, Dangers, Strategies* (Oxford, UK: Oxford University Press, 2014).

21. P. Carruthers, "Animal Mentality: Its Character, Extent, and Moral Significance," in T. L. Beauchamp and R. G. Frey (eds), *The Oxford Handbook of Animal Ethics* (Oxford, UK: Oxford University Press, 2011).
22. D. DeGrazia, "Moral Status as a Matter of Degree?" *The Southern Journal of Philosophy* 46(2), 2008, 181–198. Shelly Kagan calls the "Equal Consideration" view the "unitarian" view, and the "Unequal Consideration" view the "hierarchical" view: Kagan, *How to Count Animals, More or Less*.
23. Scientists disagree on how strongly brain volume in humans correlates with intelligence; estimates range from around 0.2 to 0.4. But all of these correlations are pretty big. G. E. Gignac and T. C. Bates, "Brain Volume and Intelligence: The Moderating Role of Intelligence Measurement Quality," *Intelligence* 64, 2017, 18–29.
24. K. Hays-Gilpin and D. S.Whitley (eds.), *Reader in Gender Archaeology* (Hove, UK: Psychology Press, 1998).
25. The theory that babies are more conscious than adults comes from one of the world's most respected developmental psychologists, Alison Gopnik. A. Gopnik, "Why Babies Are More Conscious than We Are," *Behavioral and Brain Sciences* 30(5–6), 2007, 503–504.
26. Another version of the brain-body ratio is called the "encephalization quotient." Another way is to look at the raw number of neurons in a brain. Some have argued that our brains are mostly scaled-up primate brains. S. Herculano-Houzel, "The Human Brain in Numbers: A Linearly Scaled-up Primate Brain," *Frontiers in Human Neuroscience* 3, 2009, 31.
27. "List of Animals by Number of Neurons," Wikipedia, https://en.wikipedia.org/wiki/List_of_animals_by_number_of_neurons.
28. For example, this paper argues for different brain functions implementing consciousness in ray-finned fish: M. L. Woodruff, "Consciousness in Teleosts: There Is Something It Feels Like to Be a Fish," *Animal Sentience: An Interdisciplinary Journal on Animal Feeling* 2(13), 2017, 1–21.
29. R. I. Dunbar and S. Shultz, "Evolution in the Social Brain," *Science* 317(5843), 2007, 1344–1347.
30. R. V. Kail and J. C. Cavanaugh, *Human Development: A Life-span View* (Boston: Cengage Learning, 2018), 130.
31. L. Chittka and J. Niven, "Are Bigger Brains Better?" *Current Biology* 19(21), 2009, R995–R1008.
32. Singer, *Practical Ethics*, 85.
33. C. McFarland and D. T. Miller, "Judgments of Self-Other Similarity: Just Like Other People, Only More So," *Personality and Social Psychology Bulletin* 16(3), 1990, 475–484.
34. Peter Singer says no: "We are not, of course, going to attempt to assign numerical values to the lives of different beings, or even to produce an ordered list." And I'm like, "Of course?" Singer, *Practical Ethics*, 90.
35. Levitin, *The Organized Mind*, 45.
36. C. Koch, "How to Make A Consciousness Meter," *Scientific American*, November 2017, 28–33.
37. Some disagree. For example, in reference to being able to feel pain, philosopher Shelly Kagan claims that "there is no obvious reason to assume that this capacity is one that humans have to a higher degree than animals have." Kagan, *How to Count Animals, More or Less*, 163.
38. To get this number, you divide the species-typical number of cortical (or equivalent) neurons by the number of human cortical neurons (16 billion). L. Scherer, B. Tomasik, O. Rueda, and S. Pfister, "Framework for Integrating Animal Welfare into Life Cycle Sustainability Assessment." *The International Journal of Life Cycle Assessment* 23(7), 2018, 1476–1490.
39. It is important that we use multiplication rather than division, because with division the numbers go screwy when the level is negative. If a human and a chicken both endure

a -10 event, the human's adjusted happiness would be (-10/1) = -10, but the chicken's would be (-10/0.0038) = -2631. This is the opposite of the intention, which is to mute the consciousness of simpler beings. A better way is multiplication. I'm citing Kagan here, though Kagan is discussing these points as a way to differentiate the moral value of different animals' pains, not the differences in the felt pains themselves. Kagan, *How to Count Animals, More or Less*, 89, 138.

40 For a thorough discussion, see Tomasik, "Is Brain Size Morally Relevant?"

41 Invertebrate Sentience Table, retrieved August 18, 2019, from https://www.rethinkpriorities.org/invertebrate-sentience-table.

42 Shelly Kagan believes that even if the suffering is exactly the same, we should still favor more complex animals over simpler ones. Kagan, *How to Count Animals, More or Less*.

43 S. Segal Glick, "I Made My Kids Eat Crickets," *Today's Parent*, January 22, 2018, retrieved July 7, 2019, from https://www.todaysparent.com/family/parenting/i-made-my-kids-eat-crickets/.

44 Scherer, et al., "Framework for Integrating Animal Welfare," 1476–1490.

45 T. M. Khuong, et al., "Nerve Injury Drives a Heightened State of Vigilance and Neuropathic Sensitization in Drosophila," *Science Advances* 5(7), 2019, eaaw4099.

46 D. Crummett, "The Problem of Evil and the Suffering of Creeping Things," *International Journal for Philosophy of Religion* 82(1), 2017, 71–88.

47 R. Bruns and J. Davies (under review), *Modeling Uncertainty in Animal Welfare and Ethics*.

48 The Muted Animal Theory was further broken down into what measure of brain complexity was used (these percentages sum to 62): Raw Brain Mass: 2 percent, Brain-Body Ratio: 5 percent, Encephalization Quotient: 20 percent, Neuron Count: 15 percent, Cortical Neuron Count: 20 percent.

49 For the spreadsheets that I created with economist Richard Bruns regarding animal welfare and climate change, see http://www.jimdavies.org/science-of-better/.

50 "Numbers of Insects (Species and Individuals)," retrieved July 30, 2019, from https://www.si.edu/spotlight/buginfo/bugnos.

51 Ethnobiologists created a category of "bug" to explain the predominance of the bug category in so many languages. Levitin, *The Organized Mind*, 28.

52 Scherer, et al., "Framework for Integrating Animal Welfare," 1476–1490.

53 Human (2,100,000,000 seconds) (Wolframalpha.com)
Cattle (18 months), pig (5.5 months), chicken (6 weeks) lifespans are from Aussie Abbatoirs, https://www.aussieabattoirs.com/facts/age-slaughtered.
Salmon Slaughtered at 2 years of age:
https://www.avma.org/KB/Resources/FAQs/Pages/A-Primer-on-Salmon.aspx.
Shrimp eaten at 5 months:
https://en.wikipedia.org/wiki/Neocaridina_davidi.
Crickets eaten at 2 months:
https://modernfarmer.com/2018/08/how-to-raise-crickets-for-food/.
Mealworms eaten at 1 month ("anywhere from a week to a couple of months"):
https://www.wikihow.com/Raise-Mealworms.

54 In animals, smallness of the body and population density are correlated. C. N. Johnson, "Relationships Between Body Size and Population Density of Animals: The Problem of the Scaling of Study Area in Relation to Body Size," *Oikos* 85(3), 1999, 565–569.

55 For wild animals, there are probably between 1,011 and 1,014 vertebrate land animals, at least 1,013 vertebrate sea creatures. But terrestrial and marine arthropods (bugs) are vastly more, numbering at about 1,018. There are probably 5*1,030 bacteria. B. Tomasic, "How Many Wild Animals Are There?" 2018, retrieved July 9, 2019, from https://reducing-suffering.org/how-many-wild-animals-are-there/.

56 Ten quintillion insects (a 1 with 19 zeroes) times the cricket sentience discount = 29,000,000,000,000, which is 29 trillion. Divide that by the number of humans and you get 3,851.26. 10,000,000,000,000,000,000 insects * 0.0000029 / 7.53 billion people = 3,851.26

57 Z. Groff and Y. K. Ng, "Does Suffering Dominate Enjoyment in the Animal Kingdom? An Update to Welfare Biology," *Biology & Philosophy* 34(4), 2019, 40.

58 B. Key, "Fish Do Not Feel Pain and Its Implications for Understanding Phenomenal Consciousness," *Biology & Philosophy* 30(2), 2015, 149–165.

59 "Predatory Behaviour," ALERT, January 8, 2020, retrieved January 10, 2020, from http://lionalert.org/page/predatory-behaviour.

60 Zoos vary enormously in the welfare the animals in them enjoy. There certainly are animals in zoos who have it worse than their wild counterparts, particularly in zoos of the past, where efforts for behavioral enrichment and large enclosures was not as much of a priority. Designers and workers at contemporary zoos in rich countries are very concerned with the welfare of the animals.

61 C. Wilcox, "Bambi or Bessie: Are Wild Animals Happier?" *Scientific American Blogs*, April 12, 2011, retrieved July 9, 2019, from https://blogs.scientificamerican.com/guest-blog/bambi-or-bessie-are-wild-animals-happier/.

62 This is actually a bit controversial. It could be that consciousness, which is what makes valanced mental states possible, was not directly selected for, but is rather a by-product of other adaptations. Aversion and attraction are more certainly adaptive.

63 Harari, *Sapiens*.

64 B. Tomasik, "Medicine vs Deep Ecology," October 30, 2012, retrieved July 10, 2019, from https://reducing-suffering.org/medicine-vs-deep-ecology/.

65 J. McMahan, "The Meat Eaters," *New York Times*, September 19, 2010, retrieved July 8, 2019, from https://canvas.harvard.edu/files/4295822/download?download_frd=1.

66 Singer, *Ethics in the Real World*, 44–45.

67 Singer, *Practical Ethics*, 54.

68 Pinker, *Enlightenment Now*, 19.

69 "Baby Death Parents Spared Jail," BBC News, September 14, 2001, retrieved June 4, 2019, from http://news.bbc.co.uk/2/hi/health/1542293.stm.

70 B. Tomasik, "Does Vegetarianism Make a Difference?" 2006, retrieved July 7, 2019, from https://reducing-suffering.org/does-vegetarianism-make-a-difference/.

71 Singer, *Ethics in the Real World*, 50–51.

72 Singer, *The Most Good You Can Do*, 138.

73 "The Effects of Diet Choices," Animal Charity Evaluators, March 2016, https://animalcharityevaluators.org/research/dietary-impacts/effects-of-diet-choices/.

74 Harish, "The Fish We Kill to Feed the Fish We Eat," *Counting Animals* (blog), July 10, 2011, http://www.countinganimals.com/the-fish-we-kill-to-feed-the-fish-we-eat/.

75 "The Efffects of Diet Choices," Animal Charity Evaluators.

76 Singer, *Ethics in the Real World*, 44.

77 Considering 6-ounce portions, mussels have 40.4 grams of protein, and a T-bone steak has 48; mussels have 8 grams of fat, steak has 12. "Mussel Nutrition & Health Benefits," retrieved August 22, 2019, from http://canadiancove.com/recipes/nutrition_and_health.html.

78 D. Fleischman, "The Ethical Case for Eating Oysters and Mussels," 2013, retrieved July 7, 2019, from https://sentientist.org/2013/05/20/the-ethical-case-for-eating-oysters-and-mussels/; Some think that mussels and other bivalves might feel a bit of pain, but agree that eating mussels be morally superior to eating just about any other kind of meat. B. Tomasik, "Can Bivalves Suffer?" February 6, 2017, retrieved July 10, 2019, from https://reducing-suffering.org/can-bivalves-suffer/.

79. This spreadsheet has the most complete recording of harms of eating all foods that I've seen. You can download it yourself. M. B. Budolfson, "Harm Footprint of Food," 2015, retrieved August 22, 2019, from http://www.budolfson.com/footprints.
80. It's a little more complicated than that. A reduction in demand at the store causes a reduction in prices, which increases sales. So reducing chicken consumption from, say, 10 birds to 0 causes a demand reduction of *less* than 10. How much less is called the cumulative elasticity factor. Estimates for this vary widely for meat, especially fish, but for chickens it's about 0.3. This means that eating 10 fewer chickens can be expected to result in 3 fewer chickens being farmed. Going vegetarian (compared to a typical American diet) would prevent between 2 and 21 land animals from being raised every year, and 232 fish and shellfish being caught and farmed. https://animalcharityevaluators.org/research/dietary-impacts/effects-of-diet-choices/; That said, chicken farming is a low-margin, competitive business. Some would say that in the long run, producers exit the market, creeping price back up, and eliminating this elasticity. On this view, eating one chicken causes one chicken to suffer. For more explanation of this, see https://www.youtube.com/watch?v=BZ31XUGj34.
81. N. Rott, "Decline in Hunters Threatens How U.S. Pays for Conservation," NPR, March 20, 2018, retrieved December 13, 2019, from https://www.npr.org/2018/03/20/593001800/decline-in-hunters-threatens-how-u-s-pays-for-conservation.
82. Singer, *Ethics in the Real World*, 44.
83. B. Key, "Fish Do Not Feel Pain and Its Implications for Understanding Phenomenal Consciousness," *Biology & Philosophy* 30(2), 2015, 149–165. Woodruff, "Consciousness in Teleosts," 1.
84. B. Fischer and A. Lamey, "Field Deaths in Plant Agriculture," *Journal of Agricultural and Environmental Ethics* 31(4), 2018, 409–428.
85. Ibid.
86. This study was conducted in 2017. K. Greig, "Ace Highlight: When Will There be Cost-Competitive Cultured Animal Products," May 18, 2017, retrieved July 1, 2019, from https://animalcharityevaluators.org/blog/ace-highlight-when-will-there-be-cost-competitive-cultured-animal-products/.
87. "The Humane League," December 2019, retrieved July 1, 2019, from https://animalcharityevaluators.org/charity-review/the-humane-league/.
88. 3.5 chickens saved per dollar * 365 days = 1,277.5 chickens saved per year; 1,277.5 chickens saved – 365 chickens eaten = 912.5 chickens saved.
89. Harish, "The Fish We Kill to Feed the Fish We Eat."
90. One recent study recommends eating meat. B. C. Johnston, D. Zeraatkar, M. A. Han, R. W. Vernooij, C. Valli, R. El Dib, and F. Bhatia, "Unprocessed Red Meat and Processed Meat Consumption: Dietary Guideline Recommendations from the Nutritional Recommendations (NutriRECS) Consortium," *Annals of Internal Medicine*, 171, 2019, 756–64.
91. C. Radnitz, J. Ni, D. Dennis, and B. Cerrito, "Health Benefits of a Vegan Diet: Current Insights," *Nutrition and Dietary Supplements* 12, 2020, 57.
92. The reason these studies are so hard to do is because diet is often confounded with other factors relevant to health and longevity (exercise and social interaction, for example), and because you can't control what people eat, long term, in order to study them in a more experimental setup. But some American Adventists go to churches with dietary restrictions and others do not, making a given Adventist's diet randomly assigned, while holding culture and genes more or less constant (or equally variable). These studies suggest the 3.6 years of life gained from a vegetarian diet. David G and Froolow [2019 Adversarial Collaboration Contest], "Is Eating Meat a Net Harm?" *Slate Star Codex*, December 11, 2019, retrieved March 2, 2020, from https://slatestarcodex.com/2019/12/11/acc-is-eating-meat-a-net-harm/.

93 David G and & Froolow [2019 Adversarial Collaboration Contest], "Is Eating Meat a Net Harm?"
94 Ibid.
95 Ibid.

Chapter 11. Comparing Human to Animal Suffering

1 Technically, they estimate that $1,000 saves between -6,000 and 13,000 animals. That is, it might be that the Humane League is doing more harm than good. I took the middle of this range, 7,000 animal lives saved. "Animal Charity Evaluators," *The Humane League*, 2018, retrieved August 12, 2019, from https://animalcharityevaluators.org/charity-review/the-humane-league/.
2 The most effective charities save lives for thousands of dollars. I've seen numbers ranging from $2,000 to $7,500 per life saved. R. Wiblin, "Most People Report Believing It's Incredibly Cheap to Save Lives in the Developing World," *80,000 Hours Blog*, May 9, 2017, retrieved July 2, 2019, from https://80000hours.org/2017/05/most-people-report-believing-its-incredibly-cheap-to-save-lives-in-the-developing-world/.
3 M. Capriati and H. Hillebrandt, "Against Malaria Foundation," Giving What We Can, April 25, 2018, retrieved July 18, 2019, from https://www.givingwhatwecan.org/report/against-malaria-foundation/.
4 "Aussie Abattoirs," *Age of Animals Slaughtered*, retrieved August 12, 2019, from https://www.aussieabattoirs.com/facts/age-slaughtered.
5 If Americans eat 98.6 kg per year, and that is an expected 30 animals, and Nigerians eat 5.91 kg per year, then $(5.91*30)/98.6 = 1.798$ animals eaten yearly by the average Nigerian. 1.798 animals * 20 years = 35.96, which I rounded to 36, https://en.wikipedia.org/wiki/List_of_countries_by_meat_consumption_per_capita.
6 This is ignoring the environmental impact of a saved Nigerian life (which, again, would be far less than an American life saved).

Chapter 12. Environmental Morality

1 Singer, *Practical Ethics*, 216.
2 T. Carleton, M. Delgado, M. Greenstone, T. Houser, S. Hsiang, A. Hultgren, et al., "Valuing the Global Mortality Consequences of Climate Change Accounting for Adaptation Costs and Benefits," Working Paper of the Becker Friedman Institute, 2018, retrieved August 20, 2019, from https://papers.ssrn.com/sol3/papers.cfm?abstract_id=3224365.
3 An earlier version of the following paper estimated that 72 percent of regions would be adversely affected by rising temperatures, and the rest will benefit. Carleton, et al., "Valuing the Global Mortality Consequences of Climate Change."
4 Ibid., 37.
5 Ibid.; (Table H.3, EPA Valuation, Panel A, RCP 4.5, lowest discount rate, full uncertainty, divided by $80k value of statistical life year per Broome.) R. D. Bressler, *The Mortality Cost of Carbon*, CEEP, Working paper number 11, 2020, retrieved September 10, 2020, from https://ceep.columbia.edu/sites/default/files/content/papers/n11.pdf; Broome, "How Much Harm Does Each of Us Do?" in M. Budolfson, et al., *Philosophy and Climate Change*.
6 For the spreadsheets that I created with economist Richard Bruns regarding animal welfare and climate change, see http://www.jimdavies.org/science-of-better/.
7 J. Glover and M. J. Scott-Taggart, "It Makes No Difference Whether or Not I Do It," *Proceedings of the Aristotelian Society* supplementary volumes, 49, 1975, 171–209.
8 Singer, *Practical Ethics*, 217.

9   The ranges presented are 90 percent confidence intervals. These ranges are more instructive than point estimates, due to the great uncertainty, but sometimes a point estimate is required, so here they are: the mean effect of eating a serving of beef is adding the equivalent of 0.053 days of human life, and the mean effect of eating chicken is losing 0.68 days of human life. It's tough to justify going to eat chicken at a restaurant if it means you're causing someone's life to be cut half a day shorter. These calculations take into account animal suffering now, and human life lost in the next 100 years. It does not take into account animal suffering and life lost over the next 100 years, as there are no estimates of this. Calculations are available at http://www.jimdavies.org/science-of-better/.

10 Even those trying to "de-extinct" species know very well that they are only approximating what the species was. B. Wray, *Rise of the Necrofauna: The Science, Ethics, and Risks of De-extinction* (Vancouver: Greystone Books, Ltd., 2017).

11 D. Bourn and J. Prescott, "A Comparison of the Nutritional Value, Sensory Qualities, and Food Safety of Organically and Conventionally Produced Foods," *Critical Reviews in Food Science and Nutrition* 42(1), 2002, 1–34.

12 R. Blair, *Organic Production and Food Quality: A Down to Earth Analysis* (Chichester UK: John Wiley & Sons, 2012).

13 V. Seufert, N. Ramankutty, and J. A. Foley, "Comparing the Yields of Organic and Conventional Agriculture," *Nature* 485(7397), 2012, 229.

14 M. B. Budolfson, "Consumer Ethics, Harm Footprints, and the Empirical Dimensions of Food Choices," in *Philosophy Comes to Dinner*, A. Chignell, T. Cuneo, and M. C. Halteman (eds.) (London: Routledge, 2015), 163–191.

15 H. L. Tuomisto, I. D. Hodge, P. Riordan, and D. W Macdonald, "Does Organic Farming Reduce Environmental Impacts?—A Meta-Analysis of European Research," *Journal of Environmental Management* 112, 2012, 309–320; G. van Huylenbroek, K. Mondelaers, J. Aertsens, and K. Mondelaers, "A Meta-Analysis of the Differences in Environmental Impacts Between Organic and Conventional Farming," *British Food Journal* 111(10), 2009, 1098–1119.

16 Budolfson, "Consumer Ethics, Harm Footprints," 163–191; W. Wakeland, S. Cholette, and K. Venkat, "Food Transportation Issues and Reducing Carbon Footprint," in *Green Technologies in Food Production and Processing* (Boston: Springer, 2012), 211–236; P. Desrochers and H. Shimizu, *The Locavore's Dilemma: In Praise of the 10,000-mile Diet* (New York: Public Affairs, 2012).

17 D. R. Boyd, *The Rights of Nature: A Legal Revolution that Could Save the World* (Toronto: ECW Press, 2017), 165.

18 To put some numbers to that, we can look at how many people die as a result of pollution created. For each kilowatt-hour, coal kills 387 times as many people as nuclear. Pinker, *Enlightenment Now*, 147.

19 You can find GiveWell's top recommended charities at https://www.givewell.org/charities/top-charities.

Chapter 13. Choosing Charities

1   Numbers adapted from Toby Ord's essay, as described in Singer, *The Most Good You Can Do*, 111, 119, footnote 7.

2   Singer, *The Most Good You Can Do*, 111.

3   D. Matthews, "Join Wall Street. Save the World," *Washington Post*, May 31, 2013, retrieved June 21, 2019, from https://www.washingtonpost.com/news/wonk/wp/2013/05/31/join-wall-street-save-the-world/?utm_term=.d917e0e3ed88.

4   This is brilliantly illustrated in an xkcd comic: https://xkcd.com/871/.

5   This financial planning structure was inspired by this excellent book: E. Warren and A. W. Tyagi, *All Your Worth: The Ultimate Lifetime Money Plan* (New York: Simon & Schuster, 2005). I added the charity part.
6   Effective altruist Ian Ross expressed this view, as reported in: Singer, *The Most Good You Can Do*, 45.
7   Buying your cousin's peanut butter helps your cousin, too, of course, but buying peanut butter from anybody helps *somebody*, and by a utilitarian value system your cousin isn't any more valuable than anybody else. So helping your cousin over someone else is only good to the extent that it makes you feel good, because other than that the good to the world is, we'll assume, the same. If buying the grocery store peanut butter helps really poor people, though, then the economic help done would be greater for the grocery store case.
8   Cost to prevent blindness: Singer, *The Most Good You Can Do*, 111.
9   What I mean is the production will cost $10,000 after ticket sales. Most theater productions do not recoup their expenses with ticket sales, and rely on grants and donations to keep going.
10   Singer, *The Most Good You Can Do*, 120.
11   T. Syme, "Charity vs. Revolution: Effective Altruism and the Systemic Change Objection," *Ethical Theory and Moral Practice* 22(93), 2019, 1–28.
12   Work stating that capitalism causes poverty: Syme, "Charity vs. Revolution," 1–28. Work stating that capitalism is curing poverty: Pinker, *Enlightenment Now*, 107–109.
13   J. Somé, S. Pasali, and M. Kaboine, "Exploring the Impact of Healthcare on Economic Growth in Africa," *Applied Economics and Finance* 6(3), 2019, 45–57.
14   Pinker, *Enlightenment Now*, 74.
15   Gross National Product (GDP) correlates with just about every measure of human flourishing, including peace, longevity, freedom, health, human rights, and nutrition. Pinker, *Enlightenment Now*, 96.
16   Singer, *Practical Ethics*, 208.
17   Singer, *The Most Good You Can Do*, 28.
18   K. Kushlev, D. M. Drummond, S. J. Heintzelman, and E. Diener, "Do Happy People Care About Society's Problems?" *The Journal of Positive Psychology*, 2019, DOI: 10.1080/17439760.2019.1639797.
19   Singer, *Practical Ethics*, 213.
20   For a more in-depth and nuanced discussion of fighting climate change vs. global health, see https://forum.effectivealtruism.org/posts/GEM7iJnLeMkTMRAaf/climate-change-interventions-are-generally-more-effective.
21   As mentioned earlier, it costs the average American regulatory agency about eight million dollars to save one life. If we estimate that saving one life is saving about thirty years, that comes out to about $266,666 to save one year of life. S. Ben-Achour, "How to Value a Life, Statistically Speaking," *Marketplace*, 2019, retrieved February 12, 2020, from https://www.marketplace.org/2019/03/20/how-value-life/.

Chapter 14. How to Motivate People to Be Good

1   T. Nordhaus and M. Shellenberger, "Global Warming Scare Tactics," *New York Times*, April 8, 2014, retrieved September 4, 2019, from https://www.nytimes.com/2014/04/09/opinion/global-warming-scare-tactics.html.
2   B. Ottenhoff and G. Ulrich, "More Money For More Good," 2012, retrieved July 2, 2019, from https://www.guidestar.org/ViewCmsFile.aspx?ContentID=4718.
3   Even $37,500 is quite expensive to save a life. The best charities are much more efficient.
4   Singer, *The Most Good You Can Do*, 90.
5   R. B. Cialdini, L. J. Demaine, B. J. Sagarin, D. W. Barrett, K. Rhoads, and P. L. Winter, "Managing Social Norms for Persuasive Impact," *Social Influence* 1(1), 2006, 3–15.

6. T. Rogers, N. J. Goldstein, and C. R. Fox, "Social mobilization," *Annual Review of Psychology* 69, 2018, 357–381.
7. Ibid., 357–381.
8. Ibid.
9. D. Krech and R. S. Crutchfield, "The Field and Problems of Social Psychology," 1948.
10. T. Hamm, "Buying Foods Based on Cost Per Calorie," *The Simple Dollar,* January 25, 2017, retrieved May 19, 2020, from https://www.thesimpledollar.com/buying-foods-based-on-cost-per-calorie/.
11. C. Finney, "To Eat or Not to Eat: 10 of the World's Most Controversial Foods," *Guardian,* November 20, 2019, retrieved May 20, 2020, from https://www.theguardian.com/environment/2019/nov/20/to-eat-or-not-to-eat-10-of-the-worlds-most-controversial-foods.
12. E. H. Haddad and J. S. Tanzman, "What Do Vegetarians in the United States Eat?" *The American Journal of Clinical Nutrition* 78(3), 2003, 626S–632S.
13. "Diet Change and Demographic Characteristics of Vegans, Vegetarians, Semi-Vegetarians, and Omnivores," *Humane League Labs Technical Report*, retrieved June 28, 2019, from http://www.humaneleaguelabs.org/blog/2014-04-07-large-scale-survey-vegans-vegetarians-and-meat-reducers/.
14. http://www.humaneleaguelabs.org/blog/2015-01-24-which-vegan-meals-do-omnivores-find-most-appetizing-and-accessible/.
15. Between 3.4 and 7.4 years, according to https://animalcharityevaluators.org/research/dietary-impacts/effects-of-diet-choices/.
16. "Length of Adherence to Vegetarianism," Animal Charity Evaluators, November 2017, https://animalcharityevaluators.org/research/dietary-impacts/vegetarian-recidivism/.
17. "Our Use of Cost-Effectiveness Estimates," Animal Charity Evaluators, April 2018, https://animalcharityevaluators.org/research/methodology/our-use-of-cost-effectiveness-estimates/#6.
18. "Leafleting," Animal Charity Evaluators, November 2017, retrieved July 1, 2019 from https://animalcharityevaluators.org/advocacy-interventions/interventions/leafleting/#report.
19. "Protests," Animal Charity Evaluators, March 2018, retrieved July 1, 2019 from https://animalcharityevaluators.org/advocacy-interventions/interventions/protests/#3.
20. Brian Tomasik's estimate is actually $11, but he believes the true number is closer to $100, for reasons he explains in this online article. B. Tomasik, "Donating Toward Efficient Online Veg Ads," January 28, 2012, retrieved August 29, 2019, from https://reducing-suffering.org/donating-toward-efficient-online-veg-ads/; William MacAskill also suggests the that $100 is the cost of making someone vegetarian for one year. MacAskill, *Doing Good Better.*
21. "Diet Change and Demographic Characteristics."
22. Also, people respond pretty well to documentaries and books, so marketing material should promote them. Brochures, booklets, and magazines seem to be less effective. But this effect was quite small. "Diet Change and Demographic Characteristics."
23. J. Hallam, R. G. Boswell, E. E. DeVito, and H. Kober, "Gender-related Differences in Food Craving and Obesity," *Yale Journal of Biological Medicine* 89(2), 2016, 161–173.
24. MacAskill doesn't believe that you can do meat offsets like you can do carbon offsets. I don't really understand his argument, though. MacAskill, *Doing Good Better.*

Chapter 15. When Giving Gives Back
1. E. W. Dunn, L. B. Aknin, and M. I. Norton, "Prosocial Spending and Happiness: Using Money to Benefit Others Pays Off," *Current Directions in Psychological Science,* 23(1), 2014, 41–47.

2   Ibid.
3   Ibid.
4   There was a significant relationship between giving and happiness in most of these countries. Dunn, et al., "Prosocial Spending and Happiness: Using Money to Benefit Others Pays Off."
5   Dunn, et al., "Prosocial Spending and Happiness: Using Money to Benefit Others Pays Off," 41–47.
6   J. H. Fowler and N. A. Christakis, "Cooperative Behavior Cascades in Human Social Networks," *Proceedings of the National Academy of Sciences*, 107(12), 2010, 5334–5338.
7   For a compelling argument about why being good is incredibly costly, I recommend the book P. K. Unger, *Living High and Letting Die: Our Illusion of Innocence* (Oxford, UK: Oxford University Press, 1996).
8   Tomasik, "Why Activists Should Consider Making Lots of Money."
9   V. S. Helgeson, "Relation of Agency and Communion to Well-being: Evidence and Potential Explanations," *Psychological Bulletin* 116(3), 1994, 412.
10 Oakley, et al., *Pathological Altruism*.
11 A. M. Grant, *Give and Take: A Revolutionary Approach to Success* (New York: Penguin, 2013), 159.
12 R. Schulz, et al., "Patient Suffering and Caregiver Compassion: New Opportunities for Research, Practice, and Policy," *The Gerontologist* 47(1), 2007, 4–13.
13 J. Crocker, A. Canevello, and A. A. Brown, "Social Motivation: Costs and Benefits of Selfishness and Otherishness," *Annual Review of Psychology* 68, 2017, 299–325.
14 Singer, *Ethics in the Real World*, 111.
15 Matthews and Pinkerton, "How to Pick a Career that Counts."
16 You can see recent figures at https://www.jefftk.com/donations.
17 For those interested, it's the point of marginal utility: where giving away hurts you as much as the recipients would gain. As far as I know, nobody has lived up to this standard, including Peter Singer, the world's unofficial spokesperson for utilitarianism. Singer, *The Most Good You Can Do*, 15. Peter Unger puts it well. Perfect morality would be "By sending funds to the most efficient loss-lessening programs, you must incur financial losses up to the point where going further will be unproductive, overall, in lessening serious losses." Unger, *Living High and Letting Die*, 145.
18 Syme, "Charity vs. Revolution," 1–28.
19 J. L. Bühler, R. Weidmann, J. Nikitin, and A. Grob, "A Closer Look at Life Goals Across Adulthood: Applying a Developmental Perspective to Content, Dynamics, and Outcomes of Goal Importance and Goal Attainability," *European Journal of Personality* 33, 2019, 359–384.
20 S. K. Nelson, K. Layous, S. W. Cole, and S. Lyubomirsky, "Do Unto Others or Treat Yourself? The Effects of Prosocial and Self-focused Behavior on Psychological Flourishing," *Emotion* 16(6), 2016, 850–861.
21 Ibid.
22 R. B. Cialdini and D. A. Schroeder, "Increasing Compliance by Legitimizing Paltry Contributions: When Even a Penny Helps," *Journal of Personality and Social Psychology* 34(4), 1976, 599.
23 L. B. Aknin, M. I. Norton, and E. W. Dunn, "From Wealth to Well-being? Money Matters, but Less than People Think," *The Journal of Positive Psychology* 4(6), 2009, 523–527; D. Kahneman, A. B. Krueger, D. Schkade, N. Schwarz, and A. A. Stone, "Would You Be Happier If You Were Richer? A Focusing Illusion," *Science* 312(5782), 2006, 1908–1910.

24 D. Saxbe and R. L. Repetti, "For Better or Worse? Coregulation of Couples' Cortisol Levels and Mood States," *Journal of Personality and Social Psychology* 98(1), 2010, 92; D. E. Saxbe and R. Repetti, "No Place Like Home: Home Tours Correlate with Daily Patterns of Mood and Cortisol," *Personality and Social Psychology Bulletin* 36(1), 2010, 71–81.
25 For a recent review of this literature, see W. Zhang, M. Chen, Y. Xie, and Z. Zhao, "Prosocial Spending and Subjective Well-being: The Recipient Perspective," *Journal of Happiness Studies* 19(8), 2018, 2267–2281.
26 Aknin, et al., "Prosocial Spending," 635.
27 Singer, *The Most Good You Can Do*, 100.
28 Aknin, et al., "Prosocial Spending," 635.
29 Zhang, et al., "Prosocial Spending," 2267–2281.
30 J. Haidt, "Elevation and the Positive Psychology of Morality," *Flourishing: Positive Psychology and the Life Well-lived* 275, 2003, 289.
31 A. C. Brooks, "Does Giving Make Us Prosperous?" *Journal of Economics and Finance* 31(3), 2007, 403–411.
32 J. Andreoni, W. T. Harbaugh, and L. Vesterlund, "Altruism in Experiments," *The New Palgrave Dictionary of Economics: Volume 1–8*, 2008, 134–138; W. T. Harbaugh, U. Mayr, and D. R. Burghart, "Neural Responses to Taxation and Voluntary Giving Reveal Motives for Charitable Donations," *Science* 316(5831), 2007, 1622–1625.
33 Thoits and Hewitt, "Volunteer Work and Well-being," 115–131.
34 S. Lyubomirsky, L. King, and E. Diener, "The Benefits of Frequent Positive Affect: Does Happiness Lead to Success?" *Psychological Bulletin* 131(6), 205, 803.
35 Grant, *Give and Take*, 182.
36 Lyubomirsky, "Hedonic Adaptation," 200–224.

INDEX

A

academic publishers, 90–91
Accra, Ghana, 335
achievement, 137
actions: akratic, 31; avoiding, 37–38; measuring effectiveness of, 280–283; *see also* behaviors
active reading, 93–96
acute stress, 146
adaptation: to climate change, 334, 335; to disease, 261; hacking, 131–133; to new experiences, 128–129; psychological tendencies and, 204–206; to routine, 373
addictive behaviors, 52, 119, 120
adenosine, 104
Adolescent Transitions Program, 282
adrenaline, 14
advance preparation, 17
advice, 91
Against Malaria Foundation, 331, 349, 352, 365, 371
age: climate change and, 335; happiness and, 130
aging, 266–267
agricultural revolution, 193, 314
agriculture, 323–325; *see also* factory farming
akratic actions, 31
altruism: effective, 239, 342, 346, 351, 366, 369, 373; pathological, 369; reciprocal, 187, 188

Alzheimer's disease, 260
Amazon Echo, 101
ambition, 36
American culture, 137
Amish, 160
amygdala, 181, 215
anger, 196, 197, 218
animal activism, 326–327, 361
Animal Charity Evaluators (ACE), 326, 330
animals, 284–329; consciousness of, 284–288, 291–304; death of, 308, 319–320, 323–324; domesticated, 193–194, 312–313, 316, 318–320, 327–329, 331–332; emotions of, 190, 191–192, 286; ethics of eating, 316–329; evolution of, 294–295; helping, 315–329; hunting, 322–323; Jainism and, 288–289; legal status for, 195; liminal, 195; moral considerations of, 190–195, 284; moral value and, 291–304; research on, 192; wild, 305–313, 315; zoo, 313
animal suffering: compared with human suffering, 330–332; encouraging others to reduce, 359–363; evolution and, 311–313; hunting and, 322; of livestock, 312–313, 316, 328–329, 336; reducing, 315–316; theories on, 287–291; of wild animals, 307–313
animal welfare, 326–327, 357; *see also* animal suffering

animist religion, 193
Antabuse, 56
anterior cingulate cortex, 203
anxiety, 30, 158
apathy, 114
Arbesman, Samuel, 238
arts patronage, 344, 347–350
assumptions, 235–236
atheists, 183
attention, 10–12, 134
attention-deployment, 39
attraction, 165
audiobooks, 15
Audubon Society, 322
authentic self, 270
autonomy, lack of, 145
autopilot, 42
aversion therapy, 53

B

bad behavior, 213–214
bad experiences, 128–129, 170
bad feelings, 27–28, 118, 123, 232, 233, 236, 266–268, 314, 328, 330
Bailey, Chris, 11, 79
basal ganglia, 26–29, 31, 40, 52
basic research, 279
Batson, Dan, 211
beef, 300–301, 325, 326, 327, 336–337; *see also* cows
Beethoven, Ludwig van, 65
behaviors: bad, 213–214; evolution of, 204–205; goal-directed, 40; good, 213–214; habitual, 42; motivations for, 285–286; *see also* good behavior
Benatar, David, 241
Berners-Lee, Mike, 340
better judgments, 31
biases: in-group, 188–189, 248, 256; impact, 261; memory, 124; negativity, 214–215
Bible, 183–184, 223
bicycles, 150–151
bimodal philosophy, 60
biodiversity, 337
biological clock, 163
biological pain, 290–291
bitter taste, 54
blindness, 261–263, 341, 345, 347–349
Boice, Robert, 67, 68

books: reading, 90–91; writing in, 94; *see also* reading
Boole, George, 279
boredom, 14–15, 115
Borlaug, Norman, 213–214
Bosch, Carl, 213
Boyd, David R., 196
brain: basal ganglia, 26–29, 31, 40, 52; consciousness and, 289–290; cortex, 27, 30, 296–297, 323; functions, 21, 26–30; morality and, 202–206, 215–217; neurotransmitters, 50, 52, 164, 304; prefrontal cortex, 27, 28, 29, 32, 181, 203; size, 293–296, 299; species comparison of, 286; task selection by, 26–30
brain training software, 122
breaks, 83, 85, 86
Bruns, Richard, 303
Buddhism, 36, 134, 138, 144, 290
budgeting, 343–344, 349–351
bugs. *See* insects
burnout, 84–85, 273–274, 314, 351, 369–370
busy work, 119–120
bystander effect, 223

C

caffeine, 104, 105, 106
calculations, 239
callings, 271, 272
Cambridge-Somerville Youth Study, 282
capitalism, 346
CARE, 342
care, 232
career choice, 269–279
careers, 271
caregivers, 84, 370
caterpillars, 303
cell phones. *See* mobile phones
challenges, 147
Chalmers, David, 302
charitable donations: to the arts, 344, 347–350; as Band-Aids, 346; budgeting, 343–344, 349–351; effectiveness of, 256–257, 326, 330–332; happiness from, 364–366, 374; high earnings and, 274–279; maintaining, 373–374; maximizing goodness through, 250–253; mobilizing others to give, 353–358, 375–377; percent to give, 371–372; to

reduce animal suffering, 362–363; to reduce climate change, 340; self-serving motivations for, 354–355
charities: choosing, 341–352; comparing, 350, 353; effectiveness of, 251–253, 256–257, 280–283, 326, 341–342
cheaters, 188
chickens, 300–301, 316, 318, 319, 321–322, 325, 327–329, 331, 336–337, 357, 361
children: happiness and, 162–163; meaningfulness and, 119; morality of having/not having, 240–245; overpopulation and, 348–349
chimpanzees, 195
choices, morality of, 180–183
Christians, 183–184, 223
chronic stress, 146
chronotypes, 107–109
Cialdini, Robert, 355
cingulate cortex, 291
circadian rhythm, 104
city living, 148–149
classical conditioning, 55
Clean Air Task Force (CATF), 340, 352
climate change, 320–322, 333–340, 352, 357
coffee, 106
cognitive control, 33
cognitive dissonance, 194, 223, 373
cognitive load, 203, 344
cognitive system, 27–29; habits and, 41; hacking your, 30–40
cognitive therapy, 139
collaborative projects, 77
commuting time, 18, 134–136, 145
compassion, 196
compassion fatigue, 369–370
complaining, about others, 11
complex games, 121, 122
compulsion, 29
compulsive behaviors, 119, 120
computer games, 116–123
computer technologies, 11
conceptual analysis, 217–218
concern, 196
conditioning, 22; classical, 55; habit formation and, 26, 30, 47–48; learning through, 51; punishment and, 53–54; reward system and, 53; unconscious, 53–54

consciousness: of animals, 284–304; human, 285; levels of, 291–304; suffering and, 284–285, 286
contempt, 196
control, lack of, 145
cooperation, 189
corporations, 195
cortex, 27, 30, 296–297, 323
cortical brain regions, 31
cortisol, 313
cost-effectiveness, of doing good, 248–249
coworkers, relationships with, 18
cows, 300–301, 322, 324, 325, 327, 329, 331, 336–337
creativity, 70, 120
creature consciousness, 285
crickets, 297, 298, 300, 301, 302–303, 306
cropland, 316, 323–325, 338; *see also* agriculture
cross-cultural tendencies, 206
cross-species tendencies, 207
cultural learning, 205
cultural proximity, 189
cultural transmission, 244
cultural values, 137
culture, 206, 270
cultured meat, 325–326
current events, 88–89
cycling, 150–151
Csikzentmihalyi, Mihaly, 114, 115

D

daily commitments, 80–81
Dalí, Salvador, 65
Dawkins, Richard, 216
deadlines, 71, 82–83
death, 171; age of, 263–264; animal, 308, 319–320, 323–324; from climate change, 333, 334–336
de Brigard, Filipe, 172
decision fatigue, 155
decision making, 39, 42; habits and, 47–48
deep work, 10
default mode network, 28
delegation, 48, 77, 78–79
deliberate practice, 9–10
deontology, 201–202, 226, 234, 237, 258

depression, 158, 261, 262
deprivation, 55–56
Descartes, René, 190
desire, 53
detachment, 140
detectors, 217–221
Diamond, Jared, 244
diet: encouraging others to change their, 359–363; fruitarian, 317; vegan, 319, 323–325, 327, 359–360; vegetarian, 217, 317, 359–362; *see also* meat-eating
digital technologies, 11
disability, fairness and, 262–263
Disability-Adjusted Life-Year (DALY), 259–268, 280, 331
discipline, 31, 38
discretionary spending, 343–344, 349, 351
disease, saving people from, 357–358
disgust, 196, 199–201
dispositional optimism, 35
distractions: avoiding, 59–61; effects of, 13–18; electronic, 61; external, 19; internal, 19; multitasking and, 13–17, 19, 21–23; productivity and, 10–19, 21–23; removing, 18–19; seeking, 57
distress, 198
divorce, 128, 162, 166
doctors, 274, 277
dogs, 304, 313
domesticated animals, 193–195, 312–313, 316, 318–320, 327–329, 331–332
Donaldson, Sue, 195
doodling, 14–15
dopamine, 14, 16, 52, 119, 164
dorsolateral striatum (DLS), 26, 40
dorsomedial striatum, 40
double effect principle, 209, 277–278, 354–355
dreams, 191
drug addiction, 51
drug safety, 215
Duhigg, Charles, 41
Dunn, Elizabeth, 210
dying, 306–307

E
earning to give, 274–279
Easterlin Paradox, 130

eating, 128, 132–133, 144, 316–327; *see also* diet; meat-eating
ecological validity, 125
ecosystems, 195–196, 307–311, 314, 337; *see also* environment
Ecuador, 339
education, 8, 145, 283
effective altruism, 239, 342, 346, 351, 366, 369, 373
eggs, 318
ego depletion, 39
Einstein, Albert, 157
electricity, 339
electric shocks, 53
electronic distractions, 61
elephants, 289, 294–296
elevation, 212
email, 11, 13, 61
Emmons, Robert, 151
emotional needs, 22
emotions, 4, 118, 133; of animals, 190, 191, 192, 286; judgments based on, 237; morality and, 196–201; negative, 35, 118, 196–201, 232, 233; positive, 118, 123, 154, 196, 232, 233; *see also* bad feelings; good feelings
empathy, 84, 158, 196, 237, 354
enculturation, 208, 225
endorphins, 303
endowment effect, 214
energy sources, 339
environment: climate change and, 333–340; engineering to resist temptations, 57–59; habits and, 43–44; impact of having children on, 240–241; modification of, 58–59; plant-based diet and, 331–332; *see also* ecosystems
environmental morality, 333–340
environmental protection, 195–196, 315, 339
epiphanies, 141–142
Epley, Nicholas, 184
Equal Consideration View, 293, 300, 304
eternal youth, 266–267
ethical dilemmas, 221–222
ethical purchases, 245–247
ethical theories, 217–226, 228–229
ethics, 180, 203; of animal research, 192; assumptions in, 235; changes in

field of, 239; population, 243; religion and, 226–227; rule-based, 229–231, 233–234, 368; *see also* morality
Evernote, 101
evidence, 239, 354
evolution: of animals, 294–295; of behaviors, 204–205; genetic, 204–211, 240; happiness and, 311–314; morals and, 241; of pain, 303; smartness of, 311
evolutionary competition, 127
evolutionary psychology, 205–211
exceptions, to rules, 89
executive control, 27
exercise, 7, 15, 33, 43–44, 49, 133, 141, 150–153, 170
expanding circle, 187–188, 194
experiences, 153, 155–156, 170–172
expertise, 9–10
extinction, 337–338, 339
extraordinary experiences, 153
extraverts, 157
extreme poverty, 249–250
Extreme Speciesism, 292, 303
extrinsic punishments, 56
extrinsic rewards, 54

F
face validity, 218–221
facial expressions, 286
factory farming, 194, 301, 317–318, 322, 325, 328, 329
facts, scientific, 238
fairness, 185, 187, 200–201, 211, 262–263
fair trade, 246–247
family, 158, 186–187
famines, 247, 348
fear, 196
feelings. *See* emotions
Feriss, Tim, 91
fiction writing, 20–21, 92
fight, flight, or freeze response, 146
finite resources, 244
fish, 316, 319–320, 323
fishing, 186, 193, 322, 323
Fiverr.com, 78
flow, 114–116
fluid intelligence, 17
focus, 10–12
focused attention, 60
Foldit, 123

food, 316–326; *see also* diet; eating; meat-eating
freelance work, 79
freeloaders, 188
free time, 78–80
friendships, 126, 156–160, 187–188
fruit, 289, 317
fruitarian diet, 317
fruit flies, 302–303
futility, 356–358
future, 140, 142
future lives, 328–329; value of, 264–265

G
Garavito, Luis, 277–278
Gazzaley, Adam, 17
generosity, 109–111
genetic drift, 204, 205
genetic evolution, 205–211, 240; *see also* evolution
genetics, 138–139; selfish gene theory, 186–187
geniuses, 243–244
Gill, Alexander, 160
givers, 109–110, 179, 180
GiveWell.org, 283, 341–342
Glennerster, Rachel, 281, 283
globalization, 244, 247–249
Glover, Jonathan, 335
goal-directed behavior, 40
goals, 5, 47, 76, 126–127
God, 184–185, 221, 226
God of War, 119, 120
good behavior, 213–214; feelings of futility and, 356–358; promoting in others, 353–358
good experiences, 128–129, 170–171
good feelings, 27–28, 118, 123, 232, 233, 236, 241, 266–268, 314, 328
good life, 171–175
goodness: levels of, 368; maximization of, 233–234, 236, 248–253; measuring, 280–283
Google Docs, 101
Google Home, 101
Google Scholar, 90
Google Tasks, 63, 64
Gore, Al, 369
government regulations, 255–256, 352
Grant, Adam, 110, 369

gratifications, 22, 31, 44–45, 47, 118
gratitude, 149, 151–152, 196
gratitude exercises, 134
gratitude journaling, 151–152
Grinnell, George Bird, 322
guide dogs, 341
guilt, 196–198
gut instincts, 219–220, 237

H

Haber, Fritz, 213
habitat destruction, 309, 315, 338
habitat preservation, 310–311
habits, 29; behavior and, 40–41; changing, 44–45; daily commitments and, 80; decision making and, 47–48; environment and, 43–44; formation of, 26, 30, 169; implementation intentions for, 47; long-term effects of, 42–43; morning, 46; power of, 41; replacing, 45–47; rewards and, 41–42; triggers of, 43–44; using willpower to change, 47–50
habit system, 27–29, 33; hacking your, 40–51
habituation, 128–129, 131–136, 161
Haidt, Jonathan, 139, 141, 164, 200, 215, 376
half-hours method, 60–75, 81–83, 85–86
Half Life 2, 120
happiness, 5, 36, 112–175; from within, 139–143; achievement and, 137; age and, 130; beyond, 171–175; as in-born, 138–139; children and, 162–163, 241–242; defined, 113, 118; evolution and, 311–314; external factors in, 144–145; factors that contribute to, 149–156; flow and, 114–116; from giving to others, 364–366, 374; habituation and, 128–129, 131–133; health and, 145, 261–262; hinderances to, 145–147; life satisfaction and, 118, 123–128; living location and, 148–149; marriage and, 161; meaningfulness and, 119; memory and, 166–171; money and, 77–78, 129–131, 156, 375–377; pets and, 165–166; productivity and, 112–114, 137; religion and, 154–155; retirement and, 117; science of, 137–138; self-understanding and, 133–134; set-point, 138–139; social connections and, 156–160; study of, 112–114; trade-offs and, 134–136; video games and, 116–123; volunteering and, 110–111; at work, 114–116
Harari, Yuval, 270
Harinck, Fieke, 182
harm, 232, 277, 336
Harris, Tristan, 11
Haydn, Joseph, 65
health: education and, 283; exercise and, 152; happiness and, 145, 261–262; marriage and, 161; naps and, 106
hedonic treadmill, 128
helping others, 315; on global scale, 247–249, 258; morality of, 201, 231–232; rewards of, 109–111, 364–366, 375–377; *see also* charitable donations
helplessness, 121
hereditary traits, 207
heroes, 213–214
high achievers, 157–158
home, working at, 18
hormones, 164
how-to books, 91
Humane League, 326, 330–331, 352, 362, 365
humanism, 189–190
humanity, impact of, on world, 363
human life: moral value of, 328–329; numerical value of, 254–268; saving, 357; value of, 241
human nature, 239–240
human rights, 190
human suffering, 286, 292, 298, 328, 330–332
Hume, David, 236
hunter-gatherer societies, 193, 208, 314
hunting, 322–323
hyperbolic discounting, 264

I

idea capture, 96–103; index cards for, 98–100; memory palaces for, 97–98; with phone, 100–103
ideas, processing, 103–104
identity: moral, 211–213; personal, 365–366
immediate gratification, 31, 44–45, 47, 118
immortality, 266–267
impact bias, 261
impulse control, 27, 126

INDEX

incest, 199, 200
income levels, 179–180, 375–376
incubation, 70
index cards, 98–100
induction, 236
Industrial Revolution, 246
information: search for, 22–23; timely, 87–88
in-group bias, 188, 189, 248, 256
inner experiences, 190–191
inner voice, 34–36
insects, 302–303, 305–309, 315, 324
insomnia, 104
instincts, 215
intelligence, 145, 293, 296
intentions, 33, 47
Internet, 11
interpersonal conflict, 145–146
interruptions, 19, 30, 33; *see also* distractions
intrinsic rewards, 54
introverts, 157–160
intuition, 4
intuitions, moral, 216, 217, 219–226, 232–234, 239–240, 251
Isbell, Charles, 70

J

Jahncke, Helena, 23
Jainism, 288–289
jealousy, 196
Jenner, Edward, 213
job changes, 117
jobs, 271, 272
job satisfaction, 270, 273–274
job skills, 276–277
journaling, 133, 151–152
journalist philosophy, 60
Judaism, 225
judgments, 31

K

Karnazes, Dean, 54
Kasser, Tim, 131
Kaufman, Jeff, 371
kin selection, 186–187
Kline, Michelle, 244
Knight, Craig, 23
knowledge: of current events, 88–89; lost, 337–338; reading for, 87–89; useless, 93
knowledge workers, 11

Knuth, Donald, 60
Kravinsky, Zell, 370–371
Kremer, Michael, 281, 283
Kymlicka, Will, 195

L

learning, 30, 51
legal status: of animals, 195; of ecosystems, 195–196
leisure time, 73, 114, 116; television, 124–126; video games and, 116–123
life: future, 264–265, 328–329; good, 171–175; as suffering, 305–310; *see also* human life
life changes, 45–46
life expectancy, 259, 267, 346
life satisfaction, 118, 123–128, 166, 168, 171
lifespan, 169
light, 217, 219
liking, 29–30, 51, 52
limbic system, 204
liminal animals, 195
literature reviews, 95
Livescribe, 102–103
lives saved, 254–257; of animals, 330–332; costs of, 255–257, 259; giving to charities and, 274, 276–277, 330–332; by good people, 244; number of years saved and, 258–268
livestock, 193–195, 312–313, 316, 318–320, 327–329, 331–332
living location, 148–149
living standards, 348–349
local activism, 374–377
local food, 338–340
loneliness, 159
long-term goals, 126–127
long-term rewards, 31, 45, 123–124
loss aversion, 214
love, 164–165, 196
love at first sight, 165
luck, 138–139
Luminosity, 122
lunch breaks, 86

M

MacAskill, William, 275, 277
magnitudes of good and bad, 229–231, 233–234

maintenance activities, 7
Malaria Consortium, 342
Mao Zedong, 213–214
Mark, Gloria, 16
marriage, 117, 128, 146, 161, 162, 166, 218, 219
Masicampo, E. J., 62
mastery, 114
matchers, 109, 110, 179
material possessions, 376
McFarland, Cathy, 296
meaningfulness, 119
meat demand, 321–322
meat-eating: animal suffering and, 194, 300–301, 316–326; encouraging others to reduce, 359–363; environment and, 331–332, 336–337; ethics of eating, 329; moderation in, 136; nutrition of, 327; reducing, 319, 321–322
meat substitutes, 325–326
medical testing, 215
meditation, 32, 80, 134, 139, 140, 291
memories, 153, 167, 169–170
memory: biases, 124; happiness and, 166–171; procedural, 26
memory palaces, 97–98
men: desire for children in, 163; happiness of, 125; health benefits of marriage for, 161; jobs of, 271–272; retirement and, 117
mental illness, 261–262
mental states, 144
mesolimbic system, 28, 29, 30, 52, 53
microorganisms, 307–308
Middle World, 216
mind, task selection by, 26–30
mindfulness, 32, 39, 134, 139–143, 167
mindset, changing your, 37
mind-wandering system, 28–29
mobile phones: checking of, 18–19; as distractions, 61; for idea capture, 100–103
moderation, 136
monastic philosophy, 60
money, 54; budgeting, 343–344, 349–351; giving away, 375–377; happiness and, 129–131, 156, 375–377; spending for free time, 77–78
Monte Carlo simulations, 304, 336
moral authority, 227

moral bias, 208–211
moral development, 182, 187–188, 204–205
moral dilemmas, 201, 221–222
moral dissociation, 223
moral dumbfounding, 200
moral goodness, 173–175
moral identity, 211–213
moral intuitions, 216–226, 232–233, 239–240, 251
morality: animals and, 190–195; brain systems and, 202–206, 215–217; career choice and, 272–274; deontology, 201–202; differences in, 232–233; emotions and, 196–201; environmental, 333–340; environmental protection and, 195–196; evolutionary psychology and, 205–211; good and bad behavior, 213–214; in-groups and out-groups, 188–191; of having/not having children, 240–245; helping others, 201; magnitudes of good and bad, 202, 230–231; optimization of, 228–231, 233–234; religion and, 183–185, 226–227; rule-based, 229–231, 233–234, 368; science and, 234–236; selfishness, 185–186, 201
morally deviant behaviors, 200
moral psychology, 215–217, 221, 227, 234, 251, 277, 364
moral theories, 210–211, 217–226, 228–229
moral thinking, 180–183, 215–216, 219, 234
moral value, 291–304, 328–329
morning habits, 46
Morris, Desmond, 205
motivational systems, 27–30
motor vehicle accidents, 255
Mozart, 157
multitasking, 11, 13–17, 19, 21–23, 60
murder, 357
music, 15
music piracy, 224
mussels, 320–321, 327
Muted Animal Theory, 286, 287, 290, 293–304

N

napping, 106–107
nappuccino, 106

Napster, 224
narratives, 35–36
nationalism, 194, 246
naturalistic fallacy, 267, 310–311
nature, 148, 195–196, 310–311, 314, 315, 337; *see also* ecosystems; environment
Nawijn, Jeroen, 155
negative emotions, 35, 118, 196–201, 232, 233
negative experiences, 128–129, 131–133
negativity bias, 214–215
neocortex, 202
neuroscience, of pleasure, 50–51
neurotransmitters, 50, 52, 164, 302
newborn babies, 206–207
new omnivorism, 324, 325
Newport, Cal, 10, 60, 74
news, 88–89, 189
New Year's resolutions, 47
niceness, 268, 366–367
nightly habits, 46
nociception, 295, 302
noise, 23, 145
nonfiction books, 90–91
non-productivity scenarios, 68, 71, 82
norepinephrine, 164
nostalgia, 142
notebooks, 99, 102
note-taking, 94–95, 98–103
Nozick, Robert, 172
number insensitivity, 210
numbers, use of, 239
numerical value, of human life, 254–268
nutrition, 368

O

Occam's razor, 236
octopuses, 192, 294, 298–299
office environments, 118
older adults, sleep needs of, 105–106
omission bias, 210
online social networks, 158
open-plan offices, 23
opioids, 119
opportunity cost, 13, 158
optimism, 35
optimizers, 3–4
organic food, 338–340
Oslo, Norway, 335
out-groups, 188–191

overhead myth, 252
overpopulation, 348
oxytocin, 164, 189

P

pain: of death, 308; evolution of, 303; felt by animals, 285–291, 296; felt by fish, 323; felt by insects, 302–303, 306; of nonmoving creatures, 310; of others, 296; plants and, 295, 317; purpose of, 51, 54
parsimony, 236
passions, 269–270
passive decay, 17
past, 140, 142
pathological altruism, 369
Pavlok, 53
Pavlov, Ivan, 55
pay inequities, 271–272
peanut butter, 358–359
peer-reviewed journal articles, 90
Pennebaker, James, 20, 36
perfectionism, 25
personal assistants, 48, 71–72, 78–79
personal identity, 365–366
personality, 196–201
personality tests, 157
personal productivity, 4–5
personal relationships, 351
personhood, 195
pesticides, 324, 338
Petrov, Stanislav, 214
pets, 165–166, 313
phenotypes, 204
phone calls, 13
photographs, 167–168
physical activity, 86; *see also* exercise
physical contact, 209–210
pigs, 318–320, 325, 329
Pinker, Steven, 89, 119, 154
piracy, 224
plant-based meat substitutes, 325–326
plants, 288–289, 295, 317
Plato, 226
pleasure, 144, 285; function of, 51–52; good life and, 171; neuroscience of, 50–51
pleasure-seeking system, 29–30
police force, 231
political change, 346

Pomodoro Technique, 83–84, 85
poor countries, 245–247, 256–257
Pope, Nolan, 108
population control, 348
population ethics, 243
pork, 325; *see also* pigs
Portal 2, 122
portion sizes, 40
positive emotions, 118, 123, 154, 196, 232, 233
positive experiences, 128–129, 131–133
possessions, 153
poverty, 246, 249–250, 346, 348
practice, 9–10, 21
preference satisfaction, 268
prefrontal cortex, 27, 28, 29, 32, 181, 203
prestige, 173–174, 356
pride, 196
Princip, Gavrilo, 214
prioritization, 9, 10; of daily commitments, 80–81; of projects, 63
problem solving, 121, 244, 272–273, 279
procedural memory, 26
procrastination, 23–24, 74
productivity, 3–25; choosing what to do and, 75–76; distractions and, 10–19, 21–23; happiness and, 112–114, 137; naps and, 106–107; optimizing, 5–10; personal, 4–5; procrastination and, 23–24; quantity vs. quality, 65; retirement from, 7; video games and, 116; workweek and, 38
project lists, 62–65, 75–80; curation of, 76–77
protein, 320–321
Prozac, 139, 303
psychological traits, 138–139, 204–205
psychopathy, 185–186, 237
punishment, 53–54, 56, 197
purchasing decisions, 245–247
Pychyl, Tim, 24

Q
quality, 65
Quality-Adjusted Life-Year (QALY), 263
quantity, 65
quitting, 76–77

R
rain forests, 339
rationality, 234
rationalization, 31–32, 288
reading: active, 93–96; effectively, 86–96; for knowledge, 87–89; for mind-expansion, 92; picking material for, 87; for pleasure, 92; productivity and, 5–6; when to stop, 92–93; for wisdom, 91
reality, 172, 190
reasoning, 202–203, 237
reciprocal altruism, 187, 188
reciprocity styles, 109–110, 179
Redneck Rampage, 37
reducitarians, 319, 362
relationships: bad, 145–146; personal, 351; sacrificing for work, 157–158
relatives, 186–187
relaxation, 114, 115, 155
religion: animist, 193; ethics and, 226–227; in-group bias and, 189; happiness and, 154–155; morality and, 183–185
repetition, 26, 30, 47–48
reproduction strategies, 307, 308, 310–313
reputation, 356
research, 279
retirement, 117
retribution, 185, 197
rewards: extrinsic, 54; future, 44; immediate, 44; intrinsic, 54; long-term, 45
reward system, 27–30, 50–51; function of, 51; hacking your, 51–59; punishment and, 53–54
rhythmic philosophy, 60
ripple effects, 366–367
risk aversion, 76, 93
risk taking, 255
rodents, 324
routines, 153, 169
rules: ethical, 229–231, 233–234, 368; simple, 230
rural areas, 148

S
salmon, 301, 320, 327
Same Pain Theory, 286, 287–288, 303
Scared Straight, 282
Schindler, Oskar, 277
schizophrenia, 260

INDEX

scholarship, 89–90
science: experiments in, 235–236; of morality, 234–236; progress of, 238–239, 244, 267; research in, 279
scientific psychology, 239
Scott-Taggart, M. J., 335
self-control, 27, 31, 40; fluctuations in, 47; situation-selection method of, 58
self-esteem, 35
self-help books, 91, 141–142
self-image, 36
self-interrupting behaviors, 21–22
selfish gene theory, 186–187
selfishness, 185–186, 355
self-kindness, 374
self-satisfaction, 212
self-understanding, 133–134
sensory perceptions, 216
sentience, 285–286
serotonin, 303
sex, 54, 144, 206
shame, 196, 197–198
short-term goals, 126–127
shyness, 157
Singer, Peter, 345
situation-modification, 58–59
situation-selection method, 58
six-second rule, 34
skills, improving, 7–10
slavery, 196, 293
sleep, 32; chronotype and, 107–109; how to, 105–109; importance of, 105; napping, 106–107; needs, 105
sleep deprivation, 105
sleep hygiene, 105–106
sleep pressure, 104
smartphones, 61
social activity, 86
social connections, 9, 154, 156–160
social contagion, 58
social groups, 49–50, 188–189
social interactions, 126
social isolation, 135, 144
socialization, 7
social media, 19, 89
social proof, 355, 359
speciesism, 292–293, 298–299, 304
species extinction, 337–338, 339
speed limits, 255
Stalin, Joseph, 213

starvation, 348
state consciousness, 285
stealing, 224
stegosaurus, 289, 294–295, 296
Stephenson, Neal, 60
stereotypes, 125
strangers, 186–187, 189
strategic quitting, 76–77
stress, 32, 84, 146–147, 313
subscription services, 48
success, 110
suffering: animal, 286–290, 292, 296–304, 307–313, 315–316, 322, 328, 336, 359–363; comparing human to animal, 330–332; consciousness and, 284–285; estimating, 293–304; human, 286, 292, 298, 300, 328, 330–332; by insects, 306, 308–309; life as, 305–310; moral value of, 292–293, 300–304
sugar, 54
supererogatory acts, 258
supplementary motor area, 31
Suraci, Justin, 148
survivalism, 160
survivor's guilt, 197
sweatshops, 245–246
Sword, Helen, 68
sympathy, 196, 198

T

taboos, 205
tactical bombing case, 277–278
takers, 109, 110, 179
tasks: avoiding, 37–38; choosing, 75–76; concentrating on most important, 82–83; scheduling, based on chronotype, 108–109; time estimates for, 70
task switching, 14, 18–19, 22–23
task-switching costs, 15–16, 73, 74
taxation, 201
TCOB list, 75
technology, 11, 208
television, 124, 125–126
telework, 18
temporal discounting, 264–265
temptation bundling, 55–56
temptations, 39
testosterone, 166
Tetris, 121, 170
Te Urewera National Park, 195

text messages, 13
therapy, 133, 139
time, passage of, 169
time constraints, 8–9
time management, 66–75, 80–86
Tinker Bell Theory, 286, 288–290, 303
to-do lists, 64, 75
Tomasik, Brian, 309
trachoma, 341
traits: cross-species, 207; heritable, 207; polygenic, 208
traumatic events, 35–36, 170; effects of, 147; writing about, 20
travel, 155
true love, 164

U

uncertainty, 22, 133–134
Unconscious Animal Theory, 286, 287, 303
unconscious conditioning, 53–54
Unequal Consideration View, 293, 304
Unger, Peter, 234
United Way, 342
utilitarianism, 203, 221–222, 227, 237, 258, 351

V

vacations, 153, 155–156, 168–169
valanced states, 127, 190
value conflicts, 229
values, 182
vasopressin, 164
vegan diet, 319, 323–325, 327, 359–360
vegetarianism, 317, 327, 359–362
video games, 116–123
villains, 213–214
violence, 89
volunteering, 110–111, 117, 377
voting, 88

W

waking hours, 108
Walker, Matthew, 106
Wang, Zheng, 22
wanting, 29–30, 51–53
Watership Down (Adams), 305
weakness of will, 30–32
wealth, 173, 179–180; happiness and, 129–131
Western culture, 137
Whanganui River, 196
wild animals, 195, 305–313, 315
wilderness, 337
wildlife conservation, 322–323
wildlife rehabilitation, 314
willpower, 27, 31, 32; concept of, 39; fluctuations in, 47; for habits, 47–50; increasing, 32–40; life outcomes and, 40; short-term rewards and, 126; using, 38, 40
wisdom, reading for, 91
Wise, Julia, 371
women: desire for children in, 163; happiness of, 125, 138; jobs of, 271–272; retirement and, 117
work: gratifications at, 118; happiness at, 114–115; reframing thoughts about, 85
workday, structuring, 60–75
working memory, 96
work schedule, 108–109
workspaces, 23
workweek, 38
World Bank, 249
worms, 303
writing, 67; about traumatic events, 36; in books, 94; importance of, 20–21; practice and, 21; scheduling time for, 73, 83
Wrzesniewski, Amy, 271

Y

years of life saved, 258–268, 280, 331

Z

zero-sum games, 173
zipnote, 101–102
zoos, 313